Exploring the Planets

Exploring the Planets

A Memoir

Fred Taylor
Halley Professor of Physics (Emeritus)
Clarendon Laboratory
University of Oxford, UK

Great Clarendon Street, Oxford, OX2 6DP,
United Kingdom

Oxford University Press is a department of the University of Oxford.
It furthers the University's objective of excellence in research, scholarship,
and education by publishing worldwide. Oxford is a registered trade mark of
Oxford University Press in the UK and in certain other countries

First Edition published in 2016

Impression: 1

Published in the United States of America by Oxford University Press
198 Madison Avenue, New York, NY 10016, United States of America

British Library Cataloguing in Publication Data
Data available

Library of Congress Control Number: 2015951922

ISBN 978–0–19–967159–5

Printed and bound by
CPI Group (UK) Ltd, Croydon, CR0 4YY

Dedicated to all the colleagues, friends, and family, especially Doris, who helped me through all of this.

If it is the fulfilment of man's primordial dreams to be able to fly, travel with the fish, drill our way beneath the bodies of towering mountains, send messages with godlike speed, see the invisible and hear the distant speak, hear the voices of the dead, be miraculously cured while asleep, see with our own eyes how we will look twenty years after our death, learn in flickering nights thousands of things above and below this earth no one ever knew before; if light, warmth, power, pleasure, comforts, are man's primordial dreams, then present-day research is not only science but sorcery, spells woven from the highest powers of heart and brain, forcing God to open one fold after another of his cloak; a religion whose dogma is permeated and sustained by the hard, courageous, flexible, razor-cold, razor-keen logic of mathematics.

- Robert Musil, *The Man without Qualities*, 1930.

Preface

Space, the Final Frontier

Space travel was my inspiration for over 50 years. It was still on my mind in the summer of 2011 as I contemplated my upcoming retirement from the Halley Chair of Physics at Oxford University. Before me was the mundane task of clearing out the office for my successor and moving to more modest accommodation better suited to my new status. In a lifetime of research, teaching, and administration you accumulate a lot of all kinds of stuff, much of it pre-dating electronic storage media, which means stacks of paper. Eventually time catches up with you, and something has to be done with it all.

I was not actually leaving the Clarendon Laboratory, the old name for the Physics Department which refers to the original building that it had long since outgrown, I was just shifting gears. After such long service as a professor one is usually granted an emeritus position and is free to carry on in one way or another. This brings the delightful prospect of relinquishing all routine duties, while being able to continue to work at the things that still capture one's interest. I could expect to keep some office space granted by my successor as Head of the Atmospheric, Oceanic and Planetary Physics subdepartment, however, emeritus accommodation is inevitably much smaller than the grand professorial office one is used to, and more than likely shared with others. All well and good, but you have to get rid of a lot of your accumulated books and papers.

With this in mind, I started sifting through boxes of paperwork dating back literally to the beginning of the space age with *Sputnik 1* in 1957, when I was 13 years old. I soon realized something that had never quite struck me before. Without specifically setting this as a target, I had participated as a member of one or more of the investigator teams for robotic space missions to every planet in the Solar System, out as far as Saturn, and missions to the Moon and a comet as well. I had a wealth of 'insider' data on each of these missions, all of which had made or were still making history. And I was about to throw a lot of it into the bin.

I paused for thought. If there were young folk out there who had the same sorts of interests and ambitions that I had when starting out 50 years ago, they might be interested in my archives. Just thinking about those I had met and talked to over the years made me realize there must be a lot of people, not all of them young, who are interested in the first five decades of exploring the planets, and what it was like to be part of it. I lost count of the number of times I had talked about current and historical activities over dinners, at meetings, to the media, including radio and TV, and even to politicians and historians. And yet there was no solid record of any of this except my mounds of dusty letters, memos, and notebooks.

I took the first pile of notes back out of the bin, and a week or two later spoke to the commissioning editor for Oxford University Press about an idea I had hatched. We knew each other because I had written a couple of textbooks for physics undergraduates during the part of my career I had spent on teaching students, but this was something different. I wanted to distil all of the best bits from my lifetime experience and write it up as a book. With the pearls thus preserved for whatever part of posterity might be interested, the oysters could be consigned to the bin without further trepidation.

The Backstairs of History

I knew the Press had published biographies and autobiographies of people like Einstein and various Nobel Prize winners, but they also do them for people who are not quite that famous. The criterion appears to be whether there is a strong and stimulating story line that evolves from a close involvement in something interesting and important. I launched into a description of my first-hand experience of flying scientific instruments to the planets to see what is there, to seek to understand it, and to tell the world what we found when we did. A couple of hours later I left the meeting encouraged to submit a full proposal with a detailed outline and a sample chapter. This was not a complete victory; as an existing author on the Press's books a short description might normally have sufficed, but I agreed that this new endeavour tackled a quite different genre than my ten previous books.

So I started writing, and a few months later found myself contemplating several long reviews of my plan written by no less than six anonymous colleagues whom the Press had approached for their opinions. Leafing through the comments, I could tell from their remarks that these were from people that I had met and knew very well. The responses were all very positive and full of ideas for how the material in the book should be approached. A couple of them questioned some of the minor details in the sample chapter I had prepared about NASA's *Galileo* mission to Jupiter. Obviously, like myself, they had been part of the science team for that big project. They made me realize that relying on memory can be risky when it comes to the dating of otherwise well-remembered events. My dusty notes, diaries, and documents would have to be rigorously consulted throughout the two or three years it took to write, but that was the plan anyway.

This is a memoir and not an autobiography, the difference being that it focuses on the fascinating things I have been fortunate enough to be involved with, which is essentially the entire history to date of Solar System exploration by spacecraft. My own life history is peripheral, but some autobiographical elements have been woven in since they provide essential context. My experiences and views might be of interest to people whose own lives are tangled up with the story in some way, and maybe to those youngsters who share my dreams and want to follow a similar career trajectory.

The Story so Far

The exploration of the planets using spacecraft started when we saw the first views of our Earth as it looks from space, obtained by the TIROS series of satellites in 1960. Even those blurry images caused a revolution in the way we saw our planet. For instance, artistic impressions of the Earth drawn before anyone had seen the real thing always had far too little cloud cover, something that struck me at the time and started my interest in meteorology. Far more important, however, was the way it drove home to all of us the idea that our planet is a tiny, fragile globe hanging in space and not the vast playing field stretching forever in all directions that it had always seemed to us in our everyday living.

The Earth sits in the middle of the Sun's planetary family, with its life support system part of a changeable and changing climate. We depend upon a viable ecosystem but still do not really understand how it works and we cannot control it except in a crude and very limited way. Progress is being made, however, and studies of our home planet by Earth observation satellites in space are playing a large part. This aspect of planetary science—exploring and understanding our own planet—has a special urgency since we have realized that living conditions are changing and threaten the very existence of the human race, or at least its lifestyle.

Probing and monitoring our planet and its environment remains a focus for space scientists, but now we also venture beyond into deep space to deploy satellites and landers on our neighbours Venus and Mars, and send exploratory probes as far away as the rainy lakelands of Saturn's planet-sized moon Titan. The theme of this book is the whys and wherefores of these complex voyages of discovery, from one person's perspective, and the joys and hardships of working to see them achieved. I try to describe the science in user-friendly terms and mostly discuss the 'big picture', in terms of several broad themes. The first of these is what we have learned about the formation of the planetary system, and how it evolved to give us our home world and its range of companion planets we see today. I am particularly obsessed with the conditions on each world, especially their atmospheres and their own particular versions of weather and climate. Someday we will visit them all, not just with machines and instruments, but in person. We need, therefore, to learn how well conditions suit life, either indigenous (if there is any) or ourselves when we arrive as explorers and colonists.

Playing a part and writing about it

The Space Age is itself now more than 50 years old and for those like me who were lucky enough to get in on it at the beginning, it fills a working lifetime. Today, several instruments that I helped design and build have operated on board satellites orbiting around the Earth and have studied the atmosphere and the climate. In some cases I led the project, with responsibility for big cash budgets and large teams of people. As the technology improved, my colleagues and I branched out to work on 'deep space' missions to visit the planets.

Gradually we have become familiar with all of the major bodies (including the Moon and other planetary satellites, asteroids, and comets) of the Solar System, working closely with collaborators in research centres in America and Europe, often using instruments built in our labs at Oxford. Our hardware is now scattered across the Sun's family of planets out as far as Saturn; we haven't been formally involved in flights to distant Uranus, remote Neptune, or far-off, icy Pluto, although I've worked closely with people who have and had a front row seat from which to watch their story unfold.

The goals of planetary exploration are mostly scientific, but the detailed scholarship in my work is dealt with elsewhere in 'learned' journals and books aimed at fellow scientists. A bibliography is available from me that addresses that, for those that want it. This book avoids specialized language and terms and so is for people from all backgrounds who are attracted by space exploration and interested in how and why it is done. You don't have to be a professional scientist to find the planets fascinating, and to want to know what they are like and what is going on there. Also, of course, everyone cares about planet Earth, and things that affect our daily lives, such as how well we can forecast the weather and whether the climate is really changing.

Often, very little is known to non-participants, whether scientists or not, about how the global space programme has evolved: how the motivations of researchers and politicians can merge to generate the large budgets required; how the technology is planned and developed to make various feats like landing on Mars or flying along with a comet possible; and how often it all goes badly wrong. In setting out to relate the experience, most of the narrative is from my own vantage point and based on involvements I had at the time, at the space agencies, research labs, conferences, and other places as diverse as Cape Canaveral and No. 10 Downing Street. It could be useful to students and others who want some basic background to the field for their work as well as to interested laypersons. My fellow space scientists might enjoy a different point of view, and even historians might find something of interest in these pages too.

In preparation for writing this book I read a number of other scientific memoirs, looking for examples of erudition and clarity when dealing with topics that include a lot of specialized concepts. It clearly is really difficult to write about complex ventures in space in non-technical terms that make sense to everyone. I have steered my way through, avoiding the minefield of technical terms and concepts, and tried to avoid leaving loose ends. A good model, I found, was James D. Watson's *Avoid Boring People: lessons from a life in science* (2007), in which the co-discoverer of the structure of DNA writes engagingly about his work in a field I don't understand at all, and without boring me, at least, at all. And of course *The Double Helix*, which I read in a single all-day sitting while a student in the 1960s, is a masterpiece I couldn't hope to match, but I can still use it as an inspiration. I also learned from Colin Pillinger's memoir *My Life on Mars (2010),* which overlaps with mine on the science, and John Carey's book *The Unexpected Professor (2014),* which overlaps with me on life as an academic at Oxford.

Hoping for expert guidance I turned to *The Cambridge Companion to Autobiography* (2014). In this I found lots of advice to the aspiring memoir writer, in quotes such as this: 'The "new memoir" aims at truth by shifting perspectives and doubting, fundamentally, the linguistic project of an incomplete mind engaging an infinitesimal subset of functionally infinite reality'. No, I don't understand it either. I hope I haven't done that, whatever it means, but anyone who is confused or confounded by anything I have written, and interested enough to take the trouble to write kindly to me at (fred.taylor@physics.ox.ac.uk), will receive a reply.

Contents

1
Prelude

The Cosmic Grand Tour

For as long as I can remember, and from my earliest years, I was fascinated by the idea of travelling in space and visiting the other planets. I learned to read at a tender age, partly by tackling my grandfather's library of astronomy books, attracted by the pictures, grappling with the captions, and gradually mastering the text. I started to collect my own space-based books and magazines a few years later and began to understand the grown-up material in newspapers. I learned that V2 rockets, no longer instruments of war, were being launched in the New Mexico desert to explore the upper atmosphere, and visionaries were starting to talk seriously about trips to the Moon. I wanted to be part of it. I didn't know how, but I never doubted I would.

I also wanted to travel. A book review of what used to be called the 'Grand Tour' featured in the *New York Times* recently. It began: 'Three hundred years ago, wealthy young Englishmen began taking a post-Oxbridge trek through France and Italy in search of art, culture and the roots of Western civilization.' Visits to the grand capitals of Europe no longer have quite the prestige they had three centuries ago; nowadays most of us have all been to some or even all of them, and to other great cities in the wider world, and the old concept of a character-forming 'Grand Tour' has become more mundane. At the same time, however, new vistas have been opening up, way beyond old Europe and now even beyond the planet on which we live.

The journey described in this book is the story of a modern grand tour that involved voyages through space to the planets of the Solar System, as well as to our Moon, the moons of other planets, the minor planets or asteroids, and most recently to a comet. Except for the incredibly bold Apollo astronauts who walked on the Moon, the travellers were not humans but surrogate robots in the form of specially designed spacecraft, with instruments on board to act as eyes and ears. I was destined to help conceive, design, and build many of these over a period of more than half a century.

I started out on my personal journey just as the space age was beginning, and can now look back on the data we received from these first robotic explorers in space. They sent back pictures and measurements that revealed bizarre landscapes and scorching or freezing weather that was often riven by storms. Even airless and relatively familiar bodies like the Moon hold mysteries as challenging and exciting as the jungles of Africa did for the explorers of the great Victorian era. The planets were mysterious and fascinating when I set out, but now they seem to me almost as familiar as my own back yard. Exploring them was what I wanted to do when very young, and with great good fortune it really happened, against a background of learning about life, the universe, and everything, on the way.

Published in 1951, my favourite book in my very early years was something called *Rockets, Jets, Guided Missiles, and Space Ships* by Jack Coggins, Fletcher Pratt, and Willey Ley. Profusely and imaginatively illustrated, it stoked a passion, as did its fictional counterparts, none more than *Dan Dare, Pilot of the Future*, a sophisticated comic strip in a magazine for boys called *The Eagle*. The stories were also brilliantly acted out in 15-minute episodes each week night on the pirate radio station Radio Luxembourg, which defied the BBC's monopoly by broadcasting to the UK from overseas. It made money by advertising; Dan Dare was sponsored by Horlicks, a hot drink that was claimed to promote restful sleep if you drank it before going to bed. I didn't like its malty taste, but I drank a lot of it in order to collect the labels so I could join the Dan Dare Space Club and receive the 'authentic' cap badge, which is still proudly displayed with other lifetime trophies in my garage.

The Planets of the Solar System

From Grandfather's library I soon knew all there was to know about the planets, although that wasn't much in those days before space travel. I inherited some of those old books, and now I find them full of erratic speculation on everything except the most basic data about things like size and distance from the Earth. One showed the journey times to the planets if they could be made by fast train (30 years to Venus, the closest), along with a picture of imaginary rails reaching out into space. Those that ventured to describe what conditions were like on the surface of Mars or Jupiter were completely wrong (Jupiter doesn't even have a surface).

Right or wrong, I soon had the details memorized. Six planets are visible to the naked eye and were known to the ancients: dim Mercury, close to the Sun; brilliant Venus, the morning star; our own Earth; baleful, red, intriguing Mars; enormous, colourful Jupiter; and the ringed beauty of Saturn. In addition, our forebears were familiar with the glories of the Moon, of course, and some of the smaller bodies in the Solar System, of which the most spectacular are the comets that appear from time to time, with long tails that trail across the Milky Way.

Our ancestors knew the planets intimately as lights in the sky, but they didn't know much about them as objects to visit. Still, a vast amount of writing has accumulated over the centuries describing conditions on other worlds, in terms ranging from serious, if not always accurate, scientific speculation, to the purely fantastical. Then came the 1960s, when spacecraft started to visit the Moon and the nearby planets, and their true nature began to be revealed. It would not take long, 20 years or so, before probes would be launched that would travel right out to the very edge of the Solar System.

With the large telescopes that became available in the 1800s, the outer ice-giant planets Uranus and Neptune were added to the original six planets known since the earliest times. Uranus is visible to the naked eye, but only just. Also, it moves so slowly across the sky and is so faint that it was not until 1781 that William Herschel recognized it for what it is: a planet four times bigger in diameter than the Earth. Pluto also enjoyed planetary status for a while, after its discovery in 1930, but it no longer qualifies since it is now known to be a member of a large family of small objects, the Kuiper Belt. (Later in the book, I will talk about who Kuiper was, and the time I had dinner with him).

Spacecraft have now visited all eight planets, the closer ones many times. There are all sorts of reasons for wanting to explore, starting with the thrill of the unknown and the wonder of nature, through to gaining new practical knowledge and tracing our origins and planning our future. There is a spiritual dimension to it, too. Learning about new worlds, and the old one, as physical systems, brings with it some insight into the overall meaning of things. This awareness that comes from studying the big picture leads to a kind of secular religion, not always comforting, but realistic in a way that the ancient beliefs are not.

Now that I know what it is like on each of the planets, through studying them with sophisticated devices that have actually been there, I am far from sure I would make the journey across the Solar System in person, even if I could. As a place to live, Mars is the best prospect, but even there we have found an airless, bleak, and frozen desert. Venus is like the Earth, but global warming has raised the temperature so high that some metals, such as lead or tin, melt on its surface. Mercury and the Moon are airless; the outer planets are mostly fluid and have no solid surface except deep in the core. Saturn's moon Titan resembles Earth in many ways and is a decent prospect for a landing party in fifty years' time, but it is far from the Sun and very, very cold.

For me, exciting though it was, exploring these alien environments was not the main incentive for getting involved. It started with looking at our own planet from space, initially to improve weather forecasts (which used to be much worse than they are now, believe me) and then to understand threats like ozone depletion and climate change. Mixed in with this are some very fundamental questions, like where did the Earth come from, how did it form, has it changed over the aeons, and how did it come to support life?

Answers to questions like these come more easily if you think of the Earth not as an isolated place, but as a member of the Sun's family of planets. They all originated and evolved together, more than four billion years ago, and they all carry historical clues in their atmospheres, their geological records, and their very nature, of their common evolution. Those with atmospheres have weather and climate, which is not the same as the Earth's at first sight, but which under analysis turns out to be just a different mixture of the same chemistry and physics that obeys the same laws everywhere.

Putting all of this together and starting to understand our origins and things important to our survival form the motivation for the journey that my life and career have followed, and that I set out to describe in this book.

Beginnings

Memoirs may dwell on ancestral details, often it seems with tales of city suburbs or tranquil villages where the subject started out, usually either in mansions with manicured lawns and tennis courts, followed by Eton and Oxford, or in abject northern poverty where the whole family lived in two rooms and seldom went out except to work at some back-breaking toil. Mine was something in between. I grew up near the Scottish border, on the Northumbrian coast, almost as far north as you can get and still be in England. For rural beauty and traditional seclusion this area is second to none. But my father's antecedents fished in the cold North Sea and my paternal grandfather, John Henry Taylor, broke his back down an Ashington pit.

I was born in Amble, a pretty seaport at the mouth of the River Coquet. My father's side of the family had been centred there, and at Hauxley, an old fishing village just south of Amble, for as long as anyone knows. On 24 September 1944 the Second World War was still raging. The headline in the Sunday Times for that day is about the Allied attack on the bridges at Arnhem, with the troops arriving in hundreds of gliders; the book and the movie, *A Bridge too Far*, tells the story. My father used his carpentry skills to help to build the gliders, as an invalid, having been seriously wounded and losing an eye in the Army in North Africa during the early stages of the war. It took a decade and several surgical operations before he was fully fit again, and for a time my mother was the breadwinner.

Mother was from Jesmond, a smart suburb of Newcastle upon Tyne. She, and her mother before her, had college diplomas at a time when this was rare for women, and she became an infant school teacher. When I was five years old, she was offered a headmistress position in the small village of Howick, ten miles north of Amble, on the ancient estates of the Earls Grey. In 1949, when my parents, my sister Maureen, and I arrived, the fifth Earl presided at the stately home up the hill, about a mile away from the village where my mother took over the running of the venerable little two-room school.

The Grey dynasty had been at Howick since the fourteenth century, but did not become really famous until 1801 when the Earldom was created for Charles Grey, who fought with distinction in North America and elsewhere. His son Charles, the second Earl, was a Whig politician, who became (I would argue) the greatest British Prime Minister of all time, since he led the country in 1832 when the Reform Bill was passed, and was still in charge a year later when slavery was abolished throughout the Empire. Thus he was responsible more than any other single individual for introducing modern, enlightened democracy to the country and to the world. Nowadays his name is heard more often in connection with the eponymous tea, which you can sample in the sumptuous tearoom in the former ballroom of the Hall, set amidst the wonderful gardens that you can also enjoy. I visit his tomb in the little church to pay my respects almost every time I visit Howick, which I like to do often.

I have a story about that, which nobody else seems to know. The tomb takes the form of an ornate marble chest, sitting in the nave near the lectern and the front pew used by the Grey family. It is recorded that it once had a fancy canopy, which the fifth Earl so disliked that he personally took a hammer and chisel and broke it up. What is not recorded is that he also had his grandfather's tomb moved, from its original location next to the altar to its present less exalted location. I know this because my father, still convalescing from his war injuries, worked on the estate as a labourer for a time and was part of the workforce that did the job. I remember him telling us about it when he got home in the evening; mainly he was disappointed that there was no sign of the body, which presumably lies deeper in a crypt.

My maternal grandfather was a manager at International Paints Ltd., which had a vast factory complex at Felling on the south bank of the Tyne. He would walk there every day, a round trip of seven miles. His name was Robert Burns, leading me to think (without any palpable evidence) that I might be descended from the family of the great Bard of Ayrshire. Granddad Burns was a cultured man, with a collection of art, books, and music, and his main hobby was astronomy. His beautiful books on the subject, and the extensive collection of science fiction his son, my uncle Alan, built up, including some he had written himself, caught my eye. Fact and fiction merged in my imagination with the prospects offered by the early satellites, *Sputniks* and *Explorers*, that were soon orbiting the world.

The harbour at Amble, now an attractive marina, used to serve the north Northumbrian coalfields as a rail terminus and a loading point for large ships. In the background is the historic castle in the village of Warkworth, where my parents lived after I left home for university. Picture by Anna Williams, Amble Development Trust.

When I was five we moved to Howick, a feudal village in the countryside a few miles up the coast from Amble. The School House where we lived, with the school attached, is seen to the right of Widows' Row near the centre of the picture, with Sea Houses Farm in the background on the coast. To the right, surrounded by trees, is the Old Rectory. The church it used to serve is a mile away, located close to Howick Hall for the convenience of Earl Grey, as was the railway station, now vanished. Picture courtesy of Stewart Sexton.

School Days

I hankered to be an astronomer thereafter, and eventually made it to university thanks to the local grammar school, a wonderful institution that was sadly trashed with much of the rest of the English secondary education system by dogmatic politicians some years later. Every weekday I donned my school uniform and set off along narrow country roads on the big red bus operated by United Automobile Services to the old market square in Alnwick, seven miles away. Alnwick is like a larger version of Howick in that it, too, is dominated by a large aristocratic dwelling. In its case this is Alnwick Castle, home of the Dukes of Northumberland, but now enriched by its fame as the set for Hogwarts in the Harry Potter films. You can visit there, too.

The Duke of Northumberland's School had less than 200 pupils, all boys of course (there was a Duchess's School for girls, strategically located right across the other side of the town) in seven forms from First to Upper Sixth. The buildings were historic, with lavish playing fields, and many of the teachers were wonderful. I was good at my studies and if it weren't for being pretty useless at rugby, cricket, and cross-country running, memories of those halcyon days would have been idyllic. I console myself that my lack of achievement at sport—a non-trivial deficiency at an all-boys school, although prowess in the classroom leads to grudging respect from one's peers that mostly saves the day—was mainly due to a deep-rooted lack of interest in what I saw as unproductive effort and pointless aggression.

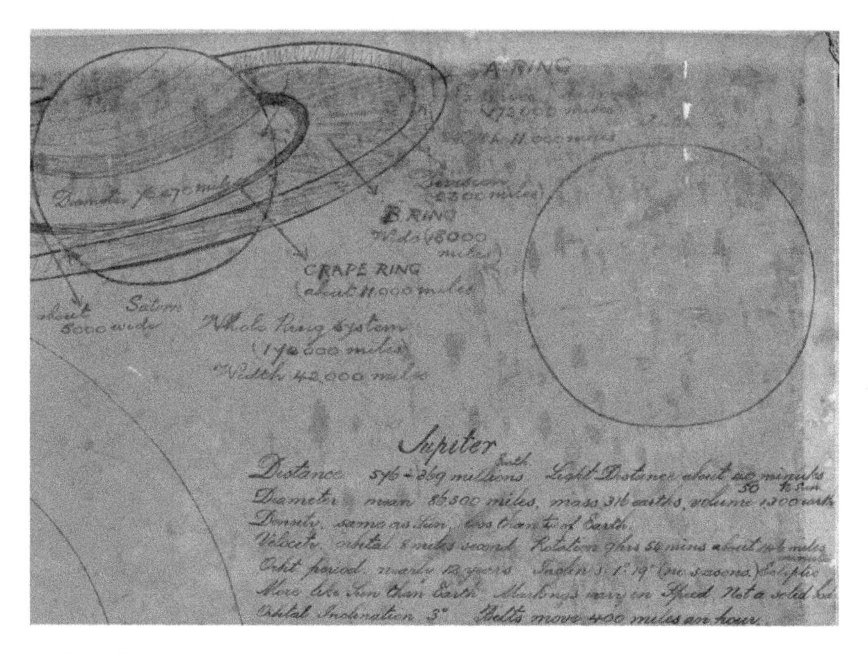

While still very young I acquired this crumbling wall chart, made more than a century ago by my grandfather Robert Burns (d. 1962), on the back of a picture of the Tyneside factory of Holzapfels Limited where he worked. The date is before 1914, because the German name was changed to International Paints Ltd when the First World War started.

At home, the local boys, about half a dozen of us, played a rudimentary version of soccer in the fields around Howick and I enjoyed that, but football wasn't allowed at school, it had to be rugby. I think this was to be consistent with schools like Eton, where soccer, once popular, had long since fallen out of favour. I thought and still think rugby is an awful game, maybe because I was tall and heavy and so assigned to play as a prop in the middle of the scrum, an uncomfortable and undignified role in an uncouth and pointless ritual. I soon broke my arm by trying too hard to get into the swing of it, which was a blessing because then I didn't have to play anymore. I took up the javelin and shot-putting instead and achieved a modicum of competence at both. I was even better at chess. Nowadays I enjoy *watching* cricket and football, in the spirit of a Roman emperor watching slaves duel to the death, especially with HD TV which reveals exactly what's going on and slow-motion replays that allow you to savour the details.

More of a success than my sporting endeavours was my role as originator and editor of the school newspaper, *The Crescent.* The name came from the school badge, which featured a crescent and fetlocks taken from a banner that an early lord of Alnwick Castle, Sir Hugh de Perce, brought back from the Crusades. My co-editor 'Tiny' Rodgerson, so called because he was large, couldn't spell crescent and so the masthead said 'The Cresent', with a large symbolic crescent moon belatedly superimposed behind and falling between the s and the second e.

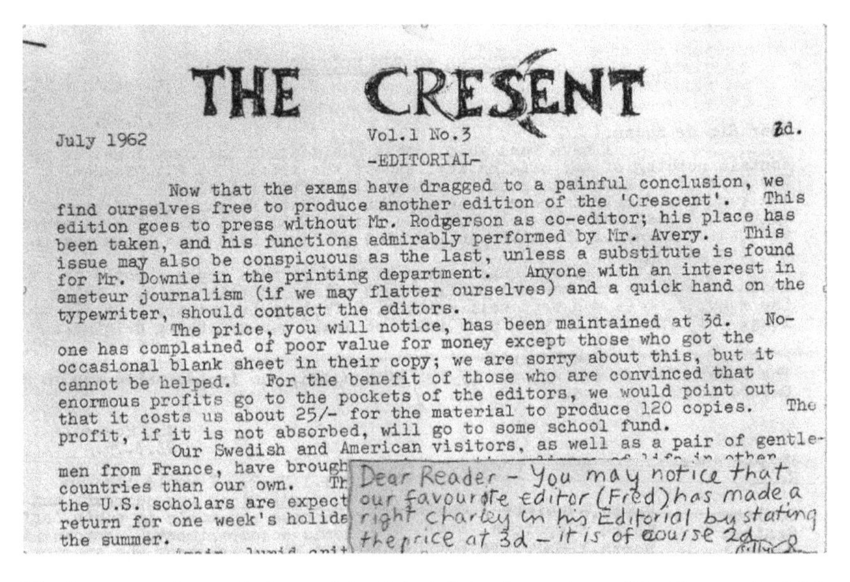

THE CRESENT

July 1962 | Vol.1 No.3 | 3d.

-EDITORIAL-

Now that the exams have dragged to a painful conclusion, we find ourselves free to produce another edition of the 'Crescent'. This edition goes to press without Mr. Rodgerson as co-editor; his place has been taken, and his functions admirably performed by Mr. Avery. This issue may also be conspicuous as the last, unless a substitute is found for Mr. Downie in the printing department. Anyone with an interest in ameteur journalism (if we may flatter ourselves) and a quick hand on the typewriter, should contact the editors.

The price, you will notice, has been maintained at 3d. No-one has complained of poor value for money except those who got the occasional blank sheet in their copy; we are sorry about this, but it cannot be helped. For the benefit of those who are convinced that enormous profits go to the pockets of the editors, we would point out that it costs us about 25/- for the material to produce 120 copies. The profit, if it is not absorbed, will go to some school fund.

Our Swedish and American visitors, as well as a pair of gentlemen from France, have brough[illegible] countries than our own. Th[illegible] the U.S. scholars are expect[illegible] return for one week's holid[illegible] the summer.

Dear Reader – You may notice that our favourite editor (Fred) has made a right charley in his Editorial by stating the price at 3d – it is of course 2d.

The front page of an early issue of the Duke's School newsletter, 'The Crescent', which I edited, and a 'note added in proof' from the back. For some reason we thought it was a rather professional production at the time. It was printed by an antiquated and very messy process called cyclostyling.

The venture went a little sour when we wrote and published a satirical play to fill the space at a time of little news. One of the subjects of the satire was the Headmaster, to whom we attributed the following lines, where he admonishes a delinquent pupil thus:

> 'Avast! Avast! You scurvy knave. Thou shallst not be the winner.
>
> I have a horrid fate for thee. I'll make thee eat school dinner.'

The problem was not so much the Head, who took this sort of thing in his stride, as the Chef, who took great exception to this slur and reportedly threatened to resign. Although I confessed to writing this piece of deathless verse, for some reason the punishment fell mainly on Tiny, who refused to take any further part in writing or publishing *The Cres(c)ent.*

The Duke's School had been founded nearly 200 years earlier as the successor to an older school in a building that is now the town library. We were old enough to have our own traditions, but our school song, *Forty Years On*, was the same as Harrow's, and we had slang terms for everything like boys at an archetypical public school do. We also had flogging, although only for serious offences like smoking in the old air raid shelters, planting fireworks on a slow fuse that went off in the drains during the French class, or attempting to manufacture dynamite in the woodland that surrounded the playing fields, to mention three examples that brought retribution on me personally. Even then it was perfunctory and relatively mild, obviously intended to administer a minor humiliation as punishment rather than actual pain.

The 2nd Form at the Duke's School, Alnwick, in 1958. The form master is Mr Alan Dodds.

English and chemistry were taught with passion and flair by Messrs Joseph Grieve and Arthur Bell respectively. Inspirational and supportive teachers like these have and always will make all the difference to young people's dreams and aspirations. I recall Mr Grieve saying 'Northumbrians are low achievers in general because they never put themselves forward. Learn to speak properly, and to be just a little bit pushy, and you'll be fine.' This I thought reminiscent of Lord Byron's comment about Edward Trelawny, 'if only we could get him to wash his hands and tell the truth, he would have the makings of a gentleman', but Joe's advice was more kindly meant. Mr Bell offered me a job as his lab assistant, which involved real responsibility and an income, both extremely modest, but novel and edifying concepts to me.

These two of my favourite teachers came together to suggest that I entered a competition that one of them had found in a magazine held in the school library. The challenge was to describe some topic in modern science using the language of Chaucer. We were deep into *The Canterbury Tales* with Mr Grieve at the time, and Mr Bell was fascinated by the new technology just arriving in the form of electronic computers. For my part, I was riveted by the serial *A for Andromeda* on BBC television at about this time, in which the plot focussed on a computer built following instructions from outer space. The story had been written by Fred Hoyle, a professor at Cambridge who was already a role model for me, and starred a stunning 19-year-old Julie Christie as a beautiful alien.

So I wrote in medieval terms about a computer, and I won a prize. A bit of passion obviously goes a long way, because the only previous time I had won anything like this was as a fan of the great children's author Enid Blyton while still at junior school. She edited her own magazine for a time, with a monthly competition in which I won an autographed copy of her latest 'Famous Five' book, which I still treasure. What I did to win that I have long forgotten, but my successful Chaucerian verse appeared in the magazine, and the final stanza ran thus:

> *To gette ye anser to a questione si fowle*
> *Would take ten menne a lifetime, were they wize as an owle!*
> *And so you must see oure terrible plite*
> *For noone doth know if our computer is rite!*

Chaucer must have spun in his grave.

Although English and Chemistry were my best and favourite subjects at school, clearly neither offered the pathway into space that I wanted. Mr Laidler's physics lessons were OK, but it takes a certain sort of brain to enjoy maths, and I didn't, which is unfortunate since a talent for maths is generally taken to be a *sine qua non* for physics, too. On the other hand, I read that Ernest Rutherford, perhaps the greatest experimental physicist of the twentieth century, was notoriously poor at maths and would fumble derivations of formulae during his lectures at Cambridge. Encouraged by this, I thought I could manage a degree in astronomy as a way into space research, especially if I could stick to the experimental side. Then I would follow my urge to join those lucky few who were paid for designing and building rockets and spacecraft.

The Headmaster of the Duke's School was a magisterial, and to the pupils terrifying, figure called Frank Mosby MA, known to us, although not to his face, as The Gaff. I recently unearthed the fact that Mosby had become Headmaster at Alnwick the month I was born, September 1944. The same internet search uncovered a large engraved silver platter, presented to him on his retirement in 1969, rather pathetically up for sale on eBay. *Sic transit gloria mundi.* But in 1962 when I meekly described my goals and plans to him, the Gaff listened carefully, and then spoke wisely. 'Forget the degree in Astronomy', he said, 'you'll end up on top of a mountain somewhere, even if you are lucky enough to get one of the few jobs that are going. Get a degree in a basic subject instead, like Physics, then specialize later.' This was excellent advice and I never regretted that I opted to follow it.

But where to apply? There were very few Astronomy degree courses, but nearly every university that existed then offered Physics. I was briefly tempted by Keele, which was then quite avant-garde in its degree subjects and had a course called Physics with English. 'No, no' said the Gaff, 'do it properly.' He wanted me to stay on at school an extra year as a one-man 3rd-year sixth form, with the express goal of taking the Oxford entrance exams. I would also have to obtain the required qualifications in Latin or Greek, then mandatory at Oxbridge even for students taking science degrees. I had done four years of Latin at school already, and a pass in the Latin exam would have been a formality, but the idea of another year at school, with all my friends gone (not to university, most of them; only a small percentage of the population took higher education in those days) was anathema.

One fine day I won a scholarship to participate in something called *the 6th International Youth Science Fortnight*. This was based in London, and was a stunning affair aimed at future scientists that involved inspirational talks and numerous visits to institutions and companies. The Duke of Edinburgh was a sponsor and I met him at an event organized for us at the National Maritime Museum in Greenwich. We also went to iconic places like Shell, Esso, English Electric, Bradwell nuclear power station, and the (slightly less ephemeral) House of Commons at Westminster. On one of the free evenings I took myself to the West End and saw the first run of the musical *Oliver*! with the fabulous Georgia Brown as Nancy. Her rendering of 'As long as he needs me' still makes my toes curl when I watch it on YouTube.

One of the daytime visits for Science Fortnight participants involved a trip to Oxford, where I unwittingly visited my future home, the Clarendon Laboratory. I was not much attracted by what we saw there, a lot of congested laboratories with benches of polished wood loaded with strange equipment, electromagnets, and early types of lasers. Small, bent, serious-looking men in three-piece suits or white coats, and sometimes both, scuttled like rats down gloomy corridors. I remember ascending to the top of the tallest building in the city, the new Biochemistry tower, which by contrast with the Clarendon was light and airy, but then still unfinished and unoccupied. As I write now, 50 years on, the tower is being demolished. I am not the first writer to remark that few things make you feel older than seeing them tear down a building that you also watched them build.

In spite of the fact that we were shown departments that included trendy subjects like Biochemistry, clustered around the Natural History Museum with its strange collections of dinosaurs and crystals, the atmosphere at Oxford seemed to me that of a place whose achievements in the humanities soared over everything else, including science. The research in physics seemed to be all about properties of matter, or nuclear physics using their own in-house particle accelerator. I did not know then, nor did anyone tell our little group of sixth-formers as we shuffled around the labs, about the pioneering work that was going on there using aircraft to study ozone and water vapour in the stratosphere. Even if I had been aware of its existence, there was then no clue that the small meteorology group in the Clarendon lab was poised to move into space research just a year or so later. Had I known that, I might have been tempted into spending another year at school to take the Oxford entrance exams. On the other hand, I did know that it was far from certain that I would get in, and then I would have spent a lonely year as the only member of the 3rd-year sixth form for nothing.

University

So, where to apply? In the end it was no contest. I narrowed it down to the five great northern universities with strong Physics departments, then quickly eliminated Newcastle because it was too close to home. Manchester had probably the best academic reputation, but this was 1963 and I was 19 years old: work and a career was one thing, but not everything, and so far as life beyond work was concerned Liverpool was the centre of the universe. The Mersey Sound was in its heyday, I knew the social life would be fantastic, and it was.

I duly applied and Liverpool took me on. The Physics Department was excellent, with really good lectures and well-equipped practical labs, and even a tutorial system which, looking back on it, was every bit as good as the renowned system at Oxford. That is, if you lived in one of the Halls of Residence. Places in Hall were scarce, and many of my fellow physics students lived in lonely digs scattered around the city, often just a spare room in somebody's house, possibly not even in Liverpool itself but across the Mersey in Birkenhead, requiring a ferry crossing every day. Hall, in contrast, was not too different from an Oxford college, offering an attractive environment with communal dining and additional support in the form of resident 'moral' tutors.

The idea of a moral tutor may sound appalling but I had every reason to be glad of mine, especially at the beginning when I was trying to find my feet, and near the end when exam pressure almost became too much. The rest of the time they didn't interfere except to occasionally dispense sherry and small talk similar to the approved Oxbridge fashion. The tutors in the Physics Department were indispensable too. Physics at Liverpool was a four-year course, but if your qualifications were good enough you were able to skip the first year and enter straight into the second, which is what I did. I suppose the assumption was that you would soon catch up on the basic material, and eventually I did, but coming on top of the trauma of being away from home for the first time it wasn't easy, especially the maths. I flunked a test for the first and last time in my life and was greatly upset, but the tutorial system kicked in and saw me through. After six months or so it was fine.

I was lucky enough to spend the whole three years in Rathbone Hall, indisputably the best of the student residences both in terms of facilities and reputation. It was located several miles from the main campus, at one end of soon-to-be legendary Penny Lane. It was single-sex, of course, despite this being Larkin's *Annus Mirabilis*, the year of the Beatles' first LP. Each resident had a decent room, not en-suite but with the luxury for the time of a wardrobe and a wash basin (which, by common consent, could double as a urinal so long as the water was left running). Set in the spacious grounds of Greenbank Hall, it provided excellent,

The residents of the top floor of D block at Rathbone Hall in 1965. L to R, Back Row: Terry Wright; Derek Senior; Dick Carver; Bob Lawless; Phil Ayres; Front Row: Fred Taylor; Dave Wilson; Geoff Crowther; Talib Al-Hadithi, Mike Maddocks. The windows behind Phil are those of the Common Room, home to the student bar, over which I had charge.

abundant meals prepared by young women training for a diploma in catering at a nearby college of further education, a laundry run by nuns, and of course unlimited fellowship. Perhaps most important of all, the Hall had its own bar, run by students, open two nights a week.

The rooms were heated but, if you wanted to be warmer or make toast, there was a small electric fire built into the wall. This ran on a meter, which took shilling coins; we soon discovered that it was a simple matter to prise open the locked coin tray and use the same coin again. Of course that was soon rectified when we foolishly killed the golden goose by leaving just one shilling in the tray at the end of term. The Bursar ordered the meters wired shut and that was that. The strange thing to me, looking back, was that we never for a second thought there was anything illegal or immoral in what we did; if we thought about it at all, we thought the electric fire should come with the rent.

A State Funeral

On 29 January 1965 I joined three of my fellow physics undergraduates on an epic and audacious journey. We had resolved to travel to London for the state funeral of Sir Winston Churchill the next day, reasoning that this was an historic event without precedent that we would always remember, and so it proved. We piled into Geoff White's elderly Ford Anglia and set off in the evening, travelling all night in the freezing cold on mostly pre-motorway roads that passed through many a town centre on the way. We arrived as planned just as dawn was breaking and found a good vantage point near St Paul's Cathedral and settled there for a long wait in near-Arctic conditions. Contemporary newspaper reports said that the temperature was –5 °C, and it felt like it. Gradually, statesmen from every nation in the world assembled, and finally the Queen arrived.

After the ceremony was over and the long cortege had gone by, with the coffin on a hand-drawn gun carriage, heading for the Thames and from the boat to Waterloo Station for the funeral train to Long Hanborough near Blenheim Palace, the crowd began to break up. Starving and freezing, we found a pub for lunch, and then wandered around Soho for several hours. Given its racy reputation, it was a disappointing and anti-climatic venue. As darkness fell we retrieved the car from the side street where we had left it, and started the long trek home, arriving back in Liverpool the next morning, tired but elated by the adventure.

Jodrell Bank

During the long vacation at the end of my first year at Liverpool I applied for a place on the Summer School that was run each year at the radio astronomy laboratory at Jodrell Bank in Cheshire. This had been founded in 1945 by Sir Bernard Lovell, a near-legendary figure in British astronomy who was still around in 1964 when I arrived there, and friendly if a little vague and remote. He, and the rest of the staff, gave us inspirational lectures, showed us over their world-class hardware, including the huge 75-metre dish for which the site was famous, and gave us a few menial tasks as an introduction to experimental research. I soldered together a power supply intended for a new receiver that would amplify the signals from space. It looked awful and resisted all my attempts to make it work; I imagine that someone with more experience than I had of assembling electronic circuits rebuilt it after I left.

I enjoyed the visit, but it put me off any idea I had of trying to get in there to do a research degree. The place was isolated and the work mostly unglamorous despite its far-sighted horizons. But worst of all, they used their fame to recruit far more graduate students than they had money, work, or room for. Around 20 were taken on each summer, bright-eyed and hopeful that they would earn a doctoral degree. However, all but two or three were shed at the end of the first year with only a master's degree to take into the outside world. Many years later, the subject of Jodrell came up when I was sitting on a committee with Chris Rapley, a highly accomplished near-contemporary of mine who later became Director of the British Antarctic Survey. It turned out that Chris had fallen victim to their semi-scrupulous recruitment approach a few years after I was there. I don't think he ever forgave them.

Running the Bar

Back in Liverpool, I completed the first two years of study and finally learned some maths. My biggest problem now was where to live; accommodation in halls of residence remained over-subscribed and the rules said that you had to move out after a maximum of two years in order to make room for new arrivals. Desperate to avoid having to live in the dreaded 'digs', I secured a rare third year in Rathbone Hall by conspiring to get elected to the student committee as Bar Secretary. In spite of the odd title (more than one of my contemporaries thought it hilarious to refer to me as a bar steward) the position involved responsibility for the actual running of the student bar, which opened two evenings a week. I ordered stock, appointed barmen, and arrange special deliveries for individual parties, all at knock-down student prices. This was my third taste of part-time management and commerce, and rather a harrowing experience for all sorts of reasons, but again excellent training in man-management (by which I mean finding out how difficult it is, not necessarily becoming good at it).

One of my dubious achievements during my year in charge of the bar was to introduce barrels of strong scrumpy, which was ludicrously cheap compared to beer, and led to cases of students being sick in inappropriate places, a crime which enraged the Warden, a normally amiable man called Donald Coult, who held me responsible (it being difficult to identify the actual perpetrator from the evidence alone). I responded by jacking up the price of scrumpy to the point where consumption fell (and the bar made a massive profit), but this so enraged the clientele that in the end we had to quietly drop the stuff from the tariff altogether.

With the post of Bar Secretary came a position on the governing body of the Hall, and I also took on the editorship of the *Rathbone*

Views of the observatory at Jodrell Bank in Cheshire, where I spent the summer vacation of 1964. Two of them are taken from high up in the structure of the 'Mk. 1' radio telescope, whose big dish can be seen in the picture at top right, behind the hut where I worked. The smaller but more modern Mk. 2 is bottom left and a small dish, used for satellite communications, is in three of the shots.

Record, the Hall's student magazine. Having been co-editor of the school newsletter at the Duke's School, I had found that sort of thing rather suited me. Glancing at a faded copy of the *Record* now, I see on the back page a report on a football match in which I played for Rathbone against a team of medics. The reporter describes me as having 'recently been turned down by Liverpool and Everton, as well as every other football club to which he applied', and turning in a performance 'reminiscent of a three-legged horse'. Still no good at sports, then.

Journalism really was my thing, however. I joined the staff of the *Panto Bulletin*, the university-wide student newspaper that focussed on the year-long build-up to what in universities elsewhere was known as Rag Week. In the climax to Panto Week, students haul huge floats through the streets of Liverpool and compete for the prize of free attendance at the grand ball that rounds out the week. Rathbone had won the previous year with a giant beetle, carrying look-alikes of the Fab Four on top and playing their music loudly. Wearing black tights borrowed from the girls' physical fitness academy at Bark Hill, I played the part of two of the insect's legs, supporting the incredibly heavy structure based on scaffolding borrowed from a building site.

In 1966 we won again, hauling a giant cannon inspired by the film *The Pride and the Passion,* set in the Napoleonic wars and starring Gary Cooper and Sophia Loren, which had been very popular a few years earlier. As we trudged down Ranelagh Street I was glad to be in the open air this time and able to see around me; also not to be deafened by huge speakers repeatedly blasting out *She Loves You* and *I Feel Fine* a few inches behind my head. Back at the *Bulletin* office, we commissioned a fund-raising extended play record featuring the Roadrunners, a popular local group once favourably compared to the Rolling Stones by no less a critic than George Harrison. The star track was their own upbeat, contemporary version of the folk song *The Leaving of Liverpool.*

Getting the work–play balance right was a bit of a challenge. There were many excellent parties, most famously at no. 9 Mossley Hill Drive, on the edge of Sefton Park. Local groups, aspiring to fame (and later making it, in a few cases) would play at your party all evening for £10 and some free beer. The Student's Union hosted many of the top groups from outside Liverpool; I remember dancing to Gene Vincent and the Bluecaps, Eric Burdon and the Animals, and even Screaming Lord Sutch. Sutch is mainly remembered these days for founding the Monster Raving Loony political party, which still contests most major elections in the UK, but he did have a chart success in the 60s with 'Jack the Ripper', despite (or perhaps because of) the fact that it was banned by the BBC, which used to regard itself as the keeper of the nation's morals. On stage, Sutch would perform a long version of his big hit with great gusto, including a lurid pantomime of the original Ripper's criminal acts.

Student Panto week at Liverpool, 1966. I am at the front of the phalanx on the right in sackcloth and a burned-cork moustache towing the prize-winning Rathbone Hall float, 'The Pride and the Passion'.

At the clubs in the centre of town you paid a small fee for membership but then got to dance to groups that were just hitting the national charts, like the Searchers. Most of them had rotating multi-faceted crystal balls on the ceiling that reflected shafts of coloured light, and ultraviolet lamps that made your shirt, and your dandruff, glow brightly in the gloom. Men still wore white shirts, jackets, and ties to go out then, and the girls nice dresses, even to grimy, sweaty, airless dives like the Cavern. This was the real Cavern—the one they take you to now is a replica, recreated where the original lies underground, filled in by bulldozers. Just as well—I'm sure the new one is more sanitary.

Looking for a Doctorate

During the build-up to our final exams, a few of us were called, one at a time, into the office of Professor Leslie Green, high in the tower of the Physics building. When my turn came he told me he thought I would do sufficiently well in the final exams to qualify to do a PhD, and would I like to think about nuclear physics research at Liverpool and the nearby government research facility at Daresbury? This vote of confidence gave me a nice lift at a time of great stress, but I had my eye on space research and the sub-atomic stuff they did at Daresbury (where

Professor Green would later become Director) did not have the same appeal. I had read about the Space Science Laboratory at the University of London, and I was targeting that.

But one day, as I left the building after a session in the practical laboratory and passed through the ground-floor lobby, now empty and echoing as it was late afternoon, I stopped to look at the student notice board. It was covered with eye-catching posters offering job opportunities to freshly minted graduates, and I almost missed the small postcard on which a few words were typed: Doctoral opportunities to work on a new Earth satellite experiment: contact Dr John Houghton, Reader in Atmospheric Physics at the Clarendon Laboratory, Oxford. So now there was space research at Oxford as well!

I wrote to Oxford and was invited to come and learn more at an interview. I motored down from Liverpool in my elderly and unreliable Austin A30. Its nickname, *Thunderball*, was discreetly stencilled on the front wings, my old air pistol was concealed under the dashboard, on which a row of switches labelled ejector seat, overdrive, and supercharger, mimicked Q's gadgets but were not actually connected to anything. The mechanic at the filling station on Smithdown Road, where I had earlier taken it for the compulsory MoT roadworthiness test, impishly told me it had failed. 'Why', I asked miserably, contemplating a swingeing repair bill. 'Some of the accessories don't work', he said.

Yes, this was the era when every young man wanted to be James Bond and I lusted after, but could not afford, an Aston Martin. *Thunderball* had hung on to life and gave good service around Liverpool, but the gearbox finally gave up the ghost on the way back from Oxford, somewhere around Wolverhampton, at 10pm on a cold March night. I flagged down a police car (the police were friendlier in those days, and even managed to be cordial about the fact that my road tax disc had expired) which led to a mechanic arriving and rigging up a temporary fix that got me home in the small hours.

This longer encounter, and the discovery that my dream subject was available, led me to completely revise the view of Oxford that I had formed on my brief schoolboy visit as part of the Schools Science Fortnight. I found the Atmospheric Physics department tucked away at the back of the main physics laboratory, the Clarendon. It was small, cramped, and intimate, but since the arrival of John Houghton as its leader it was already on the way to putting its first instrument into space on an American satellite. Previously, the group had mainly worked on observing stratospheric ozone from the ground, and stratospheric water vapour from aircraft, in both cases by making spectroscopic measurements of the atmospheric path between the instrument and the Sun. John had worked out new ways of making similar measurements from space, so that the whole planet could be observed all of the time, rather than just at a few isolated locations, and had successfully sold NASA on the idea.

The town was delightful and the university humming with life and, although it still definitely had one foot in the past, the other was now seen to be just as clearly in the future. I stayed in a guest room in Jesus College and was taken to dinner by one of Dr Houghton's current doctoral students, David Pick. 'You can order pints of beer with dinner' I wrote home excitedly, 'and they bring you breakfast in bed.' At least this was the promise, but in the event the 'scout' who would have brought it woke me up and asked if I would mind going to the hall with the students, because 'the old gentleman in the other guest room had passed away in the night, sir'.

The two guest bedrooms shared a common sitting room, through which we both had to pass to get to the staircase. My neighbour was an elderly, distinguished judge and colonial administrator, Sir John Blake-Reed OBE. He had served mainly in Egypt and became a Grand Officer of the Order of the Nile, an honour he shared with Queen Elizabeth II, Haile Selassie, and Nelson Mandela, among others. Educated at Jesus and now an honorary fellow, in addition to his legal work he published his own translation of the odes of Horace, 'capturing, in no small measure, the spirit of the original . . . where so many have failed', according to a contemporary review published in Cambridge in 1948. He was asleep (as I thought) in one of the armchairs when I came back from dinner. I tiptoed past and went to bed. In the morning, as I headed to the hall for breakfast, I passed the chair and its occupant covered by a large white sheet. Curiously, this bizarre episode did not really seem out of place among the ancient stones and halls of Oxford.

After the interview Dr Houghton said he would accept me as a doctoral student. However, I still had to gain formal admission to Oxford, and a grant to pay my fees and living expenses. This would require a good grade in my final exams at Liverpool, which were still to come. Back on Merseyside, I worked myself into the ground, getting up at 5am to study and forgoing the bar where my friends would gather in the evenings. Minutes before the barman called last orders, I would arrive just in time to gulp one pint and order another before they closed at 10.30, as required by law then. Nowadays, the student bar at the former Oxford Polytechnic, now Brookes University, which is near my home, closes at 2am.

I still had to show up at Rathbone bar, being in charge, but it only opened two evenings a week, and I had an army of helpers, some of whom were actually competent. Most evenings we went to Greenbank House, a redundant stately home in whose grounds Rathbone Hall had been built and which was now a recreation centre for students, with a lively bar that was open seven nights. Greenbank had been the home of the Rathbone family, a prominent local dynasty whose wealth derived mainly from trading with America from the port of Liverpool, from 1788 until 1944 when the house was donated to the University. Today, it stands derelict.

I had to have a back-up university to go to in the event, not unlikely I thought, that I did not make it in to Oxford. Another possibility was a job at the Space Division of the British Aircraft Corporation at Stevenage. BAC was an amalgam of several famous old aircraft companies like Vickers, Bristol, and English Electric and had several plants, but Stevenage was the centre for missiles and space. Their projects then included the British satellites *Prospero* and *Ariel*, both of which were successful and promised to be the vanguards of a brave new world of Britain in space (the politicians had other ideas, but of course I didn't know that then). I wrote to my parents: 'I visited their Physics Research and Project Planning Departments, and both offered me a job and seemed pretty keen, it's nice to be wanted!'

The physics final exams were incredibly difficult. I still have the papers, and after long experience with the more recent equivalents at Oxford I still think the papers set at Liverpool in 1966 were demanding to a point that was far beyond the pale. Probably a tired and grumpy lecturer with a headache had drawn them up late at night, without the sort of detailed cross-checking that goes on nowadays. Anyway, in spite of all my swotting, I found it impossible to answer all but a few of the questions properly, and after a detailed post-mortem calculation I decided I could not have scored more than 45% overall, which would mean a third-class degree, and bye-bye Oxford.

When, after an excruciating fortnight of waiting, I went in to the admin building to see the results being posted, I couldn't get to the notice board at first because of the scrum of anxious candidates that surrounded it. One of my class mates, whose real name I forget but who was known to us as Goldfinger, after a passing resemblance to the character in the James Bond film then doing the rounds, was just forcing his way out and he told me with a smile that I had a First. I didn't believe him, but there it was; but how? I worked out later, after discussions with my friends on the course, that nobody had done well on the physics questions. The examiners could not fail everybody, so they must have had to scale up the marks to fit a respectable range, whereupon my estimated 45% had become over 70%.

Accepted now by Oxford, I had to tell the other organizations where I had applied that I was no longer available. These included, slightly bizarrely, the University of Wyoming in Laramie, which had a programme of upper-atmosphere exploration using rockets that had attracted my attention. So had the pictures of wide open spaces, the bucking bronco on the letterhead, and the information that it was OK to wear cowboy boots in classes. They offered me a place to study for a PhD and a teaching assistant job worth $2,502 a year, a large sum by UK standards then and enough to pay my way through the course, even allowing for the cost of transatlantic travel. I sometimes wonder where I would have ended up had I gone ahead with this, as I was tempted to do.

Before I could tell the aerospace people at Stevenage that I was going to do a doctorate first and think about a job later, I came across an advert in the Sunday Times for something called the British Aircraft Corporation Open Post-Graduate Scholarship. This would support a student reading for a doctorate in an approved subject at the princely rate of £600 a year, which doesn't sound like much now but at the time was fully twice the standard government grant that I would otherwise have had. I asked my contacts at Stevenage whether they would support an application for the scholarship, and the reply was that they didn't know what I was talking about, but if I could supply the details they would see what they could do. The advert said to apply to the BAC Headquarters on Pall Mall in London, which I duly did, followed by a trip there for an interview that seemed to go well. I was the only candidate, and in due course a letter on thick embossed paper informed me that I had won the scholarship.

Thus enriched, and with a little help from Mum and Dad, I retired *Thunderball* and bought a sports car, a 1958 Triumph TR3A convertible resplendent in 'signal red'. It cost £295, but would be worth a hundred times that now had it still been in the condition it was when I bought it. In fact, although it drove like a tornado it soon started to rust badly, drank petrol from day one, and needed frequent expensive repairs, some but not all of which I did myself, illicitly using the Clarendon workshop to make the less-critical parts. Still, it was fun, especially in the summer when it wasn't cold, wet, and draughty, and the TR saw me handsomely through my first two years as a graduate student. Girls liked it, and my future wife won my heart by (inter alia) finding it hilarious when the wiring behind the dashboard burst into flames while we were driving up the Woodstock Road one afternoon. On another occasion, a radiator hose burst when I was hammering along a straight road near Wooler in Northumberland in the winter at high speed with the top up. The interior instantly filled with steam, completely blinding me. By a miracle I stayed on the road while standing on the brakes and survived unscathed.

A glimpse of the New World

With exams behind me at Liverpool by the end of June, and my arrival at Oxford not due until 1 October 1966, I had a gloriously free summer to prepare for the transition from undergraduate rote learning to post-graduate research. First-year graduate students often go to a summer school that teaches an introduction to one's research field, and my new department pointed me towards one in Space Science at Columbia University in New York that sounded wonderful. Not only was it in America, with its exciting and expanding space programme, but also it came with generous financial support. As I wrote to my parents: '. . . six weeks, all expenses paid, complete with a tour of all the major space research departments in the nation and a pocket money allowance of £20 a week!'. Again I was lucky to get not only the scholarship from

Columbia, but also a travel grant from the European Space Research Organisation that would pay for the flight from London to New York.

The course in New York was led by the distinguished astrophysicist Robert Jastrow. As well as being a professor at Columbia, Dr Jastrow was the founding director of the Goddard Institute for Space Studies near the campus. It was located on the great thoroughfare called Broadway, the same street that is home to the famous theatres, but since Broadway runs the entire length of Manhattan Island, the Institute was several miles from the footlights. Its labs, libraries, and lecture theatres were actually housed above a coffee shop on the upper West Side, near the predominantly black district of Harlem. We slept and ate in John Jay Hall, a student residence on the campus of the university. It had no air conditioning, although the Institute did, and the temperature outside was over a 100 °F during the day and not much less at night. I had never known such heat; in Northumberland people fainted if it got over 75 °F there, which it rarely did.

That summer, it was actually cooler in Florida than New York, which suited us as we were due to travel down there on a field trip. We had to go by train, stopping over in Birmingham, Alabama, as there was an airline strike at the time. This also meant that the planned tour of space centres all over the nation had to be restricted to those on the East coast. Still, it was wonderful. We saw the *Apollo* modules that would soon

Visiting NASA's launch complex at Cape Canaveral, Florida while at summer school at Columbia University in New York in 1966. The launch pad with *Lunar Orbiter 1* ready to go on an *Atlas-Agena* rocket is in the background.

be landing on the Moon under construction at Grumman Aerospace on Long Island, and we went to Cape Canaveral and witnessed the launch of the first *Lunar Orbiter* on 6 August. There were to be five of these missions altogether, designed to obtain close-up photographs of potential landing sites for the upcoming manned *Apollo* missions, which they all did successfully. Ours was the first one, and the launch was perfect. Less than 14 years later, I would be back to watch an equally flawless lift-off from the same place, this time going to Venus and carrying a payload I had designed and built myself.

2
Satellites and Spires

Gentlemen coming into residence should inform the Domestic Bursar if they may wish to bring a piano.

Letter from Jesus College,
Oxford, to the Author, June 1966

In fact, as I drove into Oxford in late September 1966, I had already learned that I was not to reside in College, with or without a piano. I was a member of what was then a rather obscure breed, the graduate student. Today, there are more graduate than undergraduate students at Oxford, but in those days men—they were nearly all men—reading for non-medical doctorates were quite rare. The colleges, of course, were all single-sex, had been for 400 years in the case of Jesus, and were to remain so for another ten years.

Rooms in College were reserved for first-year undergraduates, leaving home for the first time, so I was to find my own accommodation like the adult that I suppose, at the age of 22, I now was. I had been looking forward to continuing to be spoiled by the kind of service I had been enjoying in Liverpool in Rathbone Hall, but for me those days were over. I didn't know anybody in Oxford when I arrived, except a Welshman called Keith Webber, who had also been on the summer course in New York from which I had recently returned. We two had been the only foreign students on the course at Columbia University, and now Keith was the only other new doctoral student in Atmospheric Physics that year. Since he had been an undergraduate in Oxford he was already ensconced in digs that he liked, with a favourite landlady, and he didn't want to move.

So I took to the streets alone with a copy of the *Oxford Times* to find somewhere to live. After several disappointments—good places went fast, as the town was small and the demand high—I found a big bedsit in a Victorian house in the North Oxford suburb of Summertown. It was on the first floor, above an old-established jewellers' shop (now a Chinese restaurant), with a big bay window that looked out over the busy Banbury Road. The flat belonged to the owner of a chain of appallingly bad restaurants (Windsor Fish and Chips) that had sprung up all over

Oxford, and would later vanish just as quickly. My building had three flats managed by a couple who lived at the top of the building, an amiable Irishman called Paddy and his permanently angry wife, with whom I rubbed along quite well for a while.

This arrangement lasted only a short time before I moved into a larger flat, still in North Oxford but much nearer the centre of town, with Keith, finally prized away from his landlady, and another graduate student, Peter Norton, who was doing research in nuclear physics. Leckford Road had a long row of tall Victorian terraced houses that were nearly all divided into flats, many of them occupied by older students. Living right next door to us, although I didn't know him at the time, was Bill Clinton, the future US President, who was then a Rhodes scholar studying politics at University College.

I found Jesus College to be a friendly place and not at all snobbish. For the most part Oxford wasn't then, and it certainly isn't now. Public school ethos hung on mainly in things like the food, which was institutional rather than outright bad, featuring things like Brown Soup (officially Brown Windsor soup, but it didn't deserve the classy sobriquet) that was frequently served as a starter and then would reappear in a jug as gravy to go with the meat course. The latter, we asserted, was synthetic, and tasted pretty much the same no matter which animal the menu identified as its source. But it was much better than the fish that was served every Friday, a white unflaky mass with a most peculiar ammoniacal taste, known as the piece of cod that passeth all understanding. Another public school touch was the mass washing/bathing/toilet/laundry facility, which was all collected in a large tiled underground room of Victorian vintage known as the Palace.

The admission of women undergraduates, pioneered by Jesus and a couple of other colleges in 1975, was to change everything for the better. Nowadays the food is excellent and the bedrooms en-suite; the Palace is unrecognizable as the new student bar where today's junior members sip their subsidized wine and beer at tables located where a row of what were inelegantly called 'crappers' once stood. (Apparently the word actually appeared on them in former times; Thomas Crapper and Co. were manufacturers of sanitary fittings in the 1800s.)

My housemate Peter was on the Jesus College team for *University Challenge*, and when the time came to record the episode in which he was due to appear I went along to help provide audience support. We all piled into a coach and headed for Manchester, where the show was produced in the studios of ITV, with Bamber Gascoigne in the chair. Peter answered just one question successfully (What do the initials BALPA stand for? Answer: British Airlines Pilots' Association) and our team lost badly. During a break in the filming, I wandered off behind the set and was surprised to find myself on Coronation Street, complete with The Rover's Return; nowadays I think they film the outside scenes in a real street.

Keith, Peter, and I later moved together to the countryside to share a cottage in the village of Hardwick, near Whitney. This was ten miles from Oxford but we all had cars, and had the use of an open-fronted barn just across the road as a garage. The River Windrush meandered through the large (and, while we were there at least, unkempt) garden of our cottage. With control of our own schedules (graduate students did not attend lectures, so except for a few seminars and supervisor meetings our time was largely unstructured), the arrangement worked well. In those days the roads were usually uncrowded, except in the city centre, and free parking was available almost everywhere, even in central Oxford.

In my Triumph TR3 I would thunder down the A40, nowadays a perpetual traffic jam, at anything up to 80 miles an hour, and park by the front door of the Clarendon Laboratory. I would also drive the short distance into College for meals, and drive to the nearby River Isis (the Thames in disguise) for rowing practice, in every case without serious parking problems. I rowed in the Jesus IV boat. The 'IV' designation meant that there were three other boats that were better than we were, and our existence was recreational rather than competitive. In the first racing season at Easter, known as Torpids, our 'eight' was placed 106th, out of about 120, in the competition to become Head of the River.

Parties and Pals

Social life as a student in Oxford was dauntingly different from that at Liverpool, and at first I found it formal and stiff. Living alone had been a shock after the easy conviviality of Rathbone Hall, and whereas in the home of Merseybeat work had seemed a necessary distraction from everything else, at Oxford my contemporaries seemed to think about little else. A friend from Liverpool days, Jane Smith, who had come up at the same time as me to study at St Anne's (women's) college, threw me a lifeline by introducing me to a congenial group of hedonistic undergraduates who believed in having fun. Among them was Ken Beer, who like me was a keen Elvis fan at a time when the King's star had faded a bit. Ken and I spent many a happy afternoon sitting in a deserted cinema watching the matinee showing of movies with titles like *Speedway* and *The Trouble with Girls*, Ken wearing his silver lamé jacket for the occasion.

A larger group of us were regulars at *Heritage*, the student folk song society which met in the upstairs room of a pub in Jericho. It was more about serious traditional music than the current Beatles-dominated pop music scene, but had its hilarious moments. I once sat at the feet of a Morris dancer at one very crowded meeting while he explained that the kind of dance we were used to, in which two long lines of men in traditional dress weaved around each other clashing sticks, was only one

Rowing for Jesus IV on the Isis in 1969. It looks cool, but we rowed for fun and did not win prizes, or acclaim, or anything.

form of the traditional art. A version that was once popular, but which was almost lost in time, was represented by the solo Morris dancer such as he. To demonstrate, he fastened bells around his shins and extracted two filthy handkerchiefs from his pocket, which he proceeded to wave around in the air as he pranced and scissor-kicked his way through an ancient jig, without music, while I and the others near him tried to protect our beer from the germs and the kicks.

In search of trendy culture of a different kind, a small gang of us would put on our gowns—students attending lectures in the Humanities still wore gowns, although they had by then been done away with for lectures in the Sciences—and go weekly to lectures on Nordic mythology given by Christopher Tolkien. He was following in the footsteps of his father, the great J.R.R. Tolkien, whose *Lord of the Rings* trilogy had become a kind of secular bible. Tolkien *fils* was a much better lecturer than his father, who John Carey described as 'inaudible and incomprehensible'. We came away refreshed, and envious of the other students at the lecture who presumably could indulge in studying this intriguing subject full time.

One day, I was part of the same gang when we piled in to a packed room in the Oxford Union to listen to Robert Kennedy, whom we expected soon to be a candidate to follow his murdered brother into the White House, little realizing that soon he, too, would meet an early death at the hands of an assassin. Later, we were in the same room to hear Edward Heath who was soon to become Prime Minister of the UK. The students grilled both politicians as was the custom, with Kennedy defending the Vietnam War, saying 'I think those people [the South Vietnamese] should be able to vote for their own destiny, don't you?' and Heath illustrating his conservative principles with remarks like 'You can't have freedom and equality at the same time' (this elicited some boos). The best speaker, bar none, of the hundreds I heard at the Union was Gyles Brandreth, just a student then but who later became an MP and then a TV presenter.

The Schoolmaster

For some of the time that I was a doctoral student, I also worked as a teacher in a boys' public school, at Bloxham near Banbury. Although a private institution and not a state-funded grammar school like the Duke's School, which I had attended in Alnwick, Bloxham was superficially quite similar, with grand nineteenth-century buildings, boys wearing uniform, and vast playing fields. They had lost their sixth-form physics teacher unexpectedly and needed a temporary replacement, for which they advertised on the noticeboard in the Physics Department at Oxford. I was completely unqualified and inexperienced, but they did not mind, and I needed some extra money, as I had recently had to have some very expensive work done on my sports car. At first I travelled up to Bloxham twice a week, until the headmaster agreed to move my lessons to an afternoon and the following morning, and provided me with a nice suite of rooms to sleep in, along with dining privileges on high table. That way I only had to make the 20-mile journey once a week. The sixth form had only a dozen boys in it, and although they were not particularly keen on their studies they were invariably quiet and polite. In the evenings I would do one-on-one tutorials with the two or three who were trying for Oxbridge entrance, and with one who was struggling to qualify for army officer training at Sandhurst.

I was sorry when my year at Bloxham came to an end and the new full-time teacher arrived to take over, leaving me without a much-valued source of additional income. I consulted the notice board in the Clarendon to see if there was another such position and indeed there was, at Wallingford, south of Oxford. This time it was a grammar school, and a good one, but the contrast with Bloxham was striking. No comfortable bedroom, good dinner, or timetable adapted to my convenience. Instead I had to battle across country from Hardwick on cold dark winter mornings and get to the school by nine in the morning when the first classes started. I was driving the Noddy car—see later—by then, and was sometimes late. Secondly, the classes were much larger, around 35 boys, and included some rowdy elements on the lookout for any sign of weakness in the young and inexperienced master in charge. Thirdly, I rather foolishly took on not only the physics classes but also first year biology teaching when the real biology teacher unexpectedly left. I thought looking after 12 year olds just starting out in secondary education would be easy, but of course it isn't. In desperation I took them for nature walks along the Thames; no one seemed to mind and I don't think I ever lost anybody, at least not permanently.

The solution to keeping order in the physics classes turned out to be frequent forays away from the exam syllabus and into the realms of modern physics. When I discussed things like the nature of the universe, the boys were fascinated and asked questions intensely, with rapt attention to the answers. I had to ration these digressions or they would never have obtained the certificates they needed for their future careers and for university entrance. I like to think that I stimulated a few of them to go into the physical sciences despite probably not imparting the basics as well as a professional teacher would have done, and of course amateur teachers such as I was then have long since been outlawed. I found it stressful and did not stay the full year—the Headmaster was kind enough to say my contribution had been invaluable and he didn't know what they were going to do without me. I found a much easier job, marking exam papers for the Oxford and Cambridge schools examination board. These were for the 'A' level General Certificate of Education taken by sixth formers in schools all over the country to qualify for universities or the professions. The pay was good and I could work at home at my own speed, no more early starts or juvenile delinquents to deal with.

The members of Oxford University Atmospheric Physics Department in 1966, shortly after I arrived. Dr John Houghton, the Head of the group and my supervisor, is second from left in the front row. The Dusty Springfield look-alike in the middle between Clive Rodgers and Desmond Walshaw is Barbara, the computing assistant. I am at the right-hand end of the second row, with Keith Webber to the left and Guy Peskett behind. The workshop technicians who helped me to build my apparatus, Geoff Marshall and Martin Clarke, are third and second, respectively, from left at the back.

Natural Philosophy

There was no Physics Department as such in 1960s Oxford, just a collection of large and small quasi-autonomous groups which specialized in various segments of the physical sciences under the banner and letterhead of the Clarendon Laboratory. In the Clarendon, the older dons still addressed each other, and me, by our surnames, and apart from a few secretaries, the place contained only men. There was a single administrator, a man called Dr Croft, who was blind and went around on his wife's arm when not riding in the laboratory's venerable Bentley motor car, the ostensible purpose of which was to collect visitors from the railway station.

The main business of the Clarendon Laboratory was the study of the properties of matter, especially at very low temperatures, and some pioneering work on lasers. Meteorology, recently renamed Atmospheric Physics, was a small enclave on the top floor of the building. Most of the ground floor of the wing, where I worked in a hut on the roof, was occupied by a huge, old-fashioned electrical generator, once used to power the tram system in Manchester, but later adapted to produce the world's strongest magnetic fields which were used to probe the inner workings of crystals and semiconductors.

The Atmospheric Physics Department had its origins in the work of Frederick Lindemann, who had come from Germany to Oxford in 1919. Here he did his famous work on the upper atmosphere, leading to the discovery of the stratospheric ozone layer. As Lord Cherwell, Lindemann went on to great things in the Second World War, working with Winston Churchill as Chief Scientific Advisor, while his assistant, Gordon Dobson, carried on the ozone work at Oxford. In the 1950s, Dobson and his colleagues identified the ozone hole over the Antarctic that was later to become so notorious when it was shown to be intensifying dramatically as a result of certain types of air pollution. When Dobson retired he was succeeded by Alan Brewer, now remembered mainly for his work on the circulation of the upper atmosphere and the reasons for the extreme dryness of the stratosphere. Brewer moved to Canada in 1961, and was replaced as head of the group by John T. Houghton, who had studied the stratosphere from aircraft as a student with Brewer. John Houghton was keen to use the new platform offered by the arrival of Earth satellites to advance this work, and he was to be my research supervisor.

JTH, as we usually called him, had already developed the concept for an experiment to measure temperatures in the atmosphere from space before I arrived. He had formed a team to design a suitable instrument and written a proposal to NASA to put it on one of their new *Nimbus* series of experimental weather satellites. The device was an infrared radiometer, an optical instrument similar to a camera but working at much longer wavelengths, capable of measuring the temperature profile in the stratosphere from orbit by monitoring the intensity of the heat radiation emitted to space. If it worked it would make a key contribution to gathering global data for weather forecasting, one of the main goals of the *Nimbus* programme.

Shortly after I arrived in Oxford, I went to see John in his tiny office in the Clarendon to talk about my research plans, and found him chuckling as he scrolled through a telegram more than ten feet long that had just come off the teleprinter. This was from the American space agency confirming that his experiment had been selected for flight on *Nimbus-D*, the fourth in the series, scheduled for launch in about three years' time.

The Weather from Space

With Oxford now destined to become one of the first university groups in space, JTH was already thinking about the future beyond *Nimbus-D*. He had an idea for a larger and more complicated radiometer that would measure upper atmosphere temperature profiles more accurately and over a greater vertical range, up to as high as 70 kilometres above the ground. With this it would be possible to investigate the region of the atmosphere that lies above the stratosphere, known as the mesosphere. At the time this was *terra incognita*, despite being home to the ionosphere, a layer of charged particles that reflects radio signals, making worldwide communication possible. The erratic behaviour of the ionosphere showed that there was clearly dramatic and interesting 'weather' up there, but until now the region had been inaccessible except to a few high-altitude rocket payloads.

John suggested that I adopt his new idea for a satellite instrument, come up with a design, and build a prototype to show that it worked. This would be my doctoral project, and would also provide a sound basis for further flight proposals. Ideally, the prototype instrument would be flown on a high-altitude balloon to show it could perform as predicted in the real atmosphere. No balloon could ascend anywhere near as high as the mesosphere, but a test lower down would still verify the technique and show NASA that we meant business.

Research degrees at Oxford eschewed 'spoon feeding' and worked instead on the 'jump in the deep end and sink or swim' principle. You were given a bench to work at and an idea to get you started, there were libraries and colleagues around to consult, and you got on with it. Once I had read enough to understand the physics involved, I began to get a grasp on the various unsolved problems I would have to tackle to make a working 'pressure modulator radiometer'. This name for my new enterprise came about because the basic idea was to look down at the atmosphere from the satellite through a cell containing carbon dioxide gas. The pressure of the gas in the cell was to be cycled rapidly, about 30 times a second. The gas in the cell would absorb the infrared

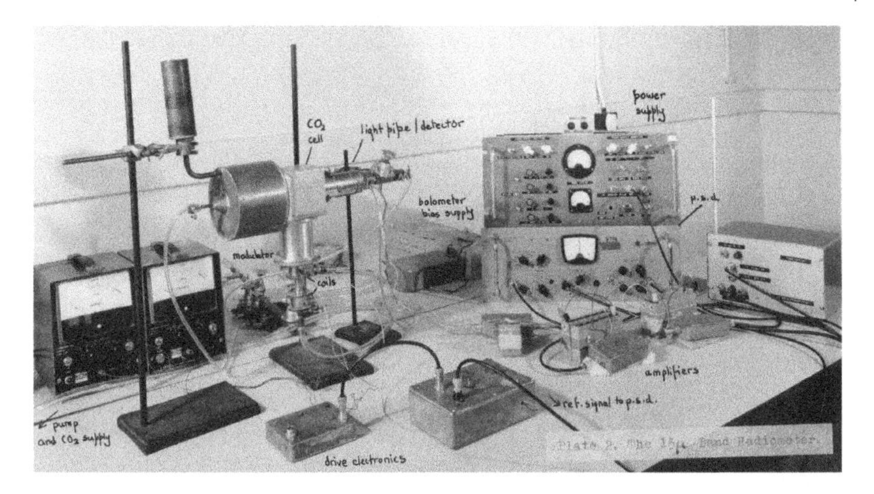

A photograph from my doctoral thesis of the 'breadboard' prototype Pressure Modulator Radiometer on the bench in the Clarendon Laboratory. Most innovative space instruments start life as individual sub-systems cobbled together like this so that the more obvious bugs can be worked out and eliminated before the compact packaging required for a flight model is attempted.

radiation being emitted from the carbon dioxide in the atmosphere, and the absorption would get weaker, then stronger, 30 times a second, modulating the radiation reaching the detector in the instrument. If the signal from the detector was read out at the frequency of modulation, the result would be a measurement that was related to the atmospheric emission, and therefore its temperature, in a known way, and the temperature profile in the atmosphere could be deduced. Simple.

The Beryllium Spider

However, it was not clear how we could modulate the pressure in the cell. The more obvious ways, such as using a piston attached to a crankshaft as in a car engine, would use too much power and create vibrations, both unacceptable in a device destined to go on a satellite. JTH's idea was to use sound waves, in a cell designed as a sort of resonant organ pipe, which was an elegant approach with few moving parts. However, when I tried it the amount of pressure modulation that resulted from the sound wave was far too small to produce the spectroscopic response we wanted. I tried making one boosted with a small bass loudspeaker (a 'woofer' from a commercial hi-fi), but the resulting device was heavy and still did not produce enough pressure variation unless a lot of power was used. Power and mass were both in short supply on those early spacecraft.

I came up with a smart idea. I made a cell with a tube coming out of the top, with a steel ball-bearing suspended in the tube by a permanent magnet. A small coil made the ball oscillate up and down the tube, using the compression of the gas as the return spring, requiring only a very small application of power to the electromagnet. When I ran it, however, the level of pressure modulation was much smaller than I expected based on a calculation of the geometry. I soon realized that this was because the ball was rolling along one side of the tube, instead of being suspended in the centre as I had naively imagined in my design. The clearance around the ball, instead of being a very fine annulus, was a much leakier crescent shape. Try as I might I could not get the ball to centralize; it was always drawn to one side or the other. Making the tube slightly smaller just resulted in the ball jamming. The problem might have been a lesser one in orbit, in zero-g, but there was no easy way to simulate that in the lab.

One of the senior members of the department, Guy Peskett, finally suggested what proved to be the solution: by replacing the ball with a more conventional piston, with a long skirt parallel to the walls to make a seal using the viscosity of the gas itself rather than a sliding piston ring. The gap had to be small and maintained very precisely for it to work. Guy's idea was to use flat 'spider' springs to locate the piston accurately in the centre of the bore. This worked under all orientations, as required, but only if built to very tight tolerances and carefully assembled. Eventually, with lots of skilled help from Geoff and Martin, the technicians in our small departmental workshop, I had one running.

Next the modulator had to be assembled into a complete radiometer, which presented several further difficulties to be overcome. These were mostly the sorts of problems that do not seem to exist when the device is drawn on paper, but which rear their ugly heads when you build something and turn it on for the first time. The one that kept me awake at night was how to get rid of the effect known as microphonics. This arises because the modulator not only cycled the gas pressure, but also vibrated its mounting and the rest of the instrument at the same frequency. The detector and optical components that focussed the radiation were all very vibration sensitive, and responded by producing an output that was mixed in with the real signal from the atmosphere, but much larger and indistinguishable from it.

The classic way of getting rid of vibration is to add lots of strength and weight to the structure, which was not an option for an instrument going into space. What I needed was a skilfully designed, lightweight structure that suppressed rather than transmitted the vibrations. Nowadays, hugely sophisticated and complicated computer programs do this sort of job: then, I messed about with a home-made aluminium frame, trying various configurations until I got something that more or less worked.

A similar seat of the pants approach produced the electronics to drive the modulator, on a home-made printed circuit board that fitted inside the modulator itself. Being inside the metal casing meant that the detector did not pick up an induced signal from the drive circuit, which, like the vibration-induced effect, would have swamped the real signal

The flight version of my balloon radiometer, fully assembled and integrated but before being enclosed in its insulating and crash-resistant housing. The pressure modulator is the cylindrical object partially exposed near the centre of the picture. The batteries that power the device, and the electronics that process the signal and transmit it to the ground station, are on the left and right, respectively, of the lower part of the package. At the top, a small motor controls the scan mirror which points the view of the instrument downwards towards the atmosphere or sideways to look at the calibration targets, one of which (full of liquid nitrogen) can be seen to the right of centre embedded in styrofoam for insulation.

from the atmosphere. None of these well-known but elusive problems with sensitive apparatus had crossed my mind before I started building the device and discovered them for myself. It was an education, and a painful one.

The decontaminated detector signal had to be amplified and digitized and provided with a transmitter and antenna for relaying the data to the ground station. Here I was lucky because this was a non-specialized job and some old but serviceable circuit boards already existed from another balloon experiment from an earlier era. This was great, except that they bristled with antiquated germanium transistors and, not having built them myself, I struggled to understand the design and had a lot of trouble diagnosing problems and fixing failures. Without Jim Williamson, who had used them on a project of his own and understood everything, I would have been lost at sea several times. More than once they stopped working late at night when I was preparing for a flight the next morning and Jim would come at once and achieve in minutes the repairs that I had failed to do with hours of struggling. I developed a paranoia about the quixotic nature of electronic circuits that my later career would do nothing to dispel.

On the trail of the R-101 airship

With all of the sub-systems working at last, the next step was to package the radiometer for the balloon flight. To simulate observations from a satellite, I had to get data from very high in the stratosphere, which required a very large, very expensive balloon and special launching facilities that were not routinely available in the UK. A firm in America called Kaysam made a special double balloon, one inside the other, that was claimed to reach very high altitudes but was easier to handle than a single, very large balloon. We bought a couple of these and arranged to launch them from Cardington airfield in Bedfordshire, where there was

A meteorological blimp being brought out of the R-101 hangar at Cardington, where I filled the balloon that would take my apparatus up into the stratosphere. The original occupant for which the hangar was built in 1930 was a large as an ocean liner and completely filled the huge shed; now it stands nearly empty.

My radiometer payload is suspended below the blimp for a low level test, prior to its planned ascent into the stratosphere on a high-altitude balloon.

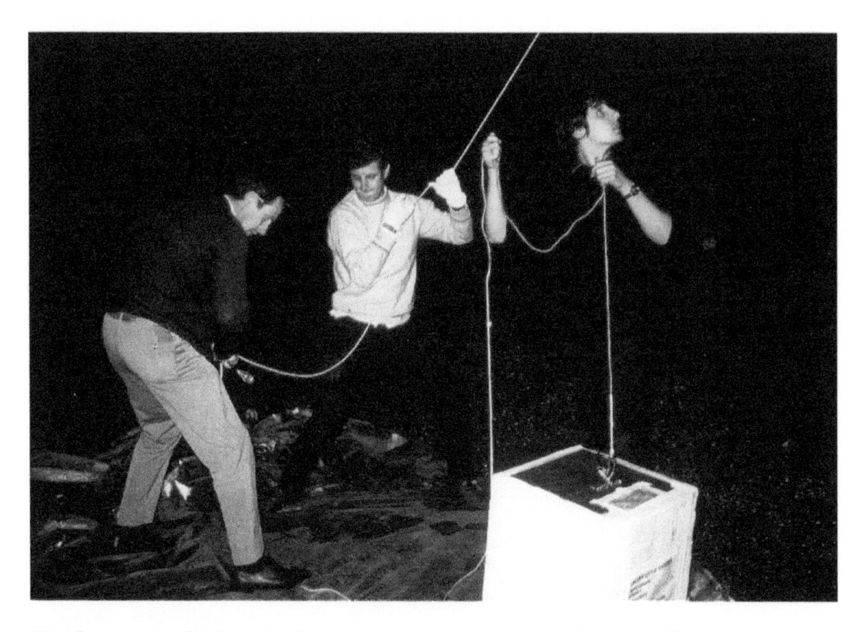

The first successful launch of the instrument took place at the Lark Hill meteorological station on Salisbury Plain at dawn on 7th August 1970. My eyes are raised towards heaven to watch the balloon but, having had several previous launch attempts fail, also in prayer that it might succeed this time.

a meteorological station that routinely launched weather balloons and a nearby radar station at Hemsby on the east coast of Norfolk for tracking the flight. The intrepid team based there was willing to have a shot at launching the Kaysams, which were still much larger than the usual meteorological balloons and required tricky handling.

There was an ample supply of hydrogen to fill our balloon: the filling took place inside the vast, almost empty hangar that once housed the ill-fated R101 airship. This eerie place now contained only a few large tethered balloons or 'blimps' that were used by the nearby meteorological station. I borrowed one of them for a while to hoist my balloon payload aloft by 100 metres or so, for a final field test before it was sent off into the wild blue yonder.

For the ascent into the stratosphere, the balloon was attached to the instrument package by a train consisting of a mesh radar reflector, a lightweight silk parachute, and 100 metres of nylon cord. The balloon was expected to lift about 15 kilograms of payload, of which ten was the instrument, including 3 kilograms of insulation and half a kilogram of liquid nitrogen. Launches were attempted when surface winds of not more than five knots prevailed. The train could then be released hand-over-hand, and a short run made with the package to stop it 'penduluming' into the ground. Tracking by the radar station would facilitate recovery of the package, which had a notice on it offering a reward of 50 pounds, and would also provide time versus height data for the analysis of the results.

Large balloons are best launched at dawn, when conditions are most likely to be calm. This means the hapless experimenter has to drive down the previous day to get set up, and after a few hours' sleep get up in the middle of the night to prepare the payload. This includes filling the reservoir of liquid nitrogen for the cold calibration target at the latest possible time, so it doesn't run out, which means that the insulation, crash protection, and other packaging also have to be finalized at the last minute.

On one visit I ran out of liquid nitrogen, fortunately during the day when there were a few people around. In a nearby hut I found two old men sitting morosely among a mass of rusty old equipment, doing nothing and looking for all the world like they had been left behind and forgotten when the giant airships left for the last time 30 years before. Perhaps they had. They pointed me towards a local supply of liquid air and seemed impressed when I told them what I was trying to do. I said I'd come back and chat to them when I had a success to report, but I was never able to find them again. Possibly they were ghosts; it's that sort of place.

The last thing one does before a launch is to carry out a final 'end-to-end' test to check that everything is working. If it isn't, dismantling the package and tracing the fault usually takes so long that the launch has to be scrubbed for the day. The launch crew, who have also got up early, would usually take this in good spirits but when it happens more times than it should have, one feels awful.

The first launch, on 26 June 1970, was a disaster. Instead of soaring up into the stratosphere, the balloon rose a few hundred feet and then levelled off and vanished over the horizon. The instrument malfunctioned as well and there was no data, not that it would have been very useful anyway so close to the ground. Instead, we were in trouble with air traffic control, because there are many air lanes around Cardington and our balloon had not followed anything like the approved trajectory. We had an inconclusive argument with the balloon manufacturer as to whether the one they sold us was faulty, or whether we had not followed the filling instructions or weighed our payload properly. At least we were able to recover the package and try again.

On about five other occasions the radiometer was painstakingly prepared for flight at Cardington, only to have the launch cancelled because the prevailing wind changed from that forecast to one which would have caused the balloon to enter the air lanes which lie to the south and west of the airfield. On other occasions the predicted trajectory passed outside the range of the Hemsby radar. After several months of delays of this kind it became clear that days on which a launch into the stratosphere could be made without infringing air traffic control or losing the vital tracking data were few indeed. This was a very low point in my fledgling career as a space scientist, but occurred during a high period in life, as I had been newly married since 28 June 1969.

My new wife was Doris Buer, an American in Oxford to read for a doctorate in the Biochemistry Department, almost next door to Physics, and a member of the oldest women's college, Lady Margaret Hall. We met when enterprising souls in both of our colleges decided to team up to provide lunches during the long vacation, when the undergraduates and most of the dons were gone, and the quads and corridors were empty except for we happy few graduate students. Those fellows of Jesus College who remained in Oxford during the depths of the long vacation had reciprocal dining rights with Exeter College, across the road, when Jesus closed for three weeks in the summer, and then returned the favour when Exeter then closed for the same length of time. That would have left us students without meals for a total of six weeks had we not made arrangements of our own. We thought we were pretty clever choosing a girls' college for this, and we were.

For our first date I took Doris to Dirty Dudley's, a pub in the countryside south of Oxford that was popular with students at the time. Its real name was the Lamb and Flag, but Dudley was omnipresent behind the bar, rotund, welcoming and not at all dirty, although if you drove past during the day you would occasionally see a seedy character handing in a brace of unplucked pheasant or duck at the back door. The menu included a lot of local game, and items such as the irreverently nicknamed 'twice poached salmon' were popular. At the end of the evening, you wrote down your own bill, added it up, and paid at the bar.

Some 15 years later, when we had moved back to Oxford from the USA, we enquired about Dudley's, to be told he had died and the pub was now part of a bland chain. However, his son Young Dudley now ran a similar business in a different pub, in Handborough, north of Oxford. We went, to find a slightly younger spitting image of Dudley behind the bar, and almost the same menu. Having eaten and drunk, we were disappointed to be presented with a bill like any other restaurant. 'What happened to making your own bill?' I asked Young Dudley. He studied me for a moment. 'You have been gone a long time, haven't you, sir?' he said.

Doris and I found a place to live in North Oxford, at the junction of Woodstock Road and First Turn to Wolvercote, in a large Victorian house occupied by a retired doctor and his wife, with their daughter and granddaughter. Our domain was the attic, which had been nicely converted into a two-bedroomed flat, with a substantial living room and kitchen. Being right under the roof, it was an odd shape and very hot in the summer and glacial in the winter, but we were pleased with it. We even had our own garage, which we shared with a guinea-pig (known as 'Hammy' because we originally took him to be a hamster) in a cage. While still living at Hardwick, I had had to retire the Truimph TR3A as the repair bills mounted, and had replaced it with a 1953 Ford Popular. This was the old 'sit up and beg' model with a six-volt electrical system, which powered lights so dim that it was very hard to see anything

Outside the Registry Office in St Giles, Oxford, on our wedding day. My mother is on the left edge and my uncle Alan behind me; father took the photo. We went from there to the Trout Inn for lunch with close friends, and then to Doris' college, Lady Margaret Hall, for a more formal reception with family, friends, and colleagues from both our research groups. Because of the large distance between our family homes, we staged a second wedding reception in Minnesota a few weeks later.

Doris by the River Coquet near my family home in Warkworth, Northumberland, 1969.

The Noddy car at Hardwick, where I lived with two other students before getting married. Doris and I painted it ourselves, and it soon became a popular sight around Oxford.

at night in the countryside, and separate, cycle-style wings. I got the Ford for nothing from someone in the Clarendon lab who was moving to the USA. It was very slow and something of an embarrassment after the Triumph, but we made light of that by painting it red with bright yellow wings to match little Noddy's car in the stories by Enid Blyton. It became known all over Oxford, and people would wave to us as we drove around.

Lift-off at Lark Hill

In August 1970 we decided to transfer the entire ballooning operation to Lark Hill, a military base just one mile away from Stonehenge on Salisbury Plain. It also had a weather balloon station, and was about as far from the nearest commercial air lane as you can get in southern England. The staff there cheerfully agreed to help us with another flight attempt. We decided to use an ordinary high-altitude balloon, and accept the fact that the data would not extend as far into the stratosphere as we had originally planned. It would also be nowhere near as high as we would have liked to simulate the planned satellite observations, but that was impossible anyway. I lugged all of my equipment down to Lark Hill and somewhat wearily set up shop again, as I had at Cardington six months earlier. The time lost there meant that I was nearing the end of my fourth year as a graduate student, close to the maximum time normally taken and in danger of the support money running out.

The first flight from Lark Hill was possible within just a few days of setting up there, but it was another failure. The balloon had a leak and after ascending to a height of one kilometre it started to come down again. I was secretly glad, because the signal from the payload vanished shortly after the balloon was released, due to, as it turned out, a failure in the transmitter electronics, and there would not have been any results anyway.

Finally, it all went well on 7 August 1970. The weather was perfect, and the air was very clear as the balloon soared into the rose-tinted dawn sky and climbed rapidly into the stratosphere. At that height the pressure is low and the balloon grows very large and reflective, stretching its skin until it is only a few molecules thick and can stretch no more. We could still see it with the naked eye more than 20 miles up in the air and still rising almost vertically above Lark Hill.

Back in the hut that was my temporary laboratory, the digital signal was coming in strongly and showing the regular patterns that corresponded to the healthy operation of the instrument. The data flowing onto the tape recorder was also displayed in red ink by a pen on a moving chart; I could see in real time the radiometer cycling between looking straight down into the atmosphere and sideways at the warm and cold calibration targets. The 'zero' signal from the cold target, basking in the liquid nitrogen I had painstakingly installed a couple of hours earlier, was drifting around all over the place as the payload swung in the wind and the atmospheric temperature and pressure fell with altitude. But if I hadn't had that data, and that from the corresponding warm target, to carry out a post-flight recalibration, the data from the atmosphere, which was also slewing around, would have been worthless.

After nearly two hours, the balloon had reached an excellent altitude of 37.8 kilometres and was still rising when it burst and the payload started to fall to Earth on its parachute. The radar at Lark Hill tracked it most of the way down, so I knew approximately where it had landed. However, the radar cannot follow the target when it gets close to the surface, and the resulting uncertainty in its final resting place is about half a mile. This may not sound like much, but on the ground it is like looking for a needle in a haystack. I drove around the area for a couple of hours and eventually gave up. A few days later a phone message from the police in Newbury said it had been found in a meadow on a farm about 18 kilometres north-west of the launch site. I drove out there and the farmer showed me the spot where it had landed, marked with a spade stuck into the ground. He pointed out a few shards of glass scattered around on the grass; I explained that these would have come from the glass dewar vessel that had held the liquid nitrogen. It was fragile and not expected to survive the impact, but I reassured him that it was not a problem as I had plenty more of them. 'Arr, but what about moi sheep?' he responded indignantly. I was able to mollify him slightly with a £50 reward and a copy of the third-party insurance policy I had taken out before being allowed to fly.

Aftermath

By the time I had usable data in my hands I was definitely in danger of running out of time and money. A D Phil is supposed officially to take three years, although this is rarely achieved, except possibly by theoreticians working entirely on their own. Building new apparatus and deploying it in the field can easily take four years, and so it was with me. After four years, you have to submit applications to the university for an extension of your status, and, much worse, the grant or scholarship covering your living costs invariably runs out. I had to make the analysis of the data short and snappy, so I argued in my thesis that, since the point of the exercise was to show that the new radiometer actually worked and offered potential for improved temperature profile measurements from satellites, I could do that by presenting the measured and calibrated measurements of infrared flux as a function of height as the balloon went up, and comparing them with the theoretical flux calculated using temperatures obtained from a nearby meteorological balloon ascent.

This was a much easier and quicker task than facing the intricacies of the 'inverse' problem, which would have involved using my measured fluxes to determine the temperature profile, and then comparing that to the radiosonde profile. This approach was fine with my supervisor and, it turned out, with my examiners also. To compute fluxes I relied on grilling Clive Rodgers, the local expert on the subject. I also pored over a textbook called *Atmospheric Radiation* by Richard Goody, a professor at Harvard University in the USA. Little did I know that I would become acquainted with Goody in the years to come, and hear directly from him the story of his own struggles to get a doctoral degree some 20 years earlier at Cambridge University.

Goody worked for his PhD just after the war ended in Europe, and had arrived at Cambridge with some wartime experience of high-flying aircraft. His doctoral project, which involved building an infrared spectrometer and flying it on a Mosquito fighter-bomber to the base of the stratosphere, where it would make measurements of water vapour concentrations, has some parallels with mine at Oxford many years later. In his own words, taken from his memoir published in 2002: 'I shall not go into the painful details. Suffice it to say that I was not provided with an airplane, and no one could tell me how to build a sensitive spectrometer for the unbelievably extreme conditions in flimsy stratospheric warplanes powered by huge reciprocating engines. My research grant was less than $500 per year, mechanical, optical, and electronic design, construction, and installation were my personal problem.' I sympathized.

A few months later, my new results had been written up and the thesis was typed and bound. Once it was officially submitted, and the wheels were turning that would eventually lead to the dreaded oral examination of its contents, known as the viva, I was free at last to take a break. Doris and I headed off in our Riley car towards the West Country. The Riley 1.5 had recently replaced the Noddy car, which was so slow it had taken 13 hours to drive from Oxford to Warkworth, so that Doris could meet my parents, including a stop by the police attracted by its distinctive appearance and erratic progress. The Ford joined my Triumph in the collection of old cars held by my friend Tony, who lived in the disused railway station at Ilderton near Wooler and had a large garage built under the canopy over the former platform.

On our post-thesis celebration (Doris had submitted hers before me, and then drawn all the figures for my thesis, including some complicated electronic diagrams that took days, working with stencils) we were bent on finding good pubs with food. We sought out the Priddy Oggy, a superior version of the Cornish pasty, to be found only in Somerset. At that time this dish was enjoying a surge of fame due to coverage in the Sunday papers of a pub restaurant called the Miners' Arms, where it had been invented by the proprietor who, it turned out, was a retired aerospace engineer. Over a dinner of Mendip snails, the aforementioned oggy, and a pudding called Miner's Delight, accompanied by an excellent (according to my diary) English wine, we discussed the options, continuing for days as we explored Glastonbury Tor, Mousehole, and the countryside all the way down to Land's End.

I was slightly less happy when I was informed by the University that my external examiner for the compulsory viva that followed was to be Professor Percival (known as Peter) Sheppard, a distinguished but elderly and fierce atmospheric physicist from Imperial College in London. The internal examiner was to be Desmond Walshaw, who used to organize our weekly departmental seminar. Desmond was the mildest of men. I only ever remember him getting angry once, when he got a letter from a meteorologist at Reading accepting his invitation to come and speak at one of our departmental seminars, addressed to

'Dr D. Walsheim'. Desmond dived for the phone and rang the hapless individual, addressing him in a shrill mock-German accent: 'Here is Doctor Valsheim. Ist zehr gut zat you accept our invitation. . . .' I wished I had been at the other end to see the man's face.

Not many people fail the viva, once their thesis has been passed for submission by an experienced supervisor, but still the experience can be harrowing. The examiners are expected to explore not just your actual work but your background knowledge, so once they had satisfied themselves that I had built a new kind of instrument and shown that it had the potential to make improved meteorological measurements from satellites, they started to ask me about things like ozone chemistry and the different types of waves that might occur high in the stratosphere.

I had swotted up on stratospheric waves, since that was the sort of thing my new instrument would be capable of observing on a global scale for the first time. As I haltingly described some fairly obscure type of wave, Professor Sheppard interrupted to say he did not think this would be important in the stratosphere, upon which I challenged him, foolishly, since I did not really know what I was talking about, by quoting a paper I had read on the subject. 'That is a bad paper, and the author is a bad scientist!' he responded, leaning slightly towards me across the table with beetled brows. 'Ah', said I, pretending to make a few enlightened notes so I could sort myself out later.

In spite of a few bad moments like this, the viva was a success and I was 'given leave to supplicate' for a doctorate, as the language goes at Oxford. The pressure modulator radiometer would fly in space about five years later on *Nimbus 6,* and it too was a great success. We patented the idea and the design, and after the *Nimbus* version had shown how temperature measurements could be extended upwards using our techniques, the Meteorological Office took it up for their operational weather satellites and paid royalties to the Oxford department. We also went on to send versions of the device to Venus and Mars, as I shall relate in due course.

Finding a job

Back in the present, it was 1970 and my mind was on what my new wife and I would do next, once we were both officially awarded our doctorates in a few months' time. We were 26 years old and it was time for a real job, and some income. But what?

There seemed to be three options. My scholarship from the British Aircraft Corporation led to job opportunities at their Stevenage plant, but they weren't at that point much involved with science; their main business seemed to be guided weapons, and that reportedly was not going well, they were actually laying people off. Technically, it was interesting stuff, that appealed to the engineer in me, but I really wanted to explore space. There was a chance to stay on at Oxford, where the work on *Nimbus 4* had finally led to a launch on 30 April, and data was flowing. Follow-on instruments for *Nimbus E* and *F*, the latter hopefully carrying a version of my pressure modulator contraption, were in the pipeline.

All of this activity generated funding for the department, which meant positions for people with backgrounds like mine (and there weren't many of us around then) to bring them to fruition. But then, out of the blue, I received an invitation to visit the main technical centre of the European Space Research Organisation, ESRO, at Noordwijk, near Amsterdam in the Netherlands, to discuss going to work there. I accepted their invitation to fly me to Holland and look around and talk about salaries. I found the laboratory behind sand dunes on a beautiful stretch of coastline famous for its harbours and sailing boats. It could hardly be nicer. While I was there they offered me a job at a salary that made me gasp.

The Oxford job would not have been permanent—such posts were and still are rare—but rather on post-doctoral contracts for a few years. Nice though that would have been, it was really just putting off the inevitable job search. I had already been in full-time education for 20 years, including seven at university, and I could not help but notice what seemed to be a very enviable existence for senior professors at Oxford. Where most companies find you less useful when you get older, and eventually unceremoniously discard you, in the groves of academia you just become more and more learned and respected. A vision of myself strolling slowly along 'the High' wearing a gown, fat, old, and bald, but content and recognized by all, became embedded in my psyche as a future earnestly to be desired.

But this was now. The ESRO job seemed to fit the bill perfectly. The agency was growing fast and had all sorts of ambitious plans for satellites around the Earth and missions to the planets. Its centre in Holland was made up of a modern complex of laboratories and offices in the middle of a tulip field, next to a sandy bay, facing England just across the sea. The salary, already high, was supplemented by 'expatriation' allowances and paid trips back to your home country, all designed to make working there appeal to a multinational workforce in the days before that became commonplace, and it did.

In 1969, Britain was not yet in the European Economic Community and the European Union we have now was still more than 20 years away. The UK had, however, been a founding member of the European Space Research Organisation in 1964. The European Space Research and Technology Centre, ESTEC, was founded in 1968 and although I didn't take their job in the end I was destined to visit there many times in future years, as part of the story which follows. It is probably ten times larger now than it was when I first saw it, and the tulips are long gone, their fields built over and turned into car parks for over 3000 employees.

In 1967, at the end of my first year as a research student, I had been to another ESRO facility, the European Space Research Institute, in Frascati near Rome. Much smaller than ESTEC, ESRIN occupied an old mansion in a beautiful location surrounded by vineyards with spectacular views over the Italian countryside. In a swelteringly hot meeting room, students from all over Europe listened to lectures on aspects of space research. These were very useful and stimulating, and the whole experience contributed to a warm feeling that Europe was in space in a serious way and ESTEC would be a great place to work.

While on the course in Frascati some of us had been drafted to take notes and write up the lectures for publication. I did this job for John Houghton's lectures on atmospheric physics, learning a lot in the process. It was a problem staying awake in the afternoon lectures, although this was not because they were uninteresting. This being Italy, there were jugs of wine on the table at lunch, and being students we partook thereof with gusto. Together with the effects of the large meal and the heat, this was too much for some of the attendees, but as one of the note takers I had to stay focussed. I discovered that the secret was to sit in the front row, finding this improves the ability to concentrate. Ever since then I have sought the front row to listen to talks and lectures. Only once did this let me down, when I had jet lag from just flying back from the USA and went straight to a small meeting which included an address by the Astronomer Royal for Scotland. My own snoring woke me up to find him staring at me from a few feet away. We both carried on as if nothing had happened.

The social life at the Frascati summer school included nightly trips into the beautiful old town to eat outside and watch the sun go down. After one of these excursions, we returned to ESRIN very late and crept through the corridors as quietly as we could. The route to the bedrooms lay through the cafeteria where we ate lunch each day, and past the food lift where we had noticed the waiter communicating with the kitchen below. One wag in our group generated much silent amusement by a stage whisper into the microphone, 'Vino! Multi vino!'. We were just moving on when the noise of a small motor split the silence of the dark house and the platform rose from below with two carafes of wine on it.

The New World beckons

While I was tidying up at Oxford and waiting for the viva, John Houghton returned from a trip to the USA and announced that a colleague of his, Dr C.B. Farmer, had a place for a young researcher to join his team at the Jet Propulsion Laboratory in California. JPL is NASA's main centre for missions to the planets, and their main project was at the time the *Viking* programme, the goal of which was to land on Mars for the first time. Dr Farmer, known to all as Barney, had started to look for someone to work on the large project planned to follow *Viking*, the *Outer Planets Grand Tour*. He needed someone who had recently graduated, with experience in the design and implementation of infrared space instruments, and I checked most of the boxes. The exception was that I was hardly an expert on the planned destination for these instruments: it wasn't going to be a satellite looking down on the Earth; this was a visit to the huge, cold, distant, and mysterious planet Jupiter and its companions in the depths of the outer Solar System. Measuring temperature profiles over the Antarctic from space still seemed exotic, so what price on doing the same for the atmospheres of Uranus and Neptune?

I was sent pictures of JPL, in which I saw a cluster of tall buildings in a leafy grove nestling in the foothills of the San Gabriel Mountains in the Californian sunshine. Its laboratories and workshops contained the most advanced space hardware in the world, some of it prototypes for imagined journeys still far in the future. Spacecraft from those very labs had already visited the Moon, Mars, and Venus, and now we were going to Jupiter and beyond. Hollywood, the mountains, and the beach, were all a short drive away, and in my mind all of southern California was a dream world. As a teenager I had been a fan of the television series *77 Sunset Strip,* and was delighted when Doris received an offer to work as a biochemist in the labs at the Children's Hospital of Los Angeles, on the eponymous boulevard near where the series was filmed.

My future boss, Barney Farmer, was himself British, starting out in Wales and working for many years for EMI Industries in Hayes, near London, a large business that made high-tech optical equipment, including infrared detectors. In fact, I had met Barney briefly very early in my doctoral project when he visited Oxford to discuss detectors for the *Nimbus* instruments. The job he had in mind for me at JPL was a Research Associate position funded by the National Research Council in Washington. This would require a competitive application, and was only for one year at first, with possible renewal for a maximum of two. Thus it was the least secure of all the opportunities I had considered so far, but I didn't care. I applied for it and got it, and soon we were heading west, into the Californian sunshine.

3
The Gateway to the Planets

I had been in Pasadena only a few days when I was stopped by the police. I was shopping for a car on Colorado Boulevard, which is where most of the automobile dealerships were to be found. Old-fashioned route markers proclaim it to be the westward end of Route 66, as that famous highway makes its way across the United States to the Pacific Coast. In Pasadena, it is the main 'drag' that runs across the centre of the city from one side to the other, and also commemorated in popular song ('She drives real fast and she drives real hard, she's the terror of the Colorado Boulevard. She's the Little Old Lady from Pasadena . . . ').

I had just crossed the road from the Ford showroom to return to my rental car parked on the other side of the street when a formidable uniformed and armed warrior pulled up on an enormous motorcycle, wearing dark aviator sunglasses and a golden helmet that glinted in the sun. This was an officer of the legendary California Highway Patrol. 'C'mere' he said, reaching for his book of citations and starting to write. 'Name?' Whether it was my English accent or the look of sheer bafflement and dismay on my face he paused and said 'Whereyafrom?' I was dumbstruck. Eventually, I managed to say 'What did I do?' He cocked his helmeted head and looked at me. 'You don't know?' I shook my head.

He put the book away and explained to me about jaywalking ('crossing a roadway other than at a signalized intersection, or without waiting for a permissive indication to be displayed.') The idea that it was a crime to cross the road anywhere I pleased astonished me, after years of waltzing through the traffic on the streets of Oxford whenever I thought I could do so and have a reasonable chance of surviving. And here I was in the land of the free being arrested for crossing the road. 'If one of them cars hits you,' said the officer, 'he can sue you for any damage to his vee-hicle if you're not in a crosswalk with a green light displayed.' 'Ah', said I, 'sorry'. The grim visage had relaxed when he heard my strange accent and I was sent on my way with a caution. This was to be the first of my many brushes with the Chippies in the next ten years as I tried to become accustomed to the idea, enshrined in law, that a car capable of over 150 miles per hour is only legal if driven at 65 mph or less, even on wide, straight, empty roads.

Wonderful Land

We arrived in California in January 1970 and, after a spell in the luxurious Huntingdon Hotel, courtesy of JPL, soon found a place to live. It was way up in the foothills near the top of Fair Oaks Avenue in Altadena, the highest part of Pasadena. We rented a cottage, old and shabby but roomy and comfortable, on the property of a Mr Allen. This was like a small farm, with a paddock containing horses and lots of trees in addition to the big house in which the Allens lived. There was a lot of wildlife as well, including, it turned out, a family of skunks who lived underneath our cottage. They didn't smell unless startled, but something must have scared one of them at 3am one night after we had been there just a couple of weeks, causing it to spray beneath the floorboards under our bed. We woke up gasping.

Fair Oaks is just a short drive away from JPL, but we needed a car. It was the only way to get around, and anyway Doris had to drop me off and drive to her new job in Hollywood at the Children's Hospital of Los Angeles on Sunset Boulevard. I looked at the current offerings from the leading brands of car and decided that a Ford Pinto was the thing to go for. The Pinto was what the Americans called a sub-compact, but to us it looked full sized and was quite roomy and economical. It hadn't yet sunk in that gas mileage was no big deal there, as the price of petrol was so low (19 cents a gallon) they virtually gave it away. I took myself down to the nearest Ford dealer, Robert H. Loud, on Colorado Boulevard of course, crossed the road on a green 'Walk' signal, and requested a test drive in a Pinto.

This went fine, but when we returned to the dealership, the salesman parked the Pinto next to a pre-owned Ford Mustang. This was a deliberate ploy—these guys are smart and he had sussed me out during the ten minutes or so that we had been driving as someone who would not be happy for long in a Pinto. The Mustang was bright mustard yellow with a black vinyl roof and, because it was a year old, the same price as the smaller, slower, and duller Pinto. A test drive of the pony car followed, and on the first bit of clear road I floored the throttle and was thrown back in my seat as the big V-8 engine kicked ass (I was learning the lingo). Soon, I was driving the Mustang home.

We didn't stay on Fair Oaks for long. By what seemed to be a very happy chance, not long after the first skunk attack one of the JPL engineers, Nat Prescott by name, came to see me at work, having heard that I had just arrived for a year. Did I have a place to live? 'Still looking', I said. He had just been posted to a prestigious job at NASA headquarters in Washington for 12 months, and was looking for someone to look after his upmarket home in the smart suburb of La Cañada (pronounced Canyada, nothing to do with the loyalists up north) for that time. At a nominal rent, would I be interested? The house was high on the hillside, overlooking Pasadena and the whole of Los Angeles, with views out to the sparkling Pacific Ocean beyond. It was new and spacious and we could never have afforded it were it not for the housekeeping deal. Soon, we had said goodbye to Mr Allen and the skunks and moved in to Big Briar Way.

Mr and Mrs Prescott insisted that we signed a contract that said we would stay for the whole year, so they would not have to find a new tenant when they were 3000 miles away on the other side of the country. This we did, and when after just four months Nat was sent home from Washington after the project he was working on got cancelled, we reminded him about the contract. We were enjoying our spectacular existence high on the Angeles Crest, in the shadow of Mount Wilson, and were in no hurry to leave. However, the Prescotts were on their way back from Washington and we knew we could not keep them out of their own house.

Until this, the only downside to life in California had been the San Fernando earthquake that struck just before dawn on 9 February 1971, while we were still in bed. The bedroom had floor-to-ceiling windows, and we used to leave the curtains open so we would be greeted by the fabulous view when we woke up. That Tuesday morning, there was a glow in the sky but the ground was still dark, and we could see big bright flashes all over Los Angeles, as if the city were being shelled or bombed, or possibly struck by a shower of meteors. All of those possibilities crossed my mind, and it was a while before I learned that the explosions were the electricity substations that were to be found on almost every block, shorting out and igniting their tar-rich insulation.

The house was tossed around and we were almost thrown out of bed, while shelves full of books and crockery came crashing down in the study and the kitchen. But we soon found that we had got off lightly: across the city, hospitals and freeways were severely damaged, and a total of 64 people died. When eventually it was deemed safe to go back to my office at JPL a few days later, the light fixtures hanging from the ceiling were being removed and modified by the maintenance staff. Several of them had shaken loose during the earthquake and crashed down onto the desks where people like me would have been working had it been later in the day. We knew people whose nerves were stretched to the point that they left the Los Angeles area for good after this. Even the less cautious would jump to their feet and get ready to run for cover when an aftershock struck, as they frequently did, for months to come.

Official government notices appeared, telling everybody how they should behave in an earthquake. You were supposed to get under a heavy piece of furniture or stand in a doorway, this being the most structurally strong part of a building. You were not to go outside, where the building might fall on you if you did not have time to get away into open space. In the corridor near my office an older notice, dating back to the days after the Cuban missile crisis, was still there alongside the earthquake precautions. This purported to list the actions to be taken in the event of

The buildings of the Jet Propulsion Laboratory nestle in the foothills of the San Gabriel mountains in the Arroyo Seco at Pasadena, California. The large building near the centre of the complex housed the Science Division where I started work in January 1970; the even larger building to the left was devoted to management and administration, with the Director's suite at the top.

In Southern California a car is essential. Shortly after arriving I bought a dashing Ford Mustang, seen here at our rented home on Big Briar Way in Pasadena, up in the foothills behind JPL, on the fringe of the treeless wilderness called the Angeles National Forest. The historic astronomical observatory on Mt Wilson could be glimpsed, high above, while in the other direction we had sweeping views over the vastness of Los Angeles to the Pacific Ocean beyond.

thermonuclear attack, still considered a real possibility in the early 1970s. The notice looked very official with a government crest and formal wording about what to do if the air raid siren went off. The advice was:

1. Do not panic,
2. Get under a strong bench or table near the center of the room,
3. Put your head between your legs,
4. Kiss your ass goodbye.

After our premature departure from Big Briar Way, we moved to a small two-storey house on Altadena Drive that had been vacated by a young JPL colleague, Dan McCleese, who was an undergraduate student on a summer placement at the laboratory. After our brief taste of modern luxury, we were back to an old (probably 1930s, ancient by California standards) and somewhat run-down building, but it was comfortable and the neighbours were friendly. We had only just stopped being students ourselves, so affordability came into it, too. The bedroom occupied the entire upper storey and had windows on all four sides, great in a sunny climate even if we did bake in the summer.

We ate out a lot, enjoying vast T-bone steaks and big jugs of very acceptable Californian Cabernet Sauvignon in restaurants with names like *Westward Ho*! and *The Velvet Turtle*. One of them served prime rib of beef with 'a side of au jus'. Another had what they described as 'St Augustine's favourite way of cooking chicken', and very good it was too, and not at all expensive. On my parents' first visit to the USA my father was astonished and delighted when he ordered a scotch in the bar of the *Turtle* and, instead of the small glass barely wetted with whisky in the bottom that he was used to in English pubs, received a huge tumbler filled nearly to the brim. My mother was bemused by the ceremonial presentation of a chilled salad fork before her plate was placed before her, piled high with fruit, nuts, and vegetables covered with tasty creamy dressing and not at all like the few leaves of wilted lettuce tainted with vinegar that she had grown up with in war-torn Britain.

Mariner Jupiter–Saturn

According to a history of the Jet Propulsion Laboratory published in 2008 (*JPL and the American Space Program*, by Peter Westwick), the Lab culture in the early 1970s, when I arrived there, 'reflected the attitudes of bright young men who were willing to work hard—and play hard'. I have already alluded to the playfulness I found when I arrived; I was soon to find that hard work was indeed very much the driving ethos. But the team spirit was so strong, and the goals so exciting, the only word to describe the way I felt about it was exhilarated. Soon I was doing long hours.

The project our erstwhile landlord from Big Briar Way had briefly worked on at NASA headquarters was the same Grand Tour of the giant planets of the outer Solar System that I had become attached to with Barney Farmer and his colleagues. Barney was the head of a team that had been selected by NASA to help the agency in planning the mission. Our main job was to evaluate the measurements that could be made at each of the outer planets from a spacecraft making a close encounter. Specifying the scientific instruments that would be required to investigate the atmospheres of bodies like Saturn, about which so little was known, was verging on inspired guesswork. Cameras for taking pictures would be essential, of course, but other possibilities included things like magnetic field measurements and remote sensing of atmospheric temperature and composition by spectroscopy and radiometry—my speciality.

Spectroscopy is about measuring how the light reflected or emitted from anything varies with wavelength. In its simplest form, when we look in visible light and see the colour of an object and deduce for example that a ripe red apple is different from a sour green apple, we are using a crude version of spectroscopy. If we actually measure how bright each colour is, this is radiometry, and the combination of the two is a very powerful tool. In our scientific work, the colours are measured at wavelengths longer (the infrared) and sometimes shorter (the ultraviolet) than visible light, but the principle is the same. At Jupiter, for instance, our study for the Grant Tour would show how we could analyse the light from the Sun after it had been reflected by the planet, or the infrared radiation emitted as heat directly from the planet itself. From these data we could determine what gases were present in the atmosphere, and in what proportions. More difficult but still potentially feasible objectives were to learn what the multi-coloured cloud layers consist of, and to probe why they are organized into bands around the equator, and studded with giant ovals that look like storms bigger than the whole Earth.

There was a real sense of exploring the unknown. So little was known about these distant atmospheres in those days, and I was having a great time filling notebooks with what I could glean from shelves full of astronomy journals, and then speculating and producing models. Atmospheric 'models' are detailed, but simplified, numerical descriptions of things like temperature profile, cloud height, and the abundance of various gases that might, or might not, be reasonable approximations to what we hope to discover. Models can be used to calculate how an instrument of a given design will perform, allowing us to define parameters such as the light-gathering power (and hence telescope size) required, and the type and sensitivity of detector needed. My notebooks from this time are full of diagrams showing how far down into the vast depth of each planet's atmosphere we could hope to peer at different wavelengths.

One of our team members, Dr Peter Stone, was from Harvard University and was an expert on the dynamical behaviour of outer planet atmospheres. He had developed theories to explain why the clouds on Jupiter form into broad bands, for instance, or what weather phenomena produce the giant ovals capped with clouds of different colours. Unlike cyclones on the Earth, which they superficially resemble, the Jovan features are not only much larger, they are also very long-lived. The Great Red Spot, the most famous, although not the only, example, seems to last forever, and has been present continuously since it was discovered shortly after the invention of the telescope in the seventeenth century.

With Peter, I learned how to speculate about the dynamical behaviour of the deep outer planet atmospheres and discuss with the team how we might investigate them with suitably designed instruments. These would have to be configured so that they could be accommodated on the Grand Tour spacecraft that was taking shape in other parts of JPL. In February 1972, I was invited by Peter to come to Harvard to give a seminar about the atmospheric science prospects for the mission. This was my first experience of the Oxford-like but distinctly American environment and ethos in Cambridge, Massachusetts, and I spent some time looking around the Grecian and Roman facades. I also ate a tough but tasty horse steak for the first and only time at lunch in the Harvard Faculty Club. One ought to have these experiences, but once is enough.

I was starting to get somewhere as a budding outer planet expert myself, and had developed a high level of enthusiasm for the ambitious mission that would put some of my ideas into practice, when the Outer Planet Grand Tour was cancelled by NASA. This derailed my plans and prospects as well as precipitating Nat Prescott's return to Pasadena and our ejection from his comfortable house. The reason for the cancellation was not so much the vast scope and expense of the project, but the fact that it had a tight and unchangeable schedule. Visiting all four gas giants required a rare cosmic alignment that would have allowed the spacecraft to swing by one planet and proceed to another, then another. This is a situation that occurs rarely, in fact only once every 175 years according to the planners at JPL. Once the project was underway, and the scale of the task of building and launching the hardware for such an ambitious undertaking had sunk in, it became clear it was impossible on the original schedule. Most missions can, and often do, absorb overruns by slipping the date of the launch, but that was not an option in this case.

All was not lost. The original mission design called for two launches in 1976 and 1977, to fly to Jupiter, Saturn, and Pluto. A further two, launched in 1979, were intended to fly to Jupiter, Uranus, and Neptune. JPL swiftly reorganized and persuaded NASA headquarters to opt for a simplified version, with just two spacecraft, aimed only at Jupiter and Saturn. Instead of the fancy new *Grand Tour* spacecraft, with its self-repairing computers and other refinements, a version of the existing *Mariner* spacecraft would be used. The *Mariner Jupiter–Saturn* mission was born, and I went back to my models, once again confident that I would eventually be able to use them to interpret the data we would acquire when the spacecraft reached Jupiter in 1979.

Jupiter through a telescope

In the meantime, I had my first taste of ground-based astronomy. Reinhard Beer, who despite his name was another Englishman, had arrived at JPL some ten years before me and had developed a powerful spectrometer that was mounted on a large telescope at McDonald Observatory in Texas. He planned to use this to observe Jupiter, and thought my theoretical work would help to analyse the spectra; I would also be able to help to make the observations. Getting to McDonald, which is in a remote area near the Mexican border, involved flying to El Paso and then driving 200 miles across the desert to the former frontier outpost of Fort Davis, where the observatory was set up on nearby Mount Locke in the 1930s.

We had good weather and our observations were a spectacular success. The superior performance of Reinhard's new spectrometer revealed the spectral lines of several species in Jupiter's atmosphere that had never been observed before. One of these was a relatively rare variant of methane, which showed the presence of heavy hydrogen, known as deuterium, on Jupiter. This was the first time deuterium had been detected anywhere outside the Earth. It isn't present in stars at all, because the high temperatures decompose it to produce ordinary hydrogen, but seawater on the Earth has a few parts per million of deuterated (heavy) water. Interestingly, the proportion of heavy to ordinary hydrogen we found on Jupiter is about ten times smaller than the terrestrial value, and we wrote several papers about the significance of this for the history of the two planets, and the way the whole Solar System had evolved.

Even more intriguing was the appearance of lines of carbon monoxide in the spectra. Jupiter's atmosphere is mostly hydrogen, and basic chemistry predicts that in such an environment carbon monoxide should quickly be converted to methane and water. How could any of it still be there? Our analysis showed, however, that the temperature of the carbon monoxide was very low, showing that it was present mainly in the cold upper atmosphere of Jupiter. Probably, it was arriving on Jupiter from space, as a flux of small, icy comet fragments. Jupiter's huge mass and position far from the Sun means it sweeps up a lot of small comets, and the occasional large one, and we know that most comets have quite a lot of frozen carbon monoxide in them, along with water and other ices. We wrote papers about that, too.

So did others, of course. Most of them were constructive contributions but we were victims of a classic manifestation of the dark side of science, which is not as rare as it ought to be. A rival astronomer denounced our results and 'proved' with his own that the carbon monoxide was in the hot, deep atmosphere of Jupiter and hence must be chemically produced somehow. Years later, the same individual published new results in which he triumphantly announced that he had discovered carbon monoxide in the cool *upper* atmosphere of Jupiter, and that therefore it must be of external origin. To make matters much worse, he deliberately ignored our original work nearly 20 years before, as well as his own criticism and conclusions at the time. We notified the editor of the journal and he made the villain issue a clarification in print, which he did only reluctantly and inaccurately, and with very poor grace.

A Golden State

After two years at JPL I had come to a fork in the road. My one-year 'associateship' had been extended to two, but that was the limit allowed by the rules. We had designed our instrument for the outer planets mission, written a formal flight proposal with Barney as the principal investigator and me on the team, and confidently expected NASA to confirm that it would fund us to build it and fly it on the spacecraft. JPL offered me a permanent job to carry on, and of course that was a great opportunity to be considered very seriously.

Doris and I fretted over where to spend our lives. California was originally meant to be an exciting interlude, not a permanent move. We thought our future lay in England, hopefully with that professorship I had dreamed about somewhere in the UK, if not actually at Oxford. But we did love California, and things were going well at work, and in the end we stayed.

Staying involved getting permanent resident status to replace my temporary visa. On a detailed inspection of the terms of the latter it turned out that I had agreed to leave the country when the visa expired and not to seek resident status, a perfectly reasonable rule that was designed to prevent immigrants from jumping the queue. There were ways around this, which I duly pursued, but the process was so long and so tedious I will draw a veil over most of it. There were long days spent queuing in the corridors of the Immigration and Naturalization Service headquarters in downtown Los Angeles for paperwork and interviews. Sometimes you had to get there before dawn to get far enough up in the line to be seen that day. Most of the others waiting seemed to be migrant workers from Mexico; at least I learned some Spanish while I waited.

Once I finally got my hands on the so-called 'green card' (which was actually blue, and later pink), we could set about buying a house. We had already found what we wanted, an enchanting Bavarian-style chalet in Rubio Canyon, which like Big Briar Way was on the mountainside high above the Los Angeles basin. Fortunately it was still available when we became able to buy, and we moved in with our cat, Daisy. The house came with ownership of two adjacent empty lots, so we had a big yard full of oak and olive trees, and a section of the canyon side covered with large cactus plants and yucca bushes. Doris set up hives and kept bees for honey, and we grew citrus fruit, oranges, lemons, grapefruit, and kumquats. Raccoons, possums, coyotes, and occasionally rattlesnakes roamed our back yard.

California is rich in history, not as ancient as Europe of course, but fascinating none the less. The garden of our new home had the ruins of an old stone adobe, in which an early resident of the area must have lived, possibly even the eponymous Jesus Rubio himself, an early settler after whom the canyon was named. A few yards away was the bed of an old railway, an amazing affair that had run from downtown Pasadena all the way up the canyon to the top of Mount Lowe, where there used to be a grand hotel with spectacular views. All of this had ceased to exist more than 50 years earlier, but there were still traces of the stations and the trackbed, which now formed a hiking trail up the mountain.

At the lower end of Pasadena, I traced the site of the original Busch Gardens, a splendid pleasure park created in a spirit of philanthropy by the founder of the great American brewery Anhauser Busch. This too has vanished, but left traces including terraces and stone stairways leading down into the nearby Arroyo Seco, the same canyon on which JPL now stands a couple of miles upstream. In its heyday, Hollywood shot movies on location there, most famously *Gone With the Wind.*

As we were settled, having bought a home, and gained a modicum of prosperity, I could now buy a seriously interesting car. The Mustang was great in many ways, but it had far too much power for its handling capability, and it used to spin if you had to brake suddenly. Other owners said that they put bags of cement in the trunk to help the rear end grip the road and reduce such behaviour, but I didn't and when I came close to spinning off the edge of one of the many mountain roads near our new home I decided it had to go.

I had my eye on an Aston Martin DB5 that was for sale in nearby Glendale. I went to see it and the owner told me he was selling because it was a magnet for the Highway Patrol, who kept pulling him over. When I sympathized with him over this unjust persecution he explained that the last straw had been an incident which occurred when he was driving recently across the desert to Las Vegas. He was travelling late one evening, in the dark, and on the long, straight road he overtook two other cars at high speed. It turned out that the hindmost of these was a police car that was chasing the car in front for speeding when the Aston went past them both. I still sympathized, but now his arrest made sense.

The DB5 became my regular transport for the next seven years. I was pulled over a lot, too, but not always for speeding. Sometimes the officers just wanted to look at the car, although I did pay a few fines as well. We also bought a Lotus Elite for Doris's daily commute to Hollywood and a Jaguar XJ6 for when we had visitors and needed four seats. For a short time, I also ran a second Aston, a DB2 convertible that I had

Most space scientists and engineers love interesting cars and I was no exception. This DB5 became my daily transportation, and I did all my own maintenance using parts that I collected from the Aston Martin factory in Newport Pagnell while on my frequent trips back to England.

I killed this 3-foot rattlesnake with my bare hands (and a long pole) when it strayed into the driveway of our house on Rubio Canyon in Altadena. My Aston Martin DB2 convertible is in the background.

spotted at a bargain price. It was in rough condition and needed work, but I was a keen if fairly inept amateur mechanic at that time and did most of my own repairs and maintenance. I became a familiar face at the parts counter at the Aston Martin factory in Newport Pagnell and brought caseloads of DB5 parts back from England on my regular trips to Oxford and Northumberland. DB2 parts were not so easily available, but could be found, mainly at a specialist garage in Dorset. However, unlike the newer model, I found the interior of the DB2 to be very cramped. It made the car difficult to drive by anyone over six feet tall, so I soon sold it on.

Doris fell out of love with her Lotus when it broke down frequently and even caught fire once. The fire was while it was in our garage, fortunately, and not on the road, and to be fair to Lotus, it was due to an inexpert piece of carburettor reassembly by myself that led to a fuel leak onto the hot exhaust manifold. The other breakdowns were mainly due to clutch failure. The clutch was cable-operated, and in US imports the arrangement was modified to deal with the left-hand-drive configuration, with the result that the cable was eight feet long, passing right around the engine compartment, rather than about eight inches as originally designed. It wasn't strong enough to handle the extra friction and would snap from time to time. One of the American car magazines said you had to drive a Lotus with one hand, so you could use the other hand to hold it together. That was a bit harsh, and when it was working it was a real flier, but when it stranded Doris in front of a notoriously seedy coffee bar, *The Four Horsemen*, frequented by fearsome-looking (but actually quite kind and helpful) Hell's Angel-type bikers, the Lotus had to go. We replaced it with a new Triumph Spitfire Mk 4, which was still fun to drive and a lot more reliable.

Et in Arcadia ego

Life at JPL was always exciting, usually but not always in a positive sense. Its flair, and much of its strength, comes from a unique combination of academic culture (as a part of Caltech) and massive government funding (as a part of NASA). All of the other NASA centres were essentially within the US civil service, with permanent posts for the staff. JPL, however, ran entirely on renewable contracts, and as a result was massively insecure. Each week, those of us that worked there had to fill in a card to tell the administration where to charge our salaries. If we didn't have a current project or grant to cover it the institution would usually bridge the gap from its reserves, but obviously there were limits to this. So, the very existence not only of one's job but of the entire laboratory depended on a culture of on-going success. When a major mission ended, failed, or was cancelled, the air was thick with angst until a new one was approved and the funds flowed again.

Fired with the optimism of youth, and because things were, for the time being at least, going well, I found this setup exhilarating rather than daunting. There was the occasional sacking or layoff among my new colleagues, even some of those I regarded as high achievers, but there were also plenty of older scientists around who had been at JPL for ages and would see out their careers there, so stability and permanence obviously could be achieved. One way to do that was to belong to, or better still be in charge of, a research programme of your own. These attracted regularly renewable chunks of funding for scientific research that was deemed to be 'underpinning' NASA's planetary missions, past, present, and future, by data analysis, support through roles like Project Scientist, and planning new concepts.

Barney Farmer had such a sinecure, called *Atmospheric Experiment Development*, from which his experiment on the *Viking* Mars mission had derived, a spectrometer called the *Mars Atmospheric Water Detector*, known as MAWD. With *Viking* at the peak of its pre-launch development, and therefore in crisis, Barney didn't have time to write the annual renewal application for the Experiment Development task, and so he turned it over to me. Although NASA has long since stopped dispensing this kind of funding, at the time it was very stable as long as you submitted regular reports that showed good work was being done, and applied annually for a renewal in painstaking detail via a long form called an RTOP (research and technology operating plan). The tedium of the RTOP was a small price to pay for a stable niche in a turbulent funding environment.

You had to deliver a steady stream of published work to show that you were productive and therefore making proper use of public funds. The length of your publications list, and the quality of the results to which the papers on it referred, were a major factor in getting renewed funding for future years, and also for getting selected as a team member on new missions. Colleagues at Oxford were amazed by this mercenary approach, by which your research tended to follow the funding, and your success was measured by counting your output. Of course, that is exactly how things work in the UK now, but it was a rather novel concept in the 'Old World' at a time when you could get through an entire career with only a handful of peer-reviewed publications, and sometimes none at all. As a member of the more modern generation, by the time I retired in 2011, I had clocked up over 300 papers in reputable journals, an average of about six a year.

I like writing books, and started early. While I was working at JPL, Cambridge University Press wrote to me asking if I would be interested in authoring a book about the planets. What they had in mind was a large-format, 'coffee table' volume with a collection of the best photographs from the close-up views obtained by cameras on spacecraft. Each picture was to have an extended caption, explaining the features that could be seen in it, and each chapter would deal with a specific planet and have a long introduction explaining what we thought we knew about it. I was keen on the idea, because of course at that time there was no internet and people had no ready access to these pictures except for glimpsing them on television as each encounter occurred, or tuning in late in the evening to watch programmes like *The Sky at Night*. Writing a book seemed like a lot of work, however, and I was worried about the fact that I was not at that time much of an expert on Mars, while the pictures then coming in from *Viking*, both from orbit and from the first landers on the planet, would be among the most spectacular and interesting in the book.

The solution to both problems was to team up with a friend and colleague who was an active member of the *Viking* imaging team, Geoffrey Briggs. Geoff did a superb job on more than half of the book, some of it after leaving JPL to take on a very high-powered job as the head of the Solar System division at NASA Headquarters in Washington, DC. We had a minor best-seller on our hands when *The Photographic Atlas of the Planets* came out in 1982. It went on to sell more than 40,000 copies, and was then translated (not by us) into German and Italian language editions, and finally taken up as a monthly special by one of the book clubs that thrived in those days. My share of the royalties paid for my third Aston Martin. Unfortunately none of the ten or so books I have authored since then have done quite so well sales-wise.

Close Encounters

Something JPL, Caltech, and Oxford had in common was a regular stream of famous names passing through. My earliest memorable encounter in Pasadena was when I shared an elevator with Carl Sagan and we had a short conversation (he was lost and I was able to direct him to his meeting location). Fred Hoyle occasionally lectured on campus, and I went along to hear him expound his latest alternative to the ‘big bang’ theory (a term he had in fact coined himself, intending it to be derogatory).

Hoyle, of course, proposed continuous creation as a model for an expanding universe with constant density on the large scale, and I believed him. Nowadays the mass of evidence that has accrued against Hoyle’s theory means it is dead, or at least there is a long way to go to get to a version that works. There is scope for that, even today; since so much of modern cosmology doesn’t add up. In spite of being one of the most imaginative thinkers of his or anyone else’s time, Fred may have been guilty of not thinking big enough. He pictured continuous creation of hydrogen atoms; now we might be tending towards believing in the continuous creation of multiple universes.

Once when leaving Caltech and heading for the car park, I found myself waiting for the light to change at the crossing on busy California Boulevard with Richard Feynman. His classic book *Lectures on Physics*, in three big red volumes that I still consult to this day, were revered as almost sacred texts by my tutors at Liverpool when I was an undergraduate. At the light, I stood in silence with the great man for what seemed like eternity (and was probably about two minutes) while the pedestrian signal remained obstinately red and the traffic streamed past. In California, crossing the road without proper permission is illegal, as I had found during my first week there, as already recounted. Eventually Feynman turned to me and said he believed the traffic light was faulty, did I agree? I nodded dumbly. That means it's OK for us to cross on red next time there's a gap in the traffic, doesn't it? Remembering the Highway Patrol, I didn't think so, but I said yes. We crossed the road and separated, and I remembered while driving home all the questions I wanted to ask him about physics. L'esprit d'escalier.

In November, 1972 I was in Flagstaff, Arizona for a conference. In those days the great Atchison, Topeka, and Santa Fe railroad ran right through the middle of Pasadena, and you could make the 450 mile journey to Flagstaff by train, travelling overnight. The line to Arizona no longer exists, but Pasadena station does, albeit reduced to serving as a stop on the LA Metro light rail line. In the olden days the car attendant would wake you in the morning at 6am, an hour out from the destination so that you had time to enjoy a cup of coffee and spectacular views of the surrounding forest and mountains.

Once at the meeting, I found myself sitting at dinner next to the great planetary astronomer Gerard Kuiper. He asked me what I did and listened politely as I told him I was working on a proposal just then for access to NASA's new airborne observatory to make planetary atmospheric observations. I gave him all the details of the big C-141 transporter plane that was being converted for the purpose. Yes, he knew a bit about it, he said. A few weeks later, shortly after NASA had turned down our proposal, they announced the schedule for the first flight of the telescope and, to my great retrospective embarrassment, baptized it the *Gerard P. Kuiper Airborne Observatory*. It was all his idea.

The best chance encounter of all had nothing to do with science. My parents came to visit and we took them on a tour of northern California in my newly acquired classic Jaguar, a Mk 2 model similar to that which Inspector Morse would later make famous. Our itinerary took us to the beautiful resort of Lake Tahoe, which includes at its centre a strip of large hotels. Being just over the border in Nevada, these are huge complexes that feature gambling casinos and floor shows, Las Vegas style. As we drove into town on 2 May 1976, the Sahara Tahoe towered in front of us, with a huge sign carrying just one word, in letters ten feet high: ELVIS.

I had been a keen Elvis Presley fan in my youth, and although maturity had calmed me down a bit I could not resist the chance to see him live. The show was that very night, and of course was sold out. After depositing Doris and my parents at the mountain cabin we had booked, overlooking the lake, I drove back to the Sahara to see if there was any chance of getting returned tickets. A friendly local heard me being rejected at the box office, took me aside and spoke confidentially. Just turn up, he said, and when the guy on the door asks for your ticket, slip him 50 bucks. He'll show you to one of the best tables.

There was time to relax between the sessions when I presented the results of my outer planet studies at a Planetary Sciences conference in Hawaii in 1972.

Shades of the Wild West are still to be found in the area around Tucson, Arizona, including Tombstone, Boot Hill and the OK Corral. I spent much time there because it is the home of the Lunar and Planetary Laboratory, a large research centre run by the University of Arizona where many of my collaborators worked and where numerous meeting and conferences relating to all aspects of the planetary sciences were held.

How I got up the courage to take my venerable mother and father on this wild and speculative adventure I will never know, but it worked exactly as he said. The four of us sat in a box at the edge of the stage, above the screaming women in the stalls below to whom Elvis threw silk scarves, donning a different one over his white, jewel-encrusted suit before each number. He would be dead just over a year later, and didn't seem completely well then, often stopping in the middle of a song, as if he had forgotten the words, but while he was singing he looked and sounded great.

At the Greek Theatre in Los Angeles we went to a concert by Sammy Davis Jr one warm summer evening. Oozing personality, he sang, danced and joked his way across the stage in the open air. He loved to play golf, he said, and someone asked him what was his handicap. I'm a one-eyed Jewish negro with a broken nose, was the reply. At the end of the show, the finale was his big hit Mr Bojangles, for which he had earned a gold record some years earlier. 'It may surprise you,' he said, 'to know that in my career I only won two gold records. I did the first one to open the show, now I'll close with the other one.' Then, *sotto voce*, turning confidentially to the front rows of the audience, shielding his mouth with his hand, 'You gotta spread them mothers out.'

Being from the Midwestern USA, Doris was wary at first of moving to the fleshpots of California. She had her own career to build, in her case in the medical field, one that is stimulating but possibly even more fraught with trials and pitfalls than space science. We have shared our ups and downs with each other and a series of cats: this one is Natasha.

Coping with a failure

The excitement of work and domestic bliss were rudely interrupted in 1973 when NASA finally announced the selection of the scientific payload for the two *Mariner* spacecraft, now renamed *Voyager*, for their mission to Jupiter and Saturn. In spite of all the work we had done under their auspices, which should have made us front-runners to become investigators on the mission, NASA decided our instrument was overdesigned and too specific for the unknown conditions to be encountered, and that a survey type instrument would be better. They selected a proposal from the rival team at the Goddard Space Flight Centre in Maryland instead of ours.

I knew the Goddard group well; one of their experiments, called IRIS, had flown alongside John Houghton's first Oxford satellite instrument on *Nimbus* 4. IRIS stood for *Infrared Interferometer Spectrometer*, a type of scanning instrument that could cover the whole infrared spectrum continuously. With all this redundant information, their team claimed, they would be better able to work out what was going on, on Jupiter and Saturn, than we would with our focussed experiment, which pre-selected a limited number of what we had worked out were the most interesting key wavelengths.

Much of this fatal specialization was down to me; in a way I had done too good a job. I was on particularly friendly terms with one of the IRIS team, Virgil Kunde, and I knew that they were planning to compete with us, but I had worked out that they couldn't gather enough infrared energy with an instrument of any reasonable size to measure a continuous spectrum at high resolution. Our pre-selection of wavelengths led to a more compact instrument with good signal strengths. Under the leadership of the experienced Dr Rudy Hanel, Goddard had trumped us by designing an interferometer with a telescope one metre in diameter, far larger than anything we had imagined possible on a spacecraft. Our JPL instrument had a ten-centimetre diameter telescope, which was not small by current standards but had to work with only 1% of the energy the super-IRIS could capture.

In a mood of near-desperation we boarded the overnight flight (the infamous 'redeye') for a debriefing at NASA Headquarters in Washington. There, confronted across the table by not only NASA officials but also our own team member Toby Owen, whom NASA had co-opted in a shrewd move as a science-savvy referee, we protested our case. The Goddard team's telescope, to avoid being impossibly heavy, had no housing around the primary mirror and no calibration system. The delicate reflecting surface would be exposed to dust and ice and other rigours of space during the whole of the long journey to Saturn and more than likely, in our view, it would be useless by the time it got there. Furthermore, when making infrared observations, which measure heat radiation from the target, conventional engineering wisdom said that it was essential to keep the instrument—which was a lot warmer than the planet—very stable in temperature, and that would be impossible.

No dice, however; it was the IRIS that flew and we didn't. It was no consolation that Virgil told me later that he and some other members of his team were not very interested in the outer planets, and would rather not have won, so that they could build another IRIS for Earth observation instead. For the record, they did a terrific job of studying all four giant planets and made a number of key findings. While JPL was denied the chance to provide the infrared instrument, it still managed and controlled the mission with its deep space tracking stations. Because of this I had the privilege to be among the first people to see the pictures and other data that was beamed back to Earth when the *Voyagers* reached not only Jupiter and Saturn but, in a neat piece of improvization, went on to Uranus and Neptune after all. By using cheaper spacecraft the launch had taken place in time to take advantage of the alignment that allowed multiple planetary encounters.

In August 1979 I was at the big party that was thrown to celebrate the successful encounter of *Voyager* with the most distant planet, big, bright blue Neptune. That was a consolation prize for our failure to be involved in designing or executing any part of the mission. It was not for want of trying, and although I was disheartened at the time, in the end the planning, modelling, and ground-based work on Jupiter did eventually come in handy for the next and even more exciting mission to Jupiter, *Galileo*, and for the fabulous trek to Saturn with *Cassini*, as will be regaled in later chapters. Meanwhile, I had been working on another idea, a mission to Venus.

4
Voyage to Venus

As a student in the 1960s I belonged to the Oxford Speculative Fiction Society, which met weekly in Magdalen College to discuss the detailed implications of what we considered to be the finest examples of that particular genre. High on the list was, of course, *The Lord of the Rings*, the full version of which had been published just a year or two before in 1965. J.R.R Tolkien was an old man then, but his son Christopher was lecturing on Nordic mythology at Oxford and some of us went regularly to listen to him. Some of our number who had learned a little Norse from this experience tried their hand at Elvish, a language which Tolkien senior had devised in elaborate detail for his books on Middle Earth.

Needless to say, the real and imaginary languages had quite a lot in common (although Welsh came in to it, too) and some of us (although not I) became quite good at translating Elvish from the books (there were no films then, of course, and indeed it was everyone's belief at the time that *The Lord of the Rings* could never be filmed successfully). The line

from *The Two Towers* translates to 'Hail Eärendel, brightest of stars!'. Eärendel is a mariner in Anglo-Saxon mythology, recorded by Cynewulf of Lindisfarne in the eighth century, who carries the Morning Star, that is Venus, across the sky. J.R.R. Tolkien also wrote a poem in 1914 entitled *The Voyage of Eärendel the Evening Star*. The Evening Star is Venus as well, of course. I took great pleasure in the way my astronomical interests met up with the literary and historical aspects of planetary studies as a result of these activities, little realizing then that about a decade later I would have a key role in a space mission to Venus.

Opportunity Knocks

It was not long after the time when I had opted for a future at JPL, and had lost the expected opportunity to explore the giant planets of the outer Solar System, that NASA decided to go ahead with a new mission to Venus. The agency was also developing a mission to follow

Voyager out to Jupiter, but this time to stay there by going into orbit. For someone like me used to thinking about weather and climate on the Earth, the inner planets Venus and Mars were actually more attractive objects of study than Jupiter or Saturn because their atmospheres are more Earthlike, as well as being closer and easier to reach.

The hot, cloudy atmosphere of Venus cried out for investigation. It was in many ways a sister planet to Earth, but one where the climate had suffered so badly from the greenhouse effect that it had long ago become uninhabitable. The Soviet Union had already sent a succession of a dozen or more spacecraft towards Venus; we never knew how many, because the Russians maintained a veil of secrecy, especially when they failed. Using their huge rocket boosters, they were able to send pressurized automatic laboratories to land on the surface. Some of these had already landed safely and sent back pictures of an arid, rocky, and hilly surface under a cloudy yellow sky. The Russians planned even more probes, and obviously thought of Venus as 'their' planet: Americans find it hard to ignore that kind of challenge, and in 1973 they threw their hat into the ring. They called their new mission *Pioneer Venus*.

Venus is the nearest planet to the Earth, a mere 25 million miles away at its closest approach. It is also very like our own world in many ways: about the same size and mass, and a similar composition with a rocky crust and a metallic core. However, Venus has a thick, hot atmosphere with yellowish, sulphurous clouds, and is bone dry with no oceans and just traces of water vapour in the atmosphere. In the 1970s, very little was known about why this should be so, or about how this alien environment behaved in terms of its winds and other weather phenomena.

To address as many puzzles as possible, *Pioneer* became two missions in one. The science goals laid down by the team that NASA formed to study the possibilities were in two separate categories: those that needed measurements inside the atmosphere, from descent probes, and those that needed global coverage, from an artificial satellite. They would be launched on separate rockets, much smaller than the Russian ones, but at about the same time so that they would arrive within a week or two of each other and operate in concert at the planet. The satellite part of the mission would make comprehensive, long-term studies of meteorological phenomena in the atmosphere, and so become the first Venusian weather satellite. Instruments would be needed that could measure temperature profiles, cloud densities, and make humidity maps under unfamiliar conditions.

Probing Venus's atmosphere

The technique I had worked on at Oxford for measuring temperature profiles in our own atmosphere remotely from space would, in principle, work even better on Venus or Mars. Their atmospheres are nearly pure carbon dioxide, and the method worked by measuring the intensity of the infrared heat radiation emitted by that gas at various altitudes. The signal measured by our instruments should be enhanced as a result, and a wider range of heights in the atmosphere could be probed from orbit. John Houghton came to visit me at JPL and suggested we write a paper about this possibility, which we duly did. Our definitive paper on Venus temperature sounding was accepted for publication by the esteemed *Journal of the Atmospheric Sciences*. It attracted attention, as it was meant to, and when NASA announced the opportunity for suitably qualified teams to participate in the new *Pioneer* mission, I started work on turning the ideas in the paper into a formal proposal.

The plan we proposed was that JPL and Oxford might jointly build an instrument similar to the Oxford *Nimbus* radiometers with a few modifications to adapt it for Venus. We would not only apply our experience of building meteorological sounders to another planet, but also we would produce compatible sets of data for Venus and Earth that could be compared to each other. We could then identify similarities and differences in their meteorological behaviour. It was an attractive proposition.

If we were selected, against what was again bound to be stiff competition, NASA would pay the cost of building the instrument, but only if the team leader—the Principal Investigator or PI—was in an American institution. If a foreigner led the team, his country would have to foot the bill for the hardware, likely to be several millions of dollars. That ruled John out as PI, and in any case both he and Barney, the other obvious candidate, were both PIs already, for their *Nimbus* and *Viking* experiments respectively. Everybody knew that the Principal Investigator role is too demanding for anyone to carry more than one project at a time. So it happened that I was nominated to lead the proposal, and head up the project itself if our bid were to be successful.

Opportunities for participation in a major space mission are rare and the selection process is fiercely competitive. The first step is to build a team. I was less than 30 years old; most Principal Investigators are more like twice that. I would have to build some experience into the team or NASA would not look seriously at the proposal. I knew who I wanted: at the National Center for Atmospheric Research in Boulder, Colorado I had met John Gille, an expert on satellite radiometry and himself an experimenter on several *Nimbus* satellites. His colleague at NCAR, Bob Dickinson, had done important theoretical work on the atmosphere of Venus at a time when very little was known about it. We could use his computer models to plan the experiment and help to analyse the data. Back in Pasadena, the atmospheric scientist who was known for some of the first studies of the possibility of a 'runaway greenhouse' on Venus, Andy Ingersoll, was just down the road at the Caltech campus. Add the senior people from Oxford and our group at JPL and we would have a team.

I approached them all, and the response was enthusiastic from everyone except Barney. The *Viking* mission was a couple of years from

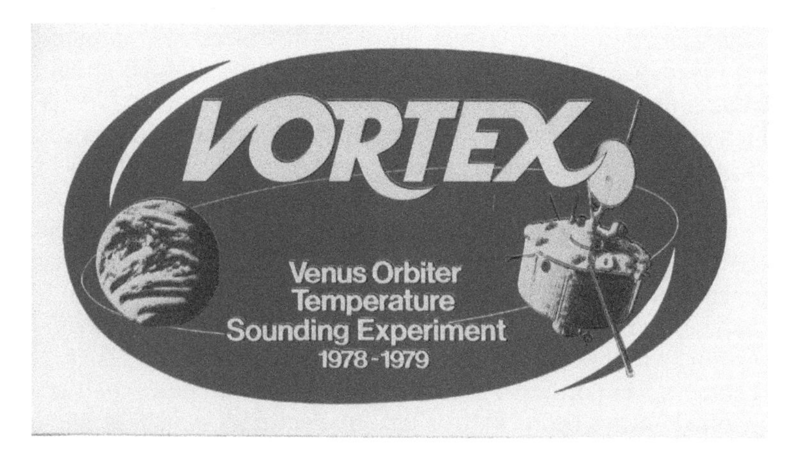

Experiments need a beguiling name, so I christened our Venus instrument VORTEX, a catchy shorthand for the full mouthful, which was Venus Orbiter Radiometric Temperature-sounding Experiment. The name had a double meaning, because one of the goals was to look for vortex-like behaviour in Venus' atmosphere, an objective that would be unexpectedly fulfilled in spectacular fashion.

launch to Mars, and had reached the stage of desperate panic that seems to grip all space projects a year or two before launch, as time and funds run out. A notice in the *Viking* project office described the six phases of a space mission as 'euphoria; excitement; pain; and panic', in that order, followed by 'persecution of the innocent' (for mistakes, failures, and cost overruns) and, after ultimate success, 'rewards for those only marginally involved'. I would become familiar with all of these in the years ahead.

Barney and the rest of the *Viking* team had just reached the pain side of panic. He had no time to think about a new project now, although he wished us luck with Venus. I persuaded him to join the team anyway, on the understanding that he would not be asked to do anything except offer advice until he might have the time at some point in the future to get more involved. I began work on the proposal, eying NASA's deadline six months away. The senior management at JPL was benign but showed very little interest; it was only later that I learned that this was because they thought I did not have a cat in hell's chance of being selected for the mission.

One of the Chosen Few

I first learned that my proposal had won the competition for participation in the project while attending the American Astronomical Society's annual Planetary Sciences meeting in April 1974. Don Hunten, who was one of the leading scientists that had been involved with designing the mission concept and subsequently led the selection committee that helped NASA to decide which instruments to fly, told me in his deep Canadian drawl 'Congratulations, you've been selected by NASA as a Principal Investigator on the Pioneer Venus mission. If you have problems at JPL go and see the Director. You're one of the chosen few.' Back at JPL, there was surprise at the successful selection of VORTEX, and on my part euphoria mixed with alarm as I rushed around an unprepared laboratory to put together the engineering support necessary to start work on building the instrument.

NASA had decided that the new Venus mission would be run by a team at its Ames Research Center, at Palo Alto near San Francisco, 300 and some miles north of Pasadena. This did not go down well at JPL, which had always been NASA's leading centre for planetary missions, but JPL was still wrestling with *Viking* and would soon be dealing with *Voyager*, and so could not have handled another big new project just then. Ames was a 'proper' NASA centre, staffed by government employees, not academics and contractors, as was the unique arrangement at JPL. At that time Ames was between projects and short of work, although they had experience with small spacecraft and many excellent people on their staff.

Once the *Pioneer Venus* project was underway I had to go up to Ames regularly for project meetings. Somewhat to my surprise, having got used to JPL management, which was very competent but a bit fierce, the *Pioneer* team at Ames were mostly charming and friendly. The location of the complex is delightful, on the southern edge of San Francisco Bay, on the airfield known as Moffett Field that used to be home to the American fleet of giant airships before the war. Like its English equivalent at Cardington, which I had come to know so well a few years earlier, it still had the huge hangars in which the American zeppelins were kept, until the accidental loss of one of them. The USS *Macon* crashed in a storm in 1935, which led to the retirement of the rest.

A big difference from the air of decay at Cardington was that in modern times Moffett Field was home to a large and very active US Naval Air Station, as well as the NASA Centre. As a foreigner, I would hardly be allowed in there in these post-9/11, security-conscious days, but in 1976 I could fly right in to the military airfield and land on the tarmac outside the building that housed the Pioneer Venus project. JPL had a light aircraft at the disposal of its high-fliers, an elite group to which as a Principal Investigator I could now claim membership, for this kind of trip. The plane, with its crew of two and space for eight passengers, was kept at the local Hollywood-Burbank airport near Pasadena. The aircraft was not only convenient, it was luxurious, and had a cocktail cabinet (another feature that is unimaginable now) that was not opened on the outgoing journey, but would be considered indispensable on the way back to help everyone to decompress once the meeting was over. On landing at Moffett Field it would pull up on the tarmac and decant us directly into the NASA buildings, while behind us US Navy Tomcats were taking off almost vertically with a shattering roar.

JPL also used to have a helicopter for the use of scientists and engineers in going to meetings. This departed from the Lab itself, from a pad on the top of the tallest building, which was adjacent to the Space Sciences building in which I had my office. It would take us to Burbank, where we could get on the JPL plane to go to Ames, or in the other direction to Los Angeles International Airport to board long-distance flights. The *Pioneer Venus* spacecraft were being built at Hughes Aircraft Corporation at Long Beach on the ocean side of Los Angeles, and we would fly straight there for meetings, landing on the lawn outside the Hughes headquarters building. The helicopter was the 'whirlybird' type, with a transparent canopy that extended right over the passenger's heads and below their feet. The pilot would skim over the Hollywood

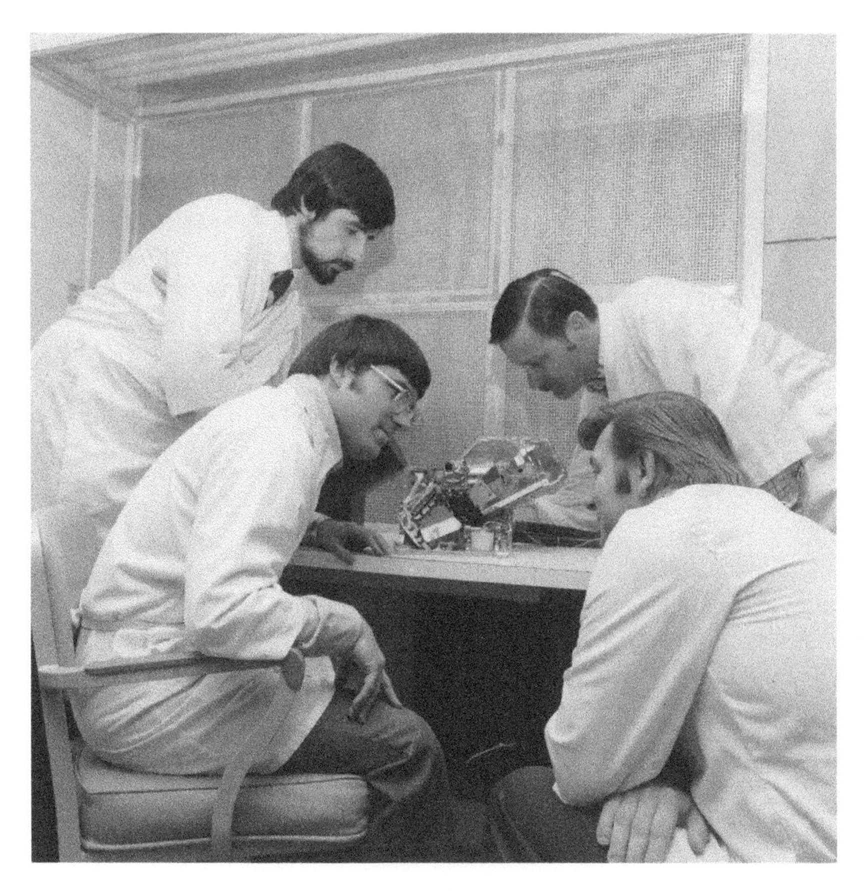

The partly-assembled VORTEX instrument for Pioneer Venus, on the bench at JPL in 1977. The author is top left (the beard did not last long); Fred Vescelus, the project manager, is below me (in glasses); top right is Paul Forney, project engineer, and below him another of the engineering team called Don.

Hills on a direct route to Long Beach and watch us all flinch as the ground rushed by a few feet below. No cocktails in this aircraft however.

Stowaway to Venus

The hardware delivered from Oxford was subjected to NASA quality control inspections after it arrived. The JPL engineers, looking for specks of dust, found a complete spider in one of the flight components and gleefully presented me with a touched-up picture, showing it wearing a bowler hat and waving a British flag.

Building a new instrument for a trip to another planet is a very demanding enterprise fraught with difficulties. One of my more experienced colleagues summed it up when he said 'things go wrong that you never dreamed of'. The three years it took to put VORTEX together were sometimes painful in the extreme, and involved cost overruns, traumatic personnel changes, all-night and Christmas Day working, and worst of all grim management reviews.

JPL's long-standing Director and founder, William Pickering, retired in the middle of this, and was replaced in April 1976 by Bruce Murray, a professor from Caltech who had worked on the early Mars missions and the *Mariner 10* Venus fly-by in 1972. Bruce was a much more aggressive character than Pickering, who was an old-fashioned gentleman, and on his arrival he saw himself as a new brush determined to sweep clean. It was my misfortune that the 'tentpole' problem at JPL when he arrived (*Viking* having left JPL for the first leg of its journey to Mars by then) was VORTEX, running late and spending way beyond its original budget. Bruce took a look at the state of my project, called me into his penthouse office, and told me to cancel the experiment. I couldn't of course, and wouldn't, but it was a bad moment. Of the words we exchanged, the only ones that I can actually remember, after the diatribe, were those when I said 'it will do great science . . . ', to which the reply was 'it's too late for that—[pause]—OK?'. It was not OK.

At this time of maximum trauma, a memo came around from our local section manager asking for candidate photographs, with captions, for a 'JPL album' that was being developed by the publicity office. Typical scenes showing the great things we do were being sought. I had just taken some pictures of the VORTEX team at work, so I submitted one of Reinhard Beer working at the controls of our test and calibration facility. This was housed in Building 11, the oldest structure on what was now a large modern site, and located right down in the Arroyo where it had housed the engineers working on early rocket motors in the days when JPL really worked on jet (or actually rocket) propulsion. It was so far from my office that I used some of my project funds to buy a bike for my frequent commutes between the two.

My colleague and team member Dr Reinhard Beer at work in the VORTEX test and calibration facility at the Jet Propulsion Laboratory.

The caption I submitted with the photo of Reinhard captures my mood at the time. It was headed JPL SCIENTIST DISCOVERS LIFE IN BUILDING 11 and went on, 'Dr Ludwig Lager, Famous JPL astronomer, is shown working with three and a half million dollars worth of government funded equipment, with which he has unambiguously detected life in the outermost reaches of JPL. Building 11 is so far away from the rest of JPL that it takes signals, travelling at the speed of light, 300,000 years to reach JPL management.'

On to the launch

We soldiered on and eventually delivered a superb instrument at a total cost of about four and a half million dollars (the original proposal had estimated a budget of not much more than half of that). In a clean bay at Cape Canaveral it was checked out on board the *Pioneer Venus Orbiter* by the technicians from Hughes Aircraft, who had brought the spacecraft, loaded with all of the instruments, down from their assembly plant in Long Beach. We had a lengthy series of tests to make sure the combination of all of the instruments and the spacecraft worked properly. For some of these I joined the other investigators and mission managers in one of those large control rooms you see on TV and in the movies, with everyone sitting in front of ranks of screens, knobs, and switches wearing headsets and microphones and saying 'check' or 'go' over and over again. During one of the last of these dry runs, just a few weeks before the spacecraft was due to be mounted onto the rocket that would send it to Venus, VORTEX stopped sending signals. 'We have LOS' said the Controller grimly (LOS is loss of signal—we talk in acronyms a lot in this business). Cue panic.

It was soon determined that this was an electronics failure. In order to make essential weight savings, we had built VORTEX using an innovative approach to packaging the components. This was a precursor to modern integrated circuits, with the semiconductor chips laid down on a single substrate instead of the usual, for then, individual transistors and resistors soldered to a printed circuit board and coated with a kind of space-approved varnish. The downside to the new hybrid circuits was that dismantling the instrument to find and fix the fault was a long and

The *Pioneer Venus Orbiter* spacecraft in the large clean room at the Hughes Aircraft facility in Long Beach, California. The VORTEX instrument can just be seen peeping out of the gold foil covering the thermal blanket, about half way from the central antenna. The scientist in the white coat is Harry Taylor, no relation to me, who was the Principal Investigator for the mass spectrometer experiment that was also on the spacecraft. We all posed for a picture similar to this; NASA mistakenly sent Harry's picture to me and the one with me in it to him; we never bothered to swop.

My first instrument heads into space: the launch of *Pioneer Venus* on an *Atlas-Centaur* rocket at Cape Canaveral on May 20th 1978. The *Pressure Modulator Radiometer* that I worked on at Oxford was launched on Nimbus 6 on June 12, 1975 but I had left that team when I moved to California.

complicated task, one that we couldn't attempt in the limited lab facilities available at the Cape, and we were very short of time to do anything at all.

Fortunately, we had a spare instrument, an exact copy of the flight hardware, fully tested and calibrated, back at JPL. We took the faulty one off the spacecraft and packed it in its special travelling case. By a cunning piece of foresight (crises like this were not unknown), this had been designed so that it just fitted on an airline passenger seat and could be strapped in on the first available flight to Los Angeles. Once back at JPL, the dysfunctional instrument was quickly swopped for the spare, which was then lugged back to Florida. When we quickly redid the most important tests with the spare instrument now sitting on *Pioneer Venus Orbiter*, there were no more problems. The spare would have to do the job at Venus as well. The instrument that should have made the trip is still, as far as I know, in a warehouse in Pasadena along with hundreds of other JPL artefacts.

Launch and arrival

Watching any launch into space is awe inspiring, but the feeling when several years of your life, and most of your aspirations as a scientist, are at stake on it is something else again. *Pioneer* was not a very large spacecraft—it weighed only about half a ton, all up, but to get to Venus it needed a two-stage rocket. On top of a large *Atlas* (developed as an intercontinental ballistic missile originally) was mounted a smaller *Agena* upper stage. Between the *Agena* and the spacecraft, sitting in its nose cone, was a small solid-fuelled rocket that would be used on arrival to slow the spacecraft down and steer it into orbit around Venus.

On the day of the launch, 20 May 1978, the team gathered early on the wooden stands, called for some reason bleachers, which were provided to give us a view of the blast-off. Although we were closer than anyone else, except the launch team in its protective bunker, safety decreed that the bleachers were actually more than a mile away from the launch pad. Disappointed by this, Paul Forney and I decided to get closer. Paul was one of the lead engineers on the VORTEX team; he was also a Commander in the US Naval reserve and used to go off on secret missions at the weekends. (One of our mutual colleagues once asked him what he did: 'Ordnance' was the one-word answer. 'Guns?' said the enquirer. 'Missiles' said Paul.)

Together we made it without interference to a vantage point closer to the pad, with a much better view of the *Atlas-Centaur*, which we could see emitting streams of vapour as its liquid oxygen tanks were topped up ready for launch. Concealed under a large, scrubby bush we sat and waited, unmolested, until the countdown started. Then, without ceremony, a security patrol drove up. 'OK guys', said the military policeman, waving his thumb back towards the rest of the team on the bleachers. They knew we were there all of the time.

When the lift-off came it was perfect, and actually quite spectacular even from that distance. After the party that always follows a successful launch we all went home and began to wait the seven-months it would take the spaceraft to get to Venus, while organizing the science team to get ready to receive and analyse the data. In December 1978 we all gathered at Ames once more to follow the arrival and orbit insertion at the mission control centre there. All went well again. After a few days to check out the spacecraft functions and determine its orbit accurately, we were soon back at Ames yet again for the turn-on of the instruments.

This was the big moment: was everything still working, would we get any data? We knew to the second when it should start to come in, allowing for the ten minutes or so it took the signal to travel at the speed of light from Venus to Earth. But for several minutes after that, there was nothing but zeros on the line printer. This is what we would expect if the cover that protected the aperture had failed to deploy when we sent the command for it to open. This was something we feared, because the cover deployment took place in zero gravity at Venus, something it was not easy to test in the laboratory on Earth. Also, it had been closed for months in the hostile space environment and could have stuck shut. Any one of a hundred other things could have gone wrong during the flight, or under the stresses experienced during launch. A cold hand gripped my heart. Then, suddenly, there was the data.

We followed it all night without sleep, and tried to do some crude data analysis so we would have something scientific to report to the press conference that was gathering the next morning. Dan McCleese noted something remarkable: the highest temperatures we were measuring seemed to be when the spacecraft was passing over the north pole. Venus spins on an axis that is nearly perpendicular to its orbit, so the Sun is always above the equator, where it is permanently midsummer. At the poles, in contrast, it is always winter. They should be colder than the equator, not warmer.

I got up at the press conference and announced that our instrument seems to be working perfectly and we had good data. I described how our goals were to study another planet for the first time with an instrument similar to those that had been revolutionizing meteorological studies and weather forecasting on the Earth for the last decade or so. Features of the early data showed things about Venus that we didn't understand yet, I said, and left it at that. The polar warming business could have been a mistake and I didn't want any headlines that I might have to withdraw later.

NASA held a press conference the morning after the first data was received from *Pioneer Venus*, May 20th 1978. My hollow-eyed look comes from being up all night checking that our instrument was working properly, and seeking early inferences from the measurements to describe to the journalists who filled the auditorium.

The mysterious Dipole

It was real, however. Venus's atmosphere really is warmer over the poles, despite much less heating by the Sun. While we were seeking to explain this, something even more remarkable appeared in the data. While *Pioneer* was once again passing over the north pole, in its 24-hour orbit, the readings were showing the warm temperatures we had now come to expect. But then they suddenly shot up to an even higher level for a short while, before falling back to the normal warm values. The scan had crossed something very hot at the level of the cloud tops. Quickly, we assembled lots of scans and determined the shape of the hot feature by making a crude image. We soon realized that what we were seeing was a big hole in the cloud cover that shrouds Venus everywhere else. The radiation VORTEX was measuring was streaming up through this gap from the hotter atmosphere below.

When you find something new like this, there is great pressure on investigators to write a short paper in an international periodical to share the discovery with colleagues worldwide. The holy grail of scientific publishing is the long-established weekly journal *Nature*, whose editors, however, are extremely picky about what they publish. They can afford to be, since they have space for less than 1% of all the papers

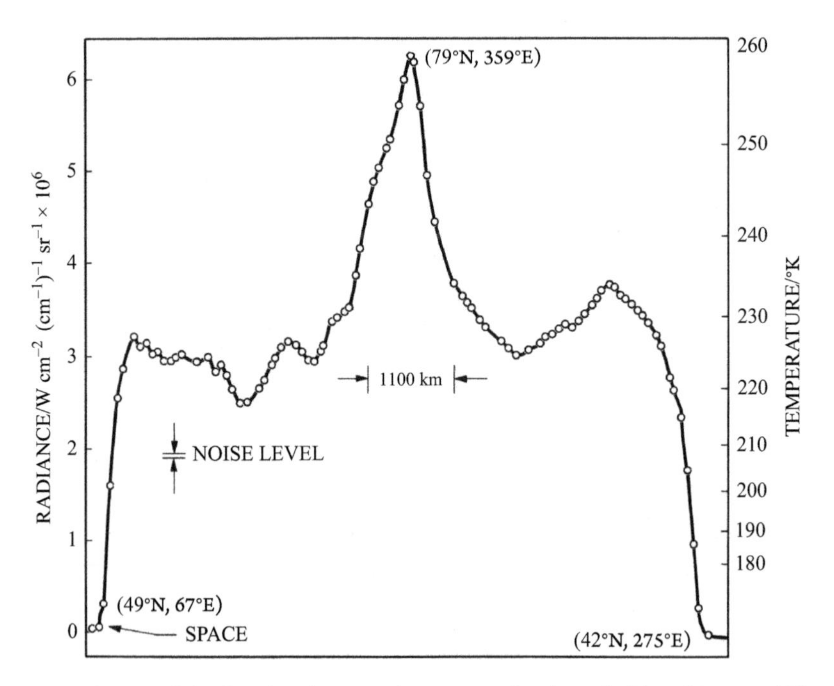

A plot of some of the first data from our instrument after its arrival in orbit around Venus revealed something remarkable. What was this huge spike in the temperature at the cloud tops that the instrument saw as it looked down on the North Pole?

they receive. In this case we had no problem—they not only accepted the paper, they decided to put one of our infrared images on the cover.

In the paper, we speculated that the hole in the cloud, and the more general polar warming, was probably a manifestation of a Hadley cell on Venus. George Hadley was a seventeenth-century meteorologist who proposed that the circulation of the Earth's atmosphere would be dominated by one large cell in each hemisphere, with warm air rising at the equator and travelling to the pole where, having cooled and become denser, it sank. Because the Earth rotates so fast, this is not exactly what happens, and although Hadley cells exist, they stop at latitudes well short of the pole. But now it looked like slowly-rotating Venus might fit Hadley's vision, not of course that he ever thought of it in that context. We argued that the descending air at the pole would heat the atmosphere by compressing it—the 'bicycle pump' effect—and this must be what produces the general polar warming. It would also push the cloud downwards to where the droplets would evaporate, and that might produce the 'hole'. Or something like that, it would take more time to tell for sure.

An early demise

As it turned out later, after much more investigation, this was more or less right, but was only the beginning of a long saga. As we built up more data and images of the hot spot in an effort to figure out what was producing it, we found it was not right at the pole as expected, but appeared to be orbiting around it, nearly 1000 miles away. That was odd, why would a Hadley cell do that? Then, we discovered that there is not one, but *two* holes, diametrically opposite each other on either side of the pole. More data showed the holes rotating around the north pole about once every four days.

Furthermore, the 'dipole', as we christened it, was surrounded by a belt of very cold air we called the polar collar. These extra discoveries suggested very complicated motions in the circulation of the atmosphere over the north pole of Venus. Presumably the south pole was similar, but we couldn't see that from *Pioneer's* orbit because the instrument was facing away from the planet when it passed over the south pole.

This is worth another paper in *Nature*, I thought. Remarkably, the editor and his advisors agreed, and not only accepted a second paper on a refined version of the same experimental result, rare for them, it also assigned us the cover again for one of our team's colourful images. Dave Diner made these by mapping the VORTEX data onto a globe representing Venus, using different colours to represent hot and cold temperatures. What a blast: it is unique in the 200-year history of the world's most prestigious scientific journal for the same author reporting the same experiment to have the cover story *twice*. And that is *ever*; we did it in a single 12-month period.

After his early inclination towards cancelling our experiment, Bruce Murray, the JPL Director, came to one of my seminars at Caltech at which I described the results we were getting from Venus and, to his great credit, stood up and said publicly that he had been mistaken about

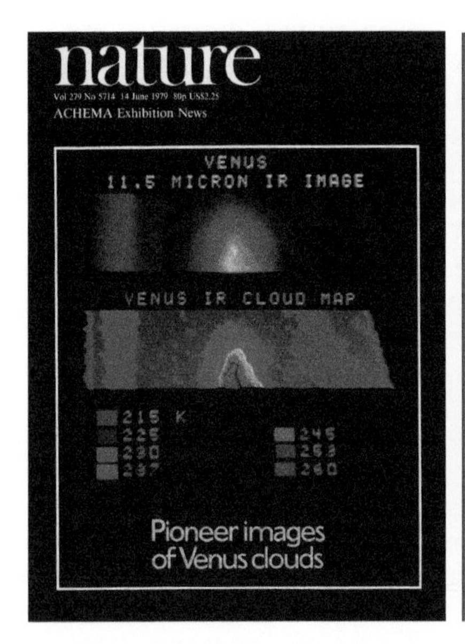

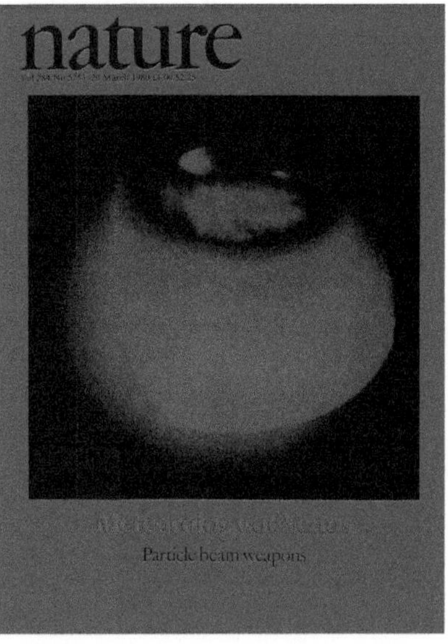

The cover of the leading scientific journal Nature for 14 June 1979 showed our detection of a hole in the Venus cloud layer near the North Pole, while the second cover published on 20 March 1980 revealed that it is part of the Venus polar dipole.

his expectations for what the experiment would achieve. He also wrote me a heart-warming letter, in which he said 'Your hard work has culminated in good science and a new understanding of Venus and the media recognition is frosting on the cake for you, your team, and JPL.' The head of NASA wrote a letter to Bruce, which he copied to me, saying '. . . in a letter from the White House, the President has expressed his congratulations to all those involved in this great achievement . . . '. Admittedly, he meant the whole *Pioneer* mission, not just VORTEX, but all this did a lot to salve the many painful wounds suffered on the way.

There were other excitements, like the time I appeared on the CBS Television evening news with its glamorous presenter Connie Chung. I arrived at the studio in Hollywood in my Aston Martin to be plastered with makeup and ushered into the studio, where the programme was already on air. I had expected a briefing or something but instead was almost immediately ushered up into the hot

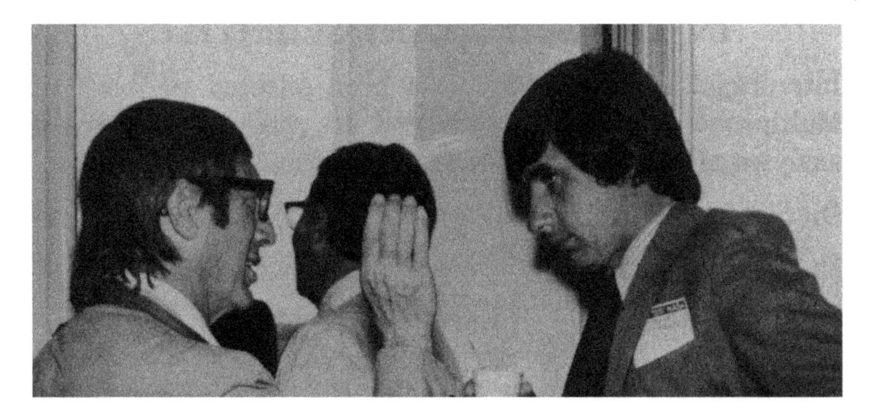

During a meeting of the Pioneer Venus science team at NASA's Ames Research Center near San Francisco, I chat with Hal Masursky of the Planetary Branch of the US Geological Survey. Hal was one of the leading experts on planetary geology at the time and a delightful colleague.

seat in front of the cameras during the first commercial break. Connie's puzzled expression during my explanations of what we were doing at Venus in response to her naive questions told me they were a bit too technical, but as it was going out live they couldn't cut it. This pleased me no end.

Things were going well for our team, and we were having a great time.

Spreading the word

The Principal Investigator's responsibilities, in return for being selected to participate in ground-breaking voyages to new worlds, is not limited to publishing the new results in scientific journals read by their colleagues, or the odd spot in the news media. They must also proclaim the findings more widely through presentations at seminars, conferences, and symposia. The larger international meetings are attended by thousands of scientists from a wide range of usually, but not always, related fields, and also by the press and by managers and politicians with an interest in the broad area covered. This might be all of space research, say, or all of astronomy, or a narrower discipline like planetary or atmospheric science. Once in a while there would be a smaller meeting devoted just to Venus.

When exciting new results come in, there is no shortage of invitations to make presentations at these meetings, near and far. After VORTEX delivered its first exciting results, and even before, since people like to hear about experiments 'in the pipeline', I travelled to Australia,

India, Africa, East Germany, and many others with my box of 35 mm slides and stacks of copies of published papers to distribute, detailing our results. The trips were not unadulterated pleasure; some of my memories are of the long boring flight to Australia, of suffering calamitous food poisoning in India, of witnessing awful politics in Africa, and of suffering the dour shabbiness and grim officialdom behind the Iron Curtain. Then there are climate extremes, hotel living, lost luggage, jet lag, and language problems that all have to be dealt with along the way. But these are all part of the essential dissemination of research findings to the world at large.

One of the largest and most famous gatherings in my field then was COSPAR, the Committee on Space Research, part of an organization based in Paris called the International Council for Space. This held a large gathering each year, moving around the world to a different venue each time, chosen to obtain a balance between geographical regions and political blocks, so that everyone got a turn to host. It was a particularly good meeting to go to with new results about Venus, because although it also dealt with things as diverse as galactic astronomy and using satellite data to improve crop yields in developing countries, it was the largest meeting with a good planetary science component that always had a good attendance from the Soviet Union. Sequestered behind the Iron Curtain, and generally prevented from discussing their progress or plans, colleagues from Russia in particular nevertheless had a vigorous and world-leading programme of Venus exploration that we in the West wanted to know about. At COSPAR they would ‘spill the beans’, as Don Hunten put it, talking somewhat out of the sides of their mouths at coffee breaks and in a more formal and discreet way in the sessions of the meeting.

That year, the COSPAR meeting was to be held in Bangalore, an attractive garden city in the southern centre of India. On 3 June 1978 I set off from Los Angeles to Delhi. I expected to find poverty, crowding, and begging there, but even so was overwhelmed by it, and found myself barely able to describe the experience in letters home. Instead I focussed on the charms of some of the buildings, and the fine hospitality of the people. I ran into a colleague from JPL, Richard Woo, bound for the same meeting, and together we toured Old Delhi and made a side trip to various attractions, including the incredible Taj Mahal.

The meeting itself went well, and, after an ill-judged farewell dinner with a small group of colleagues in a restaurant in the oldest part of town, I headed for the airport in Calcutta. The plan was to catch a flight from there to Nairobi, en route to Johannesburg where I was to visit my old friend Michael Robson, formerly of Seahouses, Northumberland. Giles, as he was for long-forgotten reasons always known at school, had been the best man at my wedding. Once again I ran into a congenial colleague from the meeting, this time Larry Soderblom from the US Geological Survey in Flagstaff, Arizona. We had already started working together as members of a team on the upcoming *Galileo* mission to Jupiter, as I discuss in the next chapter. We had a free day in Calcutta but neither of us felt like any more tourism, and in fact I was starting to feel distinctly unwell. We found a big western hotel, a Hilton I think, near the airport, and waited for our respective evening flights by the pool, with a waiter bringing us a steady supply of cold beer.

Larry's flight back to the States left before mine to Africa so I was alone by the time I made my way rather unsteadily and unhealthily to the airport. It was literally over 100 °F in the terminal, and it was crowded with people and flies. There was no sign of Kenya Airways; not even a desk, and no listing on the flight departure board. Eventually I found someone who knew something about them. He said that they kept a very loose schedule. They'll probably be here in the morning, he said. It was then about ten o'clock in the evening. My stomach at that point was in a state of complete revolt, my head was spinning from disease and beer and I could barely stand up, but there was nowhere to sit. And the toilets defied description, although I needed one urgently. In desperation, I went through a door labelled No Entry and found myself in a corridor lined with airport staff offices, where there were superior facilities. I sneaked back through that door, fortunately unchallenged, many times during a long and very hot night in which I very nearly lost the will to live. A semi-delirious note in my diary recorded it as the worst time of my life, and, looking back 45 years later, I would not change that opinion.

Thankfully, the plane did arrive in the morning, with a cheerful member of the Kenyan flight crew carrying a portable check-in desk to the centre of the concourse from where they collected our tickets and sent us to the gate to board. The plane was full of Indian men, not a single woman, and myself the only pale face (very pale, as I was still suffering from violent gastric disorders). The stewardess came around with the immigration and customs forms to be filled in and presented on arrival at Nairobi; I was pleased to see it was in English. Having filled them in, I looked around to see several of my fellow passengers staring at me and holding up their uncompleted forms; they were illiterate and wanted me to help. I tried with the nearest of them to convey his details to the form, and succeeded, with much use of sign language, until we got to the question about how much money the passenger was bringing into the country. I mimed money; he shook his head. No, no, money! I said, taking a few coins out of my pocket to illustrate the point. How much? He shook his head. After a few minutes of this someone was found who spoke a little English, and it suddenly became clear. The men were itinerant workers, going to Africa to do a construction job for several months. They would be housed and fed, and then they would be returned to Calcutta and their families clutching their pay. But for now, they had no money. None at all.

The core of the JPL VORTEX science team poses for a picture taken for the JPL newspaper 'Universe' in January 1979. From left, Martha Hanner, Fred Taylor (clutching a map of cloud top temperatures on Venus), David Diner, Paul Reichley, Dan McCleese, and Lee Elson.

We arrived in Nairobi more than 12 hours behind schedule and of course I had missed my flight connection to Johannesburg. There was another one the next day, and the airline said they would put me up in a hotel in town. The hotel looked quite nice, with a traditionally built but well-maintained building and a garden full of palm trees. The steps to my room were on the outside of the building, under the fronds of exotic trees full of lively birds. But when I got there I found that the room was already occupied. A very burly, unshaven middle-aged man sweating profusely in shorts and a Bruce Willis style dirty undershirt opened the door as I tried to use my key. I had to explain that the hotel had made a mistake and assigned us both the same room. We chatted uneasily for a moment, long enough for me to find out that he was an Israeli soldier, presumably a mercenary as there was no sign of a regiment. Nor, thankfully, were there any uniform or weapons in evidence, although we did not get into that. Back at the desk, the nice receptionist explained that the hotel was full and they expected us to share rooms. I was still ill, and this was the last straw. No way, I said. I got a room to myself.

The flight the next day was not until the evening, so I hired a Land Rover and guide, shared with a randomly encountered Australian couple on holiday, to tour around the famous game park, which was nearby. My vision of being surrounded by giraffes and tigers did not materialize although I think I saw what the guide claimed was a lion in the distance on the horizon. Back at Kenyatta International they demanded an airport tax of 20 shillings at the gate before I would be allowed to fly. I had given my last bit of local currency to the guide in the game park and only had traveller's cheques of large denomination. My new Australian friends paid for me, averting another long delay in my itinerary.

Once in Johannesburg, I had some more free time on the days when my friends had to work. Michael's wife Barbara ran a pharmacy, and soon doped me with powerful medicine that fixed my symptoms from India but left me rather woozy. In a daze, I took myself off to see the sights in my rented car, heading out to Pretoria and then to Soweto, driving with the windows up and the doors locked. At the end of the day I felt even more sheepish: despite plenty of poverty I saw no sign of danger or unrest. In fact, everybody was very nice except the police, but I mostly avoided them.

I went shopping to look for a diamond, my idea being to take home to Doris a proper engagement ring at last. Before we got married ten years earlier, all I had been able to afford was a rather pretty but undeniably commonplace turquoise gemstone called an aquamarine. As I cruised the shops in downtown Johannesburg a man with bright red veins in the whites of his eyes accosted me in one of the shop doorways. Turning his back to the street, he clandestinely produced a large and impressive diamond and showed it to me. He said I could have it for one tenth what it would cost in the shop. He scratched a deep groove in the shop window with it to show it really was a diamond. I thanked him and explained that I would be taking it back to America and would need a proper purchase certificate for the U.S. Customs. He looked at me pityingly. You don't need to show it to them, he said. Stick it up your arse! I turned and went quickly into the shop in case he decided to demonstrate.

Trouble at t' mill

Perhaps it had been going too well at Venus. VORTEX had been mapping the atmosphere for just 72 days when the data flow suddenly stopped. All of the other experiments on *Pioneer* were continuing as before, so the problem was in our instrument and not on the spacecraft or the data relay stream. When an instrument goes dead it is hard to tell what the problem is since you have no information, except that it must be in the electronics. Other common failures, in the sensitive infrared detectors, for example, still return a signal of some sort even if it is useless for science. As with the instrument that failed at the Cape, we were probably paying the price for being ahead of our time and using hybrid integrated circuits.

The instrument never revived, and its loss was a big blow, not just because we would have liked more science data, but because our contract with NASA was to build something that would work for at least one Venus year in orbit. To fall short of that was damaging to JPL's reputation as the leading space hardware centre for planetary missions. On the other hand, two and a half months is a long time and we had made millions of measurements in that time, enough to address most of our objectives. The Soviets attempted a comparable experiment in 1983, with a similar but larger infrared instrument on their *Venera-15* mission, and that failed even sooner, after just 60 days. We still found a lot to talk about when our paths crossed at international conferences, including a fierce but friendly argument about the amount and variability of water vapour in the atmosphere of Venus, something still not fully resolved today.

Because its spin axis is nearly vertical, Venus has no seasons, and for this reason planning to work for a year is an arbitrary choice. It was partly a hangover from Earth observing missions where full seasonal coverage can be critical. Having a fixed data set to work on is actually a considerable advantage when funds are limited, and we were complimented after the formal end of the mission by the chairman of the Scientific Steering Group as being the only experiment team that was on top of their data analysis task. So, we went on to analyse the data we had and published lots of results; much of it was quite amazing science and we were very happy. You can read about it in my book, *The Scientific Exploration of Venus* (Cambridge University Press, 2014).

The first British hardware to go to another planet

The pressure modulator radiometer built in Oxford for the Venus project was the first piece of equipment of any kind built in Britain that had been sent to another planet. This remarkable achievement was little noted then, and remains uncelebrated to this day, it's not the British way. When the Queen, no less, visited an exhibition set up by the Royal Meteorological Society to mark its Centenary in 1983, my co-investigator John Houghton happened to be the current President of the Society, and so he escorted her around and explained the British role in the project. John showed her a model of the *Pioneer Venus* spacecraft, which I had borrowed from NASA and shipped to England, displayed beside a large diagram of its orbit around Venus. The Queen studied this for a moment and then asked 'Does it go round and round?' I later reported to the project leaders at NASA Ames that 'our collaborators at Oxford had had the opportunity to describe the *Pioneer Venus* mission to Queen Elizabeth II, and to answer her questions about the precise nature of the orbit.'

We had helped *Pioneer* to become the first Venus weather satellite and carried out the first global investigation of the meteorology of Earth's twin planet. All sorts of interesting things showed up: the temperature increase towards the poles, contrary to the heating by the Sun; the water vapour concentrations changing sharply with time of day; the great double vortices at the poles; various cloud structure patterns, waves, and tides that mirrored the Earth, although sometimes in strange ways.

We also were able to study the energy budget of Venus, by estimating the total sunlight going in and the total heat radiation coming out, as a function of latitude. Like the Earth, there is more heating at the equator and, despite the upper-air warming, and the presence of the hot dipole, overall cooling at the poles. What is of most interest is how the atmosphere redistributes this geographical difference to produce an overall global balance. We were also able to show a crazy but vocal fringe group, of the kind that is common over all areas of science, that the heating and cooling of Venus are indeed at least approximately in balance, and the high temperatures are not due predominantly to the fact that the planet is cooling down, as they had alleged.

Behind the Berlin Wall

It was rewarding to discuss the results with the Russians and their allies, because they had such an intense interest in Venus and an ambitious programme of exploration of their own. For many years, they dispatched at least one of their huge probes in every one of the launch windows to Venus that opens up for a few weeks every 19 months. Getting detailed results from their taciturn representatives and discussing how they related to our own findings was far from easy, and involved making some forays into their territory, since they seldom ventured out. In 1975, the annual COSAR meeting was held behind the Iron Curtain in Varna, Bulgaria, and I decided to go and learn as much as I could about the *Venera* programme to help with planning our own upcoming expedition to Venus with *Pioneer*. Doris came with me as we planned to take a vacation after the working part of the trip had been concluded.

It all started badly when we got to London after the flight from Los Angeles, to be told our luggage had to be collected and transferred by hand for the next leg to Varna on Bulgarian Airlines. However, there was no sign of it among the luggage unloaded from our flight, nobody at Heathrow knew where it was, and in those days they didn't care either. Eventually we just had to get on the onward flight faced with the prospect of two weeks in Bulgaria with only the clothes we stood up in. Dinner was served on the flight and consisted of half a chicken that looked as if it had been starved to death: there was not a morsel of meat on the carcass. Tired and hungry, we arrived in Sofia to be greeted by our luggage. I don't know how it got there and this time it was me that didn't care.

We drove from Sofia to Varna, which is on the Black Sea at the other end of the country, in a rented car. There were very few cars of any kind around, and when we drove into one of the many delightfully rural and unspoiled villages the inhabitants would come out of their houses and stare at us. Communication was all but impossible, and we soon found that we needed the map the rental company had provided showing where to find one of the mere handful of filling stations that existed in Bulgaria then.

The sessions in Varna were good, but we were still building VORTEX then and I had no data of my own. That had all changed by 1987, when the Soviet Academy invited me, at their expense, to a celebration of the 30th anniversary of the first Sputnik launch. I set off to Moscow using the ticket on Aeroflot that they sent me and the visa that the letter of invitation engendered. The celebrations and the hospitality were a pleasure, but the planetary science content was not high so it was pleasing when, a short time later, there was a meeting dedicated to Venus at Potsdam in East Germany.

Getting to Potsdam involved a fascinating and very long journey, first to Berlin, and then into the underground system to cross to the East via trains and long, shuffling queues in badly lit dystopian tunnels to present credentials to the surly guards at the other end. When I finally emerged into the daylight, I was met by a car and driver who took me on a long drive of more than an hour to Potsdam. Looking at the map after I arrived, I realized that I was almost back at the airport in Berlin, but now on the other side of the Berlin Wall (actually a barbed wire fence at the Potsdam end). Years later, after the barriers had come down, I made the same journey easily by taxi in less than half an hour.

The discussions about Venus on all of these occasions were productive and the data we had acquired were pleasing and provided a lot of answers, but we had also raised a further set of questions that it would take a new mission to answer. We didn't really understand the polar vortices, for example, or why they had such a complex double shape. The massive greenhouse effect was obviously real, but the contrast with the climate on Earth was as puzzling as ever. Additional data, with more detail and some different approaches was needed. From then on I was determined to do whatever it might take to make that happen, but events were to break the chain of missions in both East and West, and many years would elapse before any of us saw Venus close up again.

Postscript

Nearly 20 years after the early demise of the VORTEX instrument, and long after I had returned to Oxford, I was at a committee meeting in Washington DC. In the evening a few of us went to decompress at a bar just off the Mall that had a reputation as a watering hole for congressmen and others in the corridors of power. It was early, and the crowds were yet to arrive, but sitting at the bar was none other than Charles F. Hall, who had been the manager in overall charge of the *Pioneer Venus* mission. He had long since retired from Ames Research Center, but still did occasional jobs for NASA. He was in Washington to sit on an internal enquiry into why the costs of space missions always escalate (although he put it in more colourful language when I asked). We talked about this for a while, and it came out that his own Venus mission was a rare counter-example, having pretty much stayed within a very tight budget. What about VORTEX, I asked, fearful of being told that this high-level body had discussed my experiment as an example of cost inflation, delivery delays, and early failure at the target. 'Hell, there's a lot worse than you' said Charlie.

5
With *Galileo* to Jupiter

It's not solid, the way you drew it, Patrick; when you see it close up it's more like the spiral sprinkler you use to water your lawn. It's not even very red.

- Author to Patrick Moore, discussing the Great Red Spot on Jupiter

One very hot Californian summer day in 1976, I was freezing in my air-conditioned office at the Jet Propulsion Laboratory in Pasadena when Torrence Johnson walked in. He took a seat and started to talk about his plans for participation in NASA's new mission to the outer Solar System, then called *Jupiter Orbiter-Probe*, or JOP for short. Missions generally had functional, descriptive names like this when they were on the drawing board. By the time they made it to the launching pad, the name had usually been changed to something more romantic and evocative. Thus, JOP became *Galileo* in February 1978.

At this time only the small 'trailblazing' spacecraft, *Pioneer 10* and *11*, had been to Jupiter. The larger and more sophisticated *Voyagers* were being developed to follow them, and *Voyager 1* reached Jupiter on 1 February 1979. It flew by like a speeding bullet at a distance of 20 million miles, much too quickly to gather the data that would be needed to unravel the activity in the atmosphere that produces prominent features like the cloud bands and the giant eddies, or to determine what produces the colours in features like the Great Red Spot. *Galileo*, in contrast, was planned to be the first artificial satellite of Jupiter, orbiting around the largest planet in the Solar System for many years. It would offer the first chance for long-term, close-up studies of Jupiter and its family of moons, and would drop a probe into the atmosphere, to make measurements down to a great depth as it descended on a parachute. This would be the first time that these things had been done for any of the four giant planets that lie out beyond Mars and the main asteroid belt.

Jupiter orbits at a distance from the Sun that is five times that of Earth and is more than ten Earth diameters across, so by volume, Jupiter is more than 1000 times larger than Earth. Between them, the two *Voyager* spacecraft would briefly visit Saturn, ten times further out and nine times larger than Earth, in 1980; Uranus, 19 times further and four

Earth diameters across in 1986; and finally Neptune, 30 times further from the Sun than Earth and about the same size as Uranus, in 1989. In NASA's long-term planning, there would eventually be *Galileo*-style orbiting missions for all four outer planets, and as we shall see in Chapter 10 this did happen for Saturn eventually, although we still face an indeterminate wait for the other two.

Destination Jupiter

Torrence was, and is, a leading expert on Jupiter's family of large icy moons, and he had spent years observing them from the Earth, using the giant Mt Wilson and Mt Palomar telescopes. He used near-infrared spectroscopy, a technique in which reflected sunlight is dispersed by a diffraction grating to produce a spectrum showing how the brightness varies with wavelength, revealing the 'colour' of the object at wavelengths longer than those the eye can detect. By analysing many such spectra, the composition of each moon can be determined, along with all of the subtle variations in the features on the surface. Now Torrence was planning a compact version of this instrument that could fit on the payload of the forthcoming *Galileo* spacecraft to study Jupiter's moons in detail from close up. He called it NIMS, for N*ear Infrared Mapping Spectrometer*.

Knowing how hard it is to win participation in any space mission, but especially a big flagship project like this, Torrence had been thinking about how to 'sell' his ideas to NASA. NIMS had already been proposed unsuccessfully for flight on the *Voyager* 2 mission to Jupiter and Uranus. The new proposal for *Galileo* was solid as far as the science was concerned, but even in orbit around Jupiter the spacecraft would only spend a few hours passing close to each of the moons, since it would take several weeks to complete each orbit of the enormous planet. This meant a lot of observing time was available for looking at Jupiter itself. Could NIMS, designed to measure the light reflected from the icy surfaces of the moons, do anything useful by way of studying Jupiter's deep and mysterious atmosphere, he wanted to know? You bet it could.

My work on 'remote sounding' of outer planet atmospheres, described in Chapter 4, originally intended for the abortive *Outer Planets Grand Tour* project and its surviving remnant, *Voyager*, was about to pay off at last. By studying how 'remote sounding' methods for measuring temperature, composition, and cloud profiles could be adapted from the Earth and Venus to the very different conditions in Jupiter's atmosphere, and publishing a couple of papers in atmospheric science journals, I had become a minor expert on this rather way-out subject. There were not many of us in those days, and it is still not exactly a mainstream subject today.

I was working on my own idea for an instrument for the same mission at the time. I called it the I*maging Infrared Radiometer*, and a

Galileo's objective: a collage of Jupiter and his family of large moons, top to bottom Io, Europa, Callisto, and the largest, Ganymede.

The *Galileo* spacecraft leaves for Jupiter on top of a solid-fuelled booster, while the space shuttle *Atlantis* that brought them both into Earth orbit attends in the background. The folding antenna intended for communications with Earth can be seen in its furled state at the top; later, when the command was given to release it, some of the spokes stuck and it never could be used.

yellowing copy of the proposal is still on my shelf. I planned to study Jupiter at long infrared wavelengths in the heat radiation from the atmosphere, to measure temperatures, and investigate the complex coloured cloud structures and their formation processes. It would need high spatial resolution to see the detail, hence the 'imaging' part. It was also going to investigate why Jupiter appears to emit more energy than it receives from the Sun; about twice as much, in fact. This remarkable fact had been discovered in 1969 by George Aumann, who was my first office mate when I arrived at JPL in 1970.

My proposal also had temperature mapping of the satellites as an important subsidiary objective. The experiment was thus highly complimentary to that of Torrence, in method and objectives, and it could easily have happened that both instruments went to Jupiter on board *Galileo*. After discussions, we agreed to help with each other's proposal, with ourselves and several other team members in common. However, in the highly competitive selection process, NASA chose NIMS but not the radiometer for flight. As I was still busy with *Pioneer Venus* at the time, this was quite satisfactory as far as I was concerned. Being a Principal Investigator is a full-time job; but the effort required to be an ordinary team member (a co-investigator in NASA-speak) is much less, and you can be on many teams at once if you are not the leader of any of them. *Galileo* was not scheduled to launch until January 1982, about three years after *Pioneer Venus*. In the event, it actually launched in October 1989, nearly eight years late and a whole decade after the Venus mission.

Saving *Galileo*

The design teams went to work on the *Galileo* spacecraft, and then the almost inevitable happened: NASA had another of its regular funding crises. These upsets were a consequence of the competitive way the government funded the space programme and not always the fault of NASA management, but the pain had to be passed on and this could be devastating for JPL. As a big, expensive project that had not seriously begun to spend yet, the Jupiter orbiter on which the laboratory depended for work and funds was in real and present danger of cancellation.

In this sort of situation, familiar to everyone at JPL from its countless recurrences over the years, the name of the game becomes to muster support for your own project and hope they cancel someone else's instead. NASA's expenditure on manned spaceflight at that time was focussed on the development of the space shuttle and looking ahead to a permanently-staffed space station. These programmes were both so expensive that a robotic planetary mission could be saved from cancellation at the expense of a relatively small reduction in the budget set aside for things like astronaut training, for instance. But first you had to convince the NASA administrators in Washington, and their political paymasters in Congress and the Senate, to go that route.

Torrence Johnson campaigned tirelessly for *Galileo* and was probably more than any other one person responsible for saving it from extinction. When NASA confirmed that the endangered mission would go ahead after all, the Director of JPL, Bruce Murray, himself a veteran of numerous space missions, rewarded Torrence with the prestigious role of *Galileo* Project Scientist. He would be responsible for the successful implementation of all of the scientific investigations on the orbiter spacecraft, providing leadership to the individual Principal Investigators in charge of each of the instruments. There was a separate Project Scientist for the atmospheric entry part of the mission. He was Richard Young from NASA Ames Research Center near San Francisco, which by then had moved on from *Pioneer Venus* to develop the atmospheric entry probe for *Galileo*.

Torrence's promotion meant that there was a vacancy for the leadership of NIMS: it would not be politically acceptable nor practical in terms of workload for the Project Scientist to also be Principal Investigator for one of the instruments. As it happened, one of the experiments on *Pioneer 10* and *11* had been developed at the University of Southern California, just across the centre of Los Angeles from JPL. One of the lead scientists for that was Robert Carlson, whom Torrence soon recruited to join JPL and, with his ready-made experience of Jupiter science and mission management, to take over as Principal Investigator for NIMS. Torrence downgraded himself to a Co-investigator role similar to mine. Bob was an inspired choice, and NIMS and *Galileo* were to flourish under his leadership, although not without having to weather many storms along the way.

The French Connection

With the future of *Galileo* secured, with NIMS on board, serious planning for the scope of our experiment could begin. Top of the list was to think ahead to the science to be done with the data, once acquired, and to design and build the hardware. As the team member charged with planning the exploration of Jupiter's complex atmosphere, I soon had big concerns on both counts. Nearly all of the data from the mission as a whole would be obtained when the instrument was looking at the atmosphere! But I was one co-investigator out of ten, and the other nine were all satellite surface scientists of one kind or another. Making proper use of such a large and novel set of data was way beyond the capabilities of one person, even with a supportive sub-team.

We needed more co-investigators, but achieving this is much harder than an outsider might think. NASA is very loath to add team members once the mission is approved and underway, for two main reasons.

The first is political—everyone with some relevant background wants to be a part of a big new mission like *Galileo*: they know it is going to lead to big breakthroughs in their field. It also means job security and respect from your peers. If NASA did allow joining-up by any other than a scrupulously fair, competitive mechanism then accusations of wheeler-dealing would be rife. The other reason is financial—if you are accepted as an official co-investigator, then your involvement is funded by NASA, including reimbursing your institution for the relevant part of your salary and overheads. For a long mission like *Galileo*, funding just one more co-investigator could easily add more than a million dollars to the overall cost.

Every problem is an opportunity in disguise. In this case, the answer came quickly. I had read a recent paper by a young French scientist, Thérèse Encrenaz, who had been studying Jupiter's infrared spectrum while visiting the Goddard Institute for Space Studies in New York, a few years after I had spent the summer of 1966 there (as discussed in Chapter 2). A paper she had written with her colleagues in GISS gave me a flying start on understanding what NIMS could expect to see when it arrived at Jupiter, and she was an obvious candidate to join the team and help with real data analysis when the time came. Not only that, but by that time she was back in France, working at the Paris Observatory, and hence could be presented to NASA as a foreign co-investigator who was not in competition with American scientists and who would cost NASA nothing. I made contact with Thérèse to confirm that she would like to take this on, and that she could raise the necessary funds from the French space agency CNES, and with Bob Carlson's help we obtained NASA's agreement as well.

Getting the instrument right

The second problem/opportunity with NIMS as it stood was that it didn't measure all the right wavelengths for atmospheric studies. The satellite scientists were only interested in reflected sunlight from the surfaces of the moons, which became too weak to measure at wavelengths longer than about 2.5 microns. (A micron or micrometre is one-millionth of a metre. Visible light ranges from about 0.3 microns (violet) to about 0.75 microns (red) in wavelength, with wavelengths longer than that termed infrared. A human hair is about ten microns across, the same as a typical mid-infrared wavelength.) However, as the work by Thérèse and others had shown, Jupiter is very bright at longer wavelengths, because of heat radiation emitted from its deep atmosphere. It happens that Jupiter's atmosphere is quite transparent at wavelengths near five microns, a so-called 'window' in the spectrum, so the emitted radiation comes from depths where temperatures are relatively high, similar to those at the surface of the Earth in fact. Radiation from these depths could shine out strongly into space in places on Jupiter where the overlying cloud cover was thin or absent; such regions had been observed with Earth-based telescopes and named 'five micron hot spots'.

Understanding the meteorological processes by which the hot spots formed, and peering through them to probe the temperature and composition in the atmosphere below the visible clouds, would be crucial to learning how Jupiter's atmosphere was constituted and how it behaved. If NIMS didn't cover the window wavelengths then its observations of the atmosphere would be restricted to altitudes around the cold, high ammonia cloud tops and most of the interesting science would be missed. Fortunately, it was not difficult to convince the rest of the team, the *Galileo* project managers, and NASA that the small extra expenditure required to re-engineer NIMS to extend its wavelength range would be worthwhile.

The science and engineering teams could now get down to work actually building hardware. This meant regular meetings, several each year, at three levels: the Project Science Group, planning things like the shape of the orbit and the number of encounters with each satellite; the Science Team, planning the mapping coverage of Jupiter and showing how it addressed our goals; and the Engineering Team, which built, tested, and calibrated the hardware with the participation of the scientists. These technical meetings could take place at the home bases of participants or contractors, or as an adjunct to a large meeting of the planetary science community where special sessions on Jupiter and *Galileo* were regularly organized.

The science team would meet mostly at the institution of the Principal Investigator, so in Pasadena in the case of NIMS, but also every so often one of the co-investigators would host a meeting. The part of the US Geological Survey that dealt with planetary geology was located in Flagstaff, Arizona and we had three team members there. As already related, in those days you could catch a train on the Santa Fe Railroad from downtown Pasadena in the evening and be woken up in your sleeper car with coffee at 6am in Flagstaff, amid spectacular scenery. The mountains near the town were home to the observatory built in 1894 by Percival Lowell to observe the canals he thought he could see so clearly on Mars.

An even more popular destination for meetings, however, was Hawaii, where Tom McCord was a professor in the university and also ran, with his wife Carol, a small company in the jungle near the centre of Oahu. It was here that we had the meetings, among the pineapple plantations. The meetings in Paris were generally in the magnificent old observatory in the city centre, rather than at its present site in Meudon, ten miles outside on the site of the old royal summer palace. The first NIMS meeting I hosted in Oxford after moving back was held in our comparatively shabby Atmospheric Physics laboratory, but was notable for coinciding with the first ever launch of a space shuttle, as related below.

A conference about Jupiter's atmosphere was held at the Goddard Institute for Space Studies in New York from 6th to 8th of May 1985, nearly twenty years after I spent six weeks there as a summer student. Here I am in the front row, nearest to the camera, asking a question of the speaker; beside me is my friend and mentor Dr Don Hunten of the University of Arizona.

Jupiter and its mysteries

When we started to work on planning and implementing the NIMS experiment, Jupiter was little explored, and even more mysterious than it is now. In our 1975 proposal to NASA for participation in the 'Jupiter Orbiter Probe 1981/82 Mission', the first five of the eight goals listed were addressed at the satellites, specifically:

- to map the compositional units on their surfaces,
- to establish the range of chemical species present,
- to identify systematic compositional trends,
- to establish the properties of geological units, and
- to understand the surface processes represented.

For the atmosphere, we said we would establish temporal and global variations in the temperature and cloud structure and mixing ratios of gases like ammonia and water vapour and provide global context for the entry probe data. Thus the probe experimenters would know how typical the region was that they studied as the probe passed through. It wasn't very typical at all, as it turned out.

Jupiter's coloured, banded appearance is enough to demonstrate that the clouds have an exotic composition and structure, and are not of a single type. In fact, at least four layers of clouds of different composition were predicted to be present within the height range probed by our instrument. We wanted to demonstrate that these multiple cloud layers actually exist, and to investigate the origins of the various colours, shed light on the morphology of the giant eddies, especially the Great Red Spot, the white plumes, and the five-micron 'hot spots'.

For studies of the composition of the atmosphere, the emphasis was less on the major constituents hydrogen and helium but rather on the variable constituents, such as the condensable vapours of water and ammonia, and the chemically reactive species like phosphine and germane. Germane is an atom of germanium with the maximum allowed number of hydrogen atoms attached, four in this case, so its chemical symbol is GeH_4. Thus it is analogous to phosphine (PH_3), ammonia (NH_3), and water (OH_2). All of the common elements, including phosphorus, nitrogen, oxygen, silicon, and germanium are in the main present as these hydrides because Jupiter's atmosphere is mostly (about 85%) made up of hydrogen. Any observed deviation from this basic mix was evidence for non-equilibrium chemistry, and the subject of much interest.

With NIMS we could measure temperatures and cloud structures to investigate the nature of the various types of eddies and their role in the general circulation of Jupiter. But to have any hope of achieving this, we had to train ourselves and our research assistants and students and write a lot of specialized data handling and analysis software, while travelling the world from Paris to Hawaii to meet each other and correlate our efforts. It was very hectic, for a while. Then the launch was delayed and various other things went wrong.

Problems

Engineering problems and cost overruns are commonplace on space projects, especially ambitious planetary missions, and *Galileo* had more than its share. The spacecraft was large and complex, weighing two and a half tons, and standing seven metres tall. It was designed with a novel (some said crazy) 'dual-spin' approach, in which one half of the main body of the spacecraft rotated with respect to the other. This was great for the science instruments, since those measuring fields and particles needed to scan through a complete range of angles, while the cameras and other optical instruments, including NIMS, required a stable pointing platform. However, these benefits were obtained at the cost of providing a large spin bearing and slip rings to transfer power and data between the two sections. As seems to always be the way, the elaborate and risky components like this worked fine, and it was one of the simpler devices that let us down.

The need for an exceptionally large antenna dish to relay the data over the vast distance between Jupiter and Earth, up to a billion kilometres, without adding too much weight, was addressed by making the dish out of gold-played metal mesh and folding it up for later deployment in space. The result looked rather like a giant umbrella nearly five metres in diameter. As a further complication arising from the great distance of Jupiter from the Sun, which made solar panels impractical, the spacecraft drew its power from two long-lasting radioisotope thermoelectric generators each containing several kilograms of radioactive plutonium.

While the engineers at JPL struggled with this monster, the smaller NIMS team in a different building was putting together a spectrometer that could cover the whole wavelength range—window and all—with a single diffraction grating. It could tilt this mechanically to scan the spectrum, while simultaneously moving one of its telescope mirrors to scan across the planet. Coverage in the perpendicular direction was achieved by moving the whole instrument using the scan platform on the spacecraft. The end result was that NIMS data came in cubes, with the three basic dimensions of latitude, longitude, and wavelength. Since different wavelengths probed different depths in the atmosphere, by scanning in and out of the strong absorption bands of the atmospheric gases, this meant our device was essentially a three-dimensional camera. Such sophistication came at a cost in money and delays, taxing the budget until it was sufficient for just one instrument instead of the usual prototype, flight, and flight spare models. Remembering how the existence of the spare had saved us on *Pioneer Venus*, I was afraid this might be false economy, but in the end we were lucky.

NASA's plans called for *Galileo* to be launched by the Space Shuttle. The Shuttle, of course, could not fly to Jupiter but could only reach low Earth orbit, but this saved the first and largest stage of what would have been a big multi-stage rocket if launched from the ground at Cape Canaveral in the usual way. The idea was to mount *Galileo* on its second stage, a liquid-hydrogen-fuelled *Centaur* rocket specially modified to fit in the Shuttle's cargo bay, and take both into orbit. There they would be released, the Shuttle would retreat to a safe distance of a few miles, and the *Centaur* would be ignited at the right place and time for it to end up on a transfer orbit to Jupiter. With the boost from the powerful *Centaur, Galileo* would be at Jupiter in just two years.

But first, the shuttle itself had to overcome numerous delays, and it was not until 12 April 1981 that the first of the fleet, *Columbia*, reached orbit. The event was televised amid great excitement, and a goodly amount of nervousness; the shuttle design had been much modified during development, including the drastic change of adding ablative tiles to the outer skin to ensure a safe re-entry. As already mentioned, there was a NIMS Science Team meeting in Oxford that week, and the whole team gathered at my house for a party to follow the flight on TV. Unsurprisingly, to those of us used to the general angst of a career in space science with its ups and downs, the loudest voice amid the cheers following lift-off was Paul Weisman exclaiming 'We've got jobs!'

Paul had the job called NIMS Experiment Representative, responsible for the interface between the NIMS instrument and the spacecraft and the technical information flow between the NIMS team and the Galileo Project. He had a further emotional experience on the same visit when I walked the team around the historic part of Oxford, including the house on New College Lane where Edmund Halley lived and worked. Paul vanished from my peripheral vision and I looked down to find him genuflecting to the house out of respect for the memory of the great astronomer and geophysicist.

But *Galileo* was still not ready to fly. By the time preparations for launch were complete it was 1986, when the loss of the shuttle *Challenger*, on the flight immediately before that which would have carried our payload, stopped the Jupiter project in its tracks. Not only was the rest of the shuttle fleet grounded, but also the liquid-fuelled *Centaur* upper stage that was to power the spacecraft on to the outer Solar System was now deemed much too dangerous to carry in the cargo bay of a manned launch vehicle. This was not paranoia: the *Centaur's* tanks full of liquid oxygen and liquid hydrogen were potentially one of the most explosive mixtures known, short of nuclear weapons, and next to these were the two plutonium-fuelled power supplies on board *Galileo* that could cause serious contamination of the Florida environment if their integrity were to be breached in an explosion.

Fortunately, an alternative to *Centaur* was available in the form of the solid-fuelled *Inertial Upper Stage* or IUS. This had been developed by Boeing for the US Air Force, to lift payloads from the low orbit of the shuttle to the geosynchronous orbit 36,000 kilometres above the Earth that is used by communications satellites. Its rather odd name came about because it was originally called the *Interim Upper Stage*, because it was intended to be replaced soon by a reusable space tug that would take over the task of moving payloads from the shuttle to their final orbits. When the tug failed to materialize, an alternative use of the 'I' was found to keep the acronym the same; 'Inertial' now referred to the guidance system (informally known among the mission controllers as 'push and pray').

The trouble was that the energy required to get to Jupiter by the original flight path was much greater than the relatively feeble IUS could deliver. The only available solution was to take much longer to get to Jupiter, and to go via multiple fly-bys of the Earth and one of Venus in order to use the gravitational 'slingshot' effect to boost the speed of *Galileo* on its trajectory. The bad news was that this would add an extra two years to the flight time, for a project that was already massively delayed. There was a sliver lining, in that the fly-by of Venus turned out to be very productive scientifically, as discussed later.

At last all was ready, and at the beginning of October we all gathered at Cape Canaveral, where the shuttle *Atlantis* was waiting on the

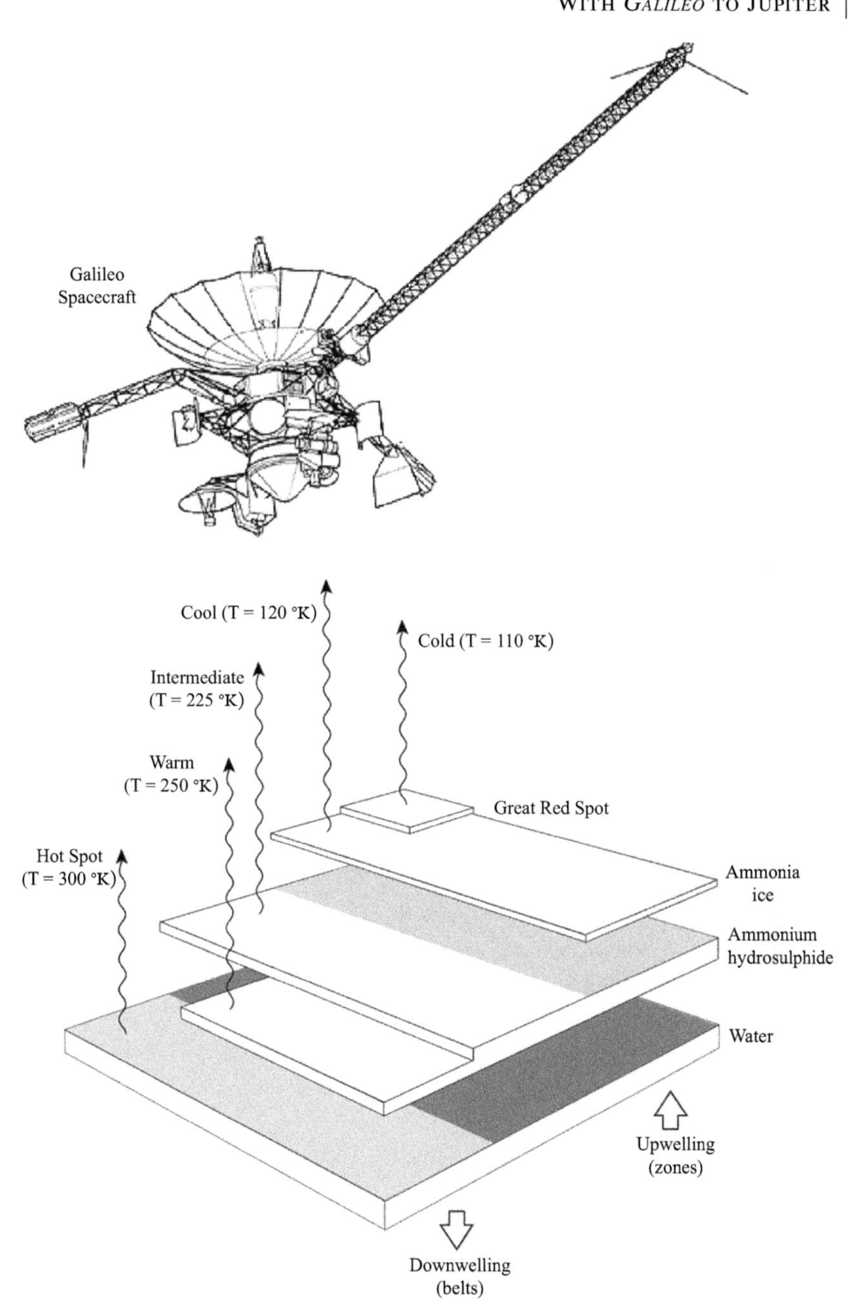

This sketch shows how the *Near Infrared Mapping Spectrometer,* NIMS, experiment worked. The instrument can be seen just to the right of centre mounted on the bottom of the *Galileo* spacecraft. The wavy lines represent infrared (heat) radiation emitted from cloud layers at different heights in the atmosphere; NIMS scans in wavelength to probe different depths. The cloud layers are shown as a simplified, averaged 'model' of what the instrument found while orbiting Jupiter.

launch pad with *Galileo*, and the IUS, aboard. After several tense days of final briefings and readiness checks we found time to relax with a football match on the beach between the NIMS and the imaging camera teams, with pelicans swooping overhead. Then the news came in that a tropical storm was approaching and the launch would have to be delayed. Many of the team headed for the airport immediately—the work was done, and although it is exciting to watch the launch in person it is not essential, and the delay due to the storm was predicted to be at least several days. Having come all the way from England, and never having seen a shuttle launch, I decided to sit it out. Now alone, for the best part of a week I sat fairly miserably by the big glass doors that opened onto the patio of my beach apartment and watched the hurricane-force winds blast the sand along the beach, and further away the crashing, foam-flecked waves under a steel-grey sky. Normally the view would have been full of pelicans and sandpipers, with the odd jogger, but not under these conditions. Even the birds were grounded.

Bored and lonely, I decided to take a trip to Disney World, which lay about an hour's drive inland towards Orlando. As I drove, the wind snatched at the car on the high bridge that links the Cape to the mainland, and the outside thermometer read something unseasonably low. It began to seem that it might be just as inclement inland as it was on the beach, and indeed it was. Disney World in Florida is nearly always warm, sunny, and very crowded, but not this time. Even the characters who usually roam around dressed up as Mickey Mouse and his chums had vanished, and I almost had the place to myself. After a few desultory rides on my own, including three trips in quick succession around 'Pirates of the Caribbean', which is mostly underground and hence weatherproof, and for which you normally had to queue for an hour or more, I gave up, drove back to Cocoa Beach and called British Airways. About another week later, on 18 October, the launch took place, flawless but unwatched by any of our team, except Torrence, who as Project Scientist had to be there to sign off on the payload. *Galileo* was on its way.

En route to Jupiter

Our troubles were not over, however. The new flight path meant first heading away from Jupiter, towards the Sun, for the impetus-imparting encounter with Venus. The big gold umbrella that would beam signals from Jupiter was not needed yet, and in any case would have got too hot so near the Sun, so it stayed furled up until after the Venus fly-by. Then, on 11 April 1991 the command was given to unfurl the antenna. It was soon clear from the signals coming back that something had gone wrong and the antenna had only partially opened. In the following weeks, the mission engineers at JPL experimented in the lab with an identical spare antenna and concluded that a few of the umbrella-type ribs had stuck because of a loss of the dry lubricant that should have helped them to slide into position. The lubricant had probably worn away due to the vibration experienced when the spacecraft was being shipped by truck between JPL and Kennedy Space Center, a trip that had to be made several times because of delays associated with the *Challenger* accident.

A 'tiger team' was formed and suggested several ingenious solutions to fix the problem. One involved turning the spacecraft first towards and then away from the Sun, in the hope that warming and cooling would free the stuck hardware through thermal expansion and contraction. When that didn't work, the antenna deployment motors were turned on and off repeatedly more than 13,000 times, sometimes with the spacecraft spun up to its fastest rotation rate of ten revolutions per minute to provide additional centrifugal force on the stubborn spokes. Nothing worked, and finally a NASA announcement grimly declared that 'after the years-long campaign to try to free the stuck hardware, the Project has determined there is no longer any significant prospect of the antenna being deployed.'

The partially unfurled high-gain antenna was completely useless, and would remain so for the entire duration of the mission. *Galileo* had a rod-like 'low-gain' antenna that could be used as a back-up, but it was only intended for sending commands to operate the spacecraft's main functions, not for relaying pictures and scientific data. Most functions on *Galileo* were automatic, under the control of an on-board computer, so the spacecraft normally only had to be told which programs to execute. The rate at which science data could be returned from Jupiter with the low gain antenna alone was painfully slow, about ten bits per second instead of the planned rate that was more than 10,000 times faster. Rather than a minute, it would take a day to send a single picture from the camera. Our NIMS instrument alone had a data rate of more than 1000 bits per second.

We had a number of fairly miserable team meetings in which we carved up the small amount of data that was left to us among our favourite science goals. We also had to fight the other teams: television imaging has an enormous data requirement, obviously, and so do 'particles and fields' experiments like the magnetometer, because they like to be on and making measurements all of the time. It reminded me of the old movie in which Charlie Chaplin and his erstwhile friends were trapped in the frozen north and reduced to eating his bowler hat. They fell out over who got the brim, considered the tastiest bit. When I recounted this to try and lighten the mood at a meeting in Pasadena, nobody laughed.

Still, NASA and JPL are brilliant under stress and moved quickly to compensate for the problem. The spacecraft had a tape recorder on board, so the instruments could take data at their normal rate and store it to be transmitted back slowly. Obviously, less data in total would reach the Earth; about a thousand times less, in fact. For short events, like the asteroid encounters that were planned, and the reception of the data from the entry probe as it descended, all of the data could be collected at maximum speed and then played back over the ensuing weeks so none was lost.

Clever improvements to the hardware at the receiving station and the installation of data compression software on the spacecraft helped to increase the downlink rate, although only to a still much-less-than-wonderful 160 bits per second. Glitches in the tape recorder performance, initially thought to be terminal but eventually manageable, did not help either. Discussions were held to think about extending the length of operations in Jupiter orbit beyond two years, to compensate for the reduced rate at which data could be acquired: in the event, we got nearly eight.

Before any data could be taken at Jupiter, *Galileo* had to fly through the asteroid belt. The earlier flights of *Pioneers* and *Voyagers* had shown that this was not as hazardous as had been feared, since it was found that the large solid bodies were well spaced out with little of the dust and debris some experts had predicted might lie between them. Furthermore, it was a chance to take our first close look at a major asteroid, a prospect that intrigued even the atmospheric scientists.

Our target was 951 Gaspra, a monkfish-shaped object about 18 kilometres long, discovered by a Russian astronomer in 1916 and named by him after a popular watering hole for writers and artists on the Black Sea. The number 951 refers to the order in which the asteroid was first discovered. Thus, the largest two are 1 Ceres (discovered in 1801) and 2 Pallas (discovered in 1802) respectively, and another 949 were identified and named in the intervening 114 years before Grigory Neujmin found Gaspra. Clearly, this was a time when observing techniques were improving rapidly, although of course the asteroids could be seen only as points of light. The cameras on *Galileo* were among the first to actually see the detail on the surface of an asteroid.

Passing close by on 29 October 1991, *Galileo* photographed the remarkable object as it rotated once every seven hours, while NIMS busily mapped out the surface at infrared wavelengths to analyse the composition. Gaspra turned out to have a smooth surface, scored in places by long grooves, and pockmarked by craters of various sizes that suggested a violent early history. It is a member of the commonest family of asteroids, the S-types, composed mainly of the rocky minerals olivine and pyroxene, which are common on the Earth. NIMS found very little compositional variation across its surface, marking it as a single integrated body rather than a conglomeration.

So Gaspra was an exploration milestone, but as a discovery its surface composition and features were a bit dull for all but the asteroid fanatics on the team. *Galileo* flew on and on 28 August 1993 encountered a larger one, 243 Ida. Discovered in 1884 and named for a nymph in Greek mythology and the mountain in Crete where she nurtured the infant Zeus, no less, Ida the planetoid is 54 kilometres long and completes a spin every four and a half hours. In order to nudge its trajectory to get nearer to Ida (about 2000 kilometres away at closest approach), *Galileo* had to burn fuel and thus shorten its operational life at Jupiter, and there was a fair bit of debate about whether this was wise.

On arrival it rather seemed it was not, since Ida turned out to be another S-type with craters and scratches rather like Gaspra, although she did have a more interesting shape rather like two potatoes joined together. But then, in February 1994 (the delay being due to the slow downloading of picture data from the spacecraft), the imaging team reported a surprise: Ida had a small moon, just a mile across, but nearly spherical, apparently orbiting its parent at a distance (probably not constant) of 90 kilometres about every 20 hours. This was named Dactyl, after the race of metalworkers and magicians who allegedly used to live on Mount Ida. The fact that Dactyl has some quite large craters, and that even a modest impact would free it from the gravitational hold of Ida, suggested that the present pairing is probably something that came about fairly recently in astronomical terms, perhaps only a few million years ago. But we found it intriguing.

The asteroid Ida and its moon Dactyl were photographed by *Galileo* in August 1993. Dactyl is just one mile across, while Ida measures about thirty miles in the longest dimension. Our NIMS instrument analysed their composition and found they are both made of rocky material similar to that commonly found in meteorites.

Meeting the Giant

By late 1995 Jupiter was looming up ahead at last. *Galileo* had to arrive on a trajectory that would allow it to enter the right orbit after a long burn of its onboard rocket motor. It also had to release the probe at the right moment, so it could enter the atmosphere at a shallow angle which would slow it down without burning up, but not so shallow that there was a risk of skipping off like a flat stone at the seashore. The details were far from certain, not least since the properties of Jupiter's atmosphere were poorly known and the speed of arrival was enormous—more than a 100,000 miles per hour. The heat shield would reach more than 10,000 °C as it burned away due to the friction needed to slow the small craft down. In the event, this went well and everything on the probe worked. After the initial violent deceleration, it drifted down on its parachute through the colourful cloud layers and reached a depth where the pressure was 20 times that on Earth, handsomely exceeding the goal of ten times. There it found ambient temperatures as high as 150 °C, hotter than the boiling point of water.

In order to make sure the orbiter did not follow the probe into Jupiter's atmosphere, but instead went on to circle the planet, the two spacecraft actually had to separate five months before the probe made its dramatic descent. At that time, we on the NIMS team were focussed on our own risky approach and orbit insertion. Our instrument was turned off, as the entire capacity of the spacecraft's tape recorder was given over to capturing the precious, short-lived data from the probe. We were passengers at this point, and it would be weeks before the slow downlink could empty the tape recorder so we could switch NIMS on and start to spool our own data.

Still, there was plenty to think about. The tight navigational requirements dictated the need for a flawless performance by the orbit insertion motor. This had been provided by the German Space Agency and built by Daimler Benz Aerospace at their space propulsion facility near Stuttgart. It used most of its fuel, nearly a tonne of hydrazine and nitrogen tetroxide, in the 49-minute burn which put *Galileo* into Jupiter orbit. The spacecraft, its communications with Earth all but cut, had a very limited tolerance for the intense radiation trapped in Jupiter's radiation belts. These are the Jovian equivalent of the Earth's van Allen belts, a plasma of charged particles, mostly very energetic electrons travelling at close to the speed of light, trapped in the planetary magnetic field. Jupiter's field is so strong that the density and energy of the radiation in the belts can easily penetrate the spacecraft's hull and then it almost literally fries the equipment inside. Had there been an astronaut in there, he would be dead in minutes, and machines do not last very much longer, their vital electronics being the first to fail.

We wanted to take data on the innermost of the large moons, Io, when the spacecraft went past just 600 kilometres above the surface. Io's orbit is deep inside the deadly radiation belts, so although *Galileo* had no choice but to brave them briefly on approach, it wouldn't be coming back for a while. The operational orbit had to be much further away from Jupiter to stop its systems from being rapidly damaged and destroyed. Io's gravity was a key component of getting into orbit, so the risk was taken, but the rest of the mission would be spent well outside the belts. In spite of the interest in Io—the same size as Earth's Moon, but hot and volcanically active, unlike its icy Jovian siblings—the answer from the mission director was no, we could not observe at this crucial time. We would have to be content with much more distant views of Io until later in the mission, when most of the primary goals had been completed, and some high-risk close encounters could be attempted.

The arrival of the orbiter and probe at Jupiter had been a complete success, but since my original conversation with Torrence nearly 20 years had passed. Characteristically, he had the words for it: 'It's a good job we were all boy wonders back then', he remarked. And, of course, scientifically speaking this was only the beginning. We had to anticipate another ten years before we collected and analysed all of the data we could now confidently expect—a whole generation of us growing old on this one project. On another occasion, when we were frustrated by the antenna data rate problem, I grumpily complained about the American pronunciation of 'skedule', schedule being a word we used a lot on space missions. Without hesitation, Torrence responded 'Well, we didn't go to the same 'shool' as you guys'.

During the mission

The first orbit lasted about seven months, with another rocket burn at the farthest point to keep *Galileo* from passing too close to the planet on later orbits and re-entering Jupiter's intense radiation belts. This impetus also shortened the time taken to complete each orbit to about two months, with successive circuits of Jupiter deliberately aligned so that *Galileo* would pass close by a different satellite each time. During the encounter, the moon's gravitational field was used to help retarget the spacecraft, so each orbit was different from the others. This often involved flying extremely close—as little as 86 miles from the surface of Callisto in one case.

The primary mission included four fly-bys of Ganymede, three of Callisto, and three of Europa, at distances that were typically a thousand times closer than those of the *Voyagers* during their fly-past of Jupiter in 1979. *Galileo's* instruments sampled the space around each moon for charged particles and magnetic fields, and mapped each surface at high resolution. After about a week of this, the tape recorder was full of data and the spacecraft spent the time until the next encounter transmitting the information back to Earth.

A fraction of the recorder memory was reserved to allow some Jupiter atmosphere data to be gathered, but nothing like as much as we wanted. We continued to have many a long discussion at every level about how to apportion the available data, still some 10,000 times less than originally expected despite the efforts made to increase it. The satellite science nearly always won priority, but I argued at planning meetings that NIMS was in the business of exploring the Jovian atmosphere, too. After much hassle we won the right to take four full spectra of different parts of the globe, at the highest possible resolution, on the first orbit. Although we had good coverage of the whole planet in the end, these early data were the best we got during the whole eight years the orbiter lived.

As we had planned, the capability of NIMS to obtain spatial and spectral information simultaneously, although originally conceived for mapping the satellites, proved ideal for investigating the composition, vertical layering, density, and particle sizes in Jupiter's multilevel and inhomogeneous cloud layers. Because of its low spectral resolution, we knew that NIMS was not likely to tell us what gives rise to the bright colours in the Jovian clouds, including for example the red pigment in the Great Red Spot or the various yellows and browns. But we did have hopes of making progress on the main materials that make up the clouds in which the coloured 'chromophores' are embedded.

The early data from *Galileo* confirmed that the face of Jupiter that we see through a telescope or in a television picture from a camera on a spacecraft is made up of four layers of cloud each with a different composition. There was never much doubt about the composition of the main cloud layer, since we know that ammonia is present in Jupiter's atmosphere and it is an easy calculation to show that it condenses at the temperature and pressure where the cloud occurs. The ammonia gas changes to solid crystals at a pressure level of around 25% of the mean pressure at the surface of the Earth. This suggests an analogy with terrestrial cirrus clouds, which are composed of (water) ice particles and found at similar pressure levels in our own atmosphere, but the Jovian ammonia cirrus layer seems to have much smaller particle sizes, by a factor of around a hundred times.

Pure ammonia ice clouds are white, and the fact that they have a yellowish tinge on Jupiter is mainly due to the fact that a thin yellow-orange haze lies over them at higher levels in the atmosphere. This haze is made up of liquid hydrocarbon droplets akin to the light oil many of us use to lubricate hinges and other household items, and seems to be produced by the action of solar ultraviolet radiation on the methane in Jupiter's atmosphere. Something similar occurs in Titan's hazy upper atmosphere, which is even more methane rich, explaining why Saturn's large moon appears in telescopes as a fuzzy, featureless orange disc.

There is another cloud layer below both the haze and the ammonia cloud on Jupiter, which has the right properties to be the ammonium hydrosulphide cloud which theory predicts will form when hydrogen sulphide reacts with ammonia. In the lab, this simple compound of ammonia and hydrogen sulphide is also white, but on Jupiter it apparently contains impurities (probably including sulphur itself) that give it the observed yellow and brown colours that are darker than those due to the overlying haze.

Below the sulphur-containing cloud, at pressures of three to five bars and temperature close to zero centigrade, is another cloud of familiar composition, water. At such a depth it is difficult to observe, but columns forming huge anvils, like those we sometimes see in thunderstorms on Earth, occasionally appear at higher levels in major updrafts, confirming that the water cloud lies where theory says it should. Deeper still in the atmosphere, where there is no data at all but only models to rely on, we expect more layers of increasingly exotic clouds as the temperature gets higher. These include, in theory at least but probably in reality also, one of pure metallic silver near the 1000 °C temperature level. The imagination is fired by the prospect of future astronauts in a submarine-style spacecraft navigating through this with spectacular views in their floodlights.

The *Galileo* probe confirmed that the composition of Jupiter's atmosphere is mainly hydrogen, as expected, with about 10% of helium and a number of minor constituents, the most abundant of which were measured and mapped by NIMS from orbit. Weather on Earth depends a lot on the condensation and evaporation of water; on Jupiter three other common species, ammonia, hydrogen sulphide, and phosphine, as well as water vapour, can evaporate and condense in the upper regions we observe, making for a remarkably complicated climate. The NIMS data track this behaviour and have shown that the water, in particular, is very variable. The probe measured a very low water abundance, but the orbiter data showed that it just happened to enter an untypically dry region.

I was in a pensive mood at this Galileo NIMS Science Team meeting held in Hawaii in 1979. These meetings were a lot work, despite the exotic surroundings, and often included difficult negotiations and hard bargaining between ourselves, with other teams, and with NASA, as we planned the details of a complex mission.

Data at last: this is an example of what Jupiter looks like through the infrared 'eyes' of the NIMS imaging spectrometer. The different colours are due to differences in the height and composition of the cloud layers.

The most conspicuous feature of Jupiter's disc, along with the colours, is the banded pattern formed by the clouds, and the large number of very big, very long-lived storm systems that stud the dark belts between the cloudy bands. The Great Red Spot, the largest example of this family of giant eddies, was revealed as having a most remarkable structure in the new observations. Instead of the deep 'plug' of cloud that most of us expected, the red spot has a 'spiral arm' structure of clouds, with gaps between where NIMS could glimpse through to the deep, relatively clear atmosphere below. Furthermore, the cloud structure is higher in the centre by more than ten kilometres and tilted towards one side, so it looks something like a crooked spiral staircase.

What seems to be happening is that 'wet' air from the deep atmosphere is rising rapidly in a relatively narrow region in the centre of the spot, and then spraying out while rotating like a giant garden sprinkler above the tops of the ammonia clouds. This is somewhat analogous to what happens in a terrestrial hurricane, but the Jovian version is bigger than the whole Earth. This dynamical analogy suggests that the material in the centre is drawn up from great depths in the atmosphere, where the high temperatures and pressures cook up some exotic constituents that happen to be red when they condense. Perhaps it is something relatively simple, like elemental phosphorus, which has the right colour. Most of the phosphorus on Jupiter at the levels we can observe is combined with hydrogen and shows up as phosphine gas, but in the hot regions deep in the planet this will be dissociated. Then, in a region of fast rising motions, phosphorus could be raised to the cloud-top level before it can combine to form phosphine again.

However, a similar argument could be made about all sorts of materials that might be present in Jupiter's pressure cooker interior or in its lightning-racked clouds. Maybe the red pigment is a much more complicated molecule than phosphorus, some sort of red organic dye. It is interesting that the other giant eddies, smaller than the red spot but still huge, comparable in size to the whole of our Earth, are not usually red, but rather white or brown. Possibly their vertical extent is less, so they import different species from different depths and end up with some other mix of condensed, coloured substances contaminating the ammonia clouds.

Or finally, as some argue, perhaps the spot isn't really red. Many of the colours reportedly observed on Jupiter by Earth-based astronomers have turned out to be part of an optical illusion. Father Godfrey Sill (he belonged to a religious order but worked as a scientist at the Lunar and Planetary Laboratory in Arizona), a noted expert on the Jovian clouds, wrote in 1976 '. . . it has been known for centuries that Jupiter has various shades of color: red (or pink, red-orange); brown (or red-brown and tan); blues (or blue-gray, purple-gray) grays; yellows (or yellow-brown, ochre, cream, greenish yellow); and perhaps even green.' My former JPL colleague Andrew Young, on the other hand, did a photometric analysis in 1985 and concluded that only various shades of yellow are present. The bluish colours sometimes seen on Jupiter are thought to originate rather like the blue sky on Earth, i.e. due to scattering in the atmosphere above very deep clouds, rather than actually blue material.

The colour filters in the *Galileo* camera should have settled this long-standing argument at last, and they did, but with a rather weak compromise. The new data did prove that the cloud material in the spot is somewhat, but not very, genuinely red; the team called it 'a sort of brickish tint'. It remains the case that there are not many materials that are red or even brickish in colour and at the same time plausible candidates as cloud constituents. Definitive answers remain a long way off, since like the clouds of purest silver a proper analysis of the Great Red Spot may require a floating 'submarine' type craft to analyse samples of the clouds and sniff around in the hidden depths of Jupiter's atmosphere.

Carl Sagan even suggested once that there might be life down there, floating like jellyfish in the warm, dense, nutritionally rich air. The *Galileo* probe could have already travelled through shoals of them, like a depth charge, and we would never have known, since the probe had no cameras on board. Carl was a member of the *Galileo* science team, but during the long unwinding of the project he became famous through his books and television series and we didn't see much of him at team meetings. I did once share an elevator with him at JPL on the way to an early meeting, and he couldn't remember which floor he was going to. 'I'm not a morning person', he said.

Mission extensions

By 1997 the 'nominal' *Galileo* mission was over, but having been forced to operate at a low data rate, and with everything on the spacecraft still healthy, including the fuel supply, everyone was keen to contemplate a mission extension. Getting NASA to agree to extend a mission that has completed its planned operations is not trivial, since even though there is no new hardware to build, the cost of paying for the teams and the use of large ground stations for tracking and data relay remains substantial. However, the science arguments were strong. Io had been found to be studded with active volcanoes, but had not been observed properly, because as we have seen its orbit lies inside Jupiter's radiation belts where the spacecraft controllers feared to go. Even more exciting was the growing evidence for a warm water ocean beneath a relatively thin crust of ice on Europa. And we didn't have nearly enough data on Jupiter's atmosphere to nail down what was happening meteorologically, although it was clearly fascinating.

So, a two-year extension began in December 1997, which included eight consecutive close encounters of Europa and four fly-bys of Callisto. It was also time to risk some close fly-bys of Io, skimming less than 100 kilometres above the surface to try to pass through some of the volcanic plumes. These 'kamikaze' flights tested the skill of the navigation team to the full. Tracking the spacecraft from Earth is not good enough to achieve the required precision; *Galileo's* camera had to be used to photograph the target against the stellar background and the pictures analysed to determine the exact flight path. This of course was made even harder by the low data rate, since the pictures were needed in near real time. Special software had to be developed to edit the pictures on board the Orbiter, to reduce the amount of data to be transmitted to Earth.

The *Galileo* Millennium Mission added another year of operations, including more fly-bys of Io and Ganymede. By now, I was part of another team which was operating an experiment on the *Cassini* spacecraft, which passed Jupiter in December 2000 for a gravity assist on its voyage to Saturn. We were able to make some interesting observations simultaneously from both platforms. As described in the next chapter, our new *Cassini* instrument was a lot fancier than NIMS—bigger, and ten years more modern—but it only had a few days when it was close enough to Jupiter to make useful observations before racing past on its quest towards the ringed giant.

A final three-year mission extension took *Galileo* to the point at which the fuel supply it carried was about to run out, which meant that control of the spacecraft trajectory would no longer be possible, and it would not be able to point its antenna towards Earth. The spacecraft electronics had already endured more than four times the cumulative dose of harmful particle radiation that they were designed to withstand, performance anomalies were being reported more and more frequently, and it could not be expected to survive much longer in any case. Clearly, the mission was almost over.

Galileo ended its saga with a final plunge into the planet's atmosphere, on Sunday, 21 September 2003, having circled Jupiter just 34 times in eight years. Rather than being switched off, and left to drift in space, the usual fate for redundant spacecraft, it was purposely destroyed to eliminate any chance of an impact in the future with one of the big moons, which might introduce biological contamination from the Earth. On everyone's minds was Europa, which was already being targeted for future missions as a possible habitat for life, because of the warm water ocean *Galileo* had discovered below its icy surface. We all knew that hardware from the JPL clean rooms would still have a lot of Californian bugs on it despite having been carefully sanitized.

Anatomy of a success

NASA held a poll of the hundreds of scientists that had been, in one way or another, involved in the mission and in analysing and interpreting the data, to find out what they thought were the top ten scientific discoveries made by *Galileo*. We all voted for our favourites, and in the end NASA listed the official winners as follows (I've paraphrased them slightly, as well as adding my own explanation):

1. The descent probe found that common elements such as carbon, nitrogen, and sulphur had different relative abundances than on the Sun.

This finding scotched the long-standing idea that Jupiter was simply a piece of the Sun that had somehow broken off and condensed separately. We now think that Jupiter formed out of 'planetesimals' typically a few kilometres across in size, which themselves accumulated out of a cloud of rocky and icy particles surrounding the young Sun. Jupiter grew fastest of all the planets, because it formed in the densest part of the dust and gas cloud that was far enough away from the hot centre so that water could freeze. With ice as well as rock available, the young Jupiter's gravity soon became large enough to trap huge amounts of gases, including even the very light hydrogen and helium, just as the Sun did, but with a different proportion of the relatively heavy elements.

2. *Galileo* made a first observation of ammonia clouds in another planet's atmosphere.

It is a little simplistic to single out just one type of cloud among all the things that *Galileo* saw in Jupiter's atmosphere, and this wasn't really a discovery as we knew, long before *Galileo* set out, that the most conspicuous bands of cloud on Jupiter, those that appear predominantly white, are frozen ammonia. On the other hand, the atmosphere turns out to be so dynamic and complex that it is hard to come up with a good one-liner for what actually was achieved. '*Galileo* showed that Jupiter's atmosphere has many different kinds of clouds, and revealed many complicated weather patterns that we don't understand yet' would be more like it.

We were disappointed that none of the other cloud layers, although they are definitely there, could be positively identified in terms of their composition. And the coloured chromophores remain as mysterious as ever, with all but the simplest aspects of atmospheric chemistry on Jupiter remaining to be understood. Even the ammonia layer has some variations in colour that have not been explained, and we still do not have much more than guesswork to explain why the Great Red Spot is red. All in all, what the headline is really saying is that we are still on the first or maybe the second rung of the ladder where understanding weather and climate on Jupiter are concerned.

3. Io's extensive volcanic activity may be one hundred times greater than that found on Earth. The heat and frequency of eruption are reminiscent of early Earth.

Io is close to Jupiter and relatively small compared to Ganymede, Calisto and Europa (Io is actually about the same size as Earth's moon). The result is that Io is tugged by its companions' gravity fields, in an effect that is similar to the forces producing the tides in Earth's oceans, into a slightly irregular orbit, which produces tidal flexing of Io's rocky body. The heat released by friction has not only boiled off any water and ice that Io once had, it melts the rock into lava throughout the body, except for a thin solid crust where the surface is exposed directly to cold space. Through vents and cracks in this crust, sulphurous gases and dust escape into space, producing spectacular plumes that were seen by the *Galileo* camera as fountain-like structures sometimes hundreds of kilometres in height. NIMS was able to map Io at longer wavelengths and measure the temperature of the eruptions by their brightness in infrared emission, finding some that were hotter than any of the eruptions on Earth.

4. Complex plasma interactions in Io's atmosphere include support for currents and coupling to Jupiter's atmosphere.

The volcanoes on Io throw around a ton per second of material into space (unlike volcanoes on Earth, where gravity is higher and the atmosphere gets in the way). There, it becomes dissociated and ionized and channelled by Jupiter's strong magnetic field. One result of this is the formation of a doughnut-shaped ring of ionized sulphur, oxygen, sodium, and chlorine around Jupiter, known as the Io plasma torus. Another is a stream of charged particles of material originally from Io that travels down the field lines and produces bright auroral emissions in Jupiter's polar regions. *Galileo* also found a stream of solid dust particles that apparently originate on Io before leaving the Jovian system altogether, at high speed.

5. Evidence supports a theory that liquid oceans exist under Europa's icy surface.

This claim is rather cautiously worded since we haven't actually *seen* the ocean. Its presence is inferred mainly from the appearance of the icy surface with its many cracks and floes that have obviously moved; this is consistent with a thin, mobile ice layer floating on liquid. The heat to keep the water in liquid form comes mainly from the fact that Europa is flexed by Jupiter's gravitational tides, like Io but not as much because it is further from Jupiter. The heat released inside the body melts Europa's mantle of water ice everywhere except near the surface, where it is very cold because of exposure to space. The patterns formed by the cracks visible on the surface suggest that the ice crust is only a few, perhaps ten, kilometres thick. With NIMS we saw coloured deposits around the cracks that seem to be soluble compounds ('salts') that were dissolved in water that escaped from the crack at some time in the past.

As a discovery, many people would rate this number one, because the existence of a subterranean ocean has obvious biological implications. Warm water spells *habitat*, and there has been much speculation about what, if anything, might live on Europa. It is quite a big 'might', because of course Europa's ocean probably never had the exposure to direct sunlight or atmospheric phenomena like lightning that are believed to have been important for producing life on Earth. Finding the truth is quite difficult because of the intervening miles of ice, but JPL has produced preliminary designs for a submarine explorer on Europa that melts or drills its way down and then explores the dark abyss beneath.

6. Ganymede is the first satellite known to possess a magnetic field.

A weak one by planetary standards—the field *Galileo* detected originating inside Ganymede is only about a millionth of Earth's, which itself is ten times weaker than Jupiter's. However, it is remarkable that an icy satellite should have any field at all, and the others don't have one strong enough to be detected. Measurements of the magnetic and gravitational fields around Ganymede give clues about what is going on deep inside the body, information which is hard to come by any other way and therefore remains in short supply. The favoured interpretation, although it seems dubious to me, is that Ganymede, which is the largest satellite in the Solar System, has an iron-rich core, and it must be convective, like Earth's. The source of the interior heating may again be tidal flexing in Jupiter's strong gravity field, although this effect is smaller for Ganymede than for Io. The same processes that produce the molten core and magnetic field on Earth are probably also involved, namely the decay of radioactive elements like uranium, and the retention of some primordial heat that has yet to escape from the interior.

7. *Galileo* magnetic data provide evidence that Ganymede and Callisto, as well as Europa, have a liquid-saltwater layer.

Europa's watery ocean is probably the nearest to the surface and potentially the most accessible, but it may not be unique. Measurements of perturbations to Jupiter's magnetic field near Ganymede and Callisto obtained during close *Galileo* fly-bys suggest that there is a layer of something more electrically conductive than solid ice below the surface. As on Europa, this is probably salty water, although on Ganymede it appears to be below a solid ice crust about 200 kilometres thick. The cracks, and the surface salt deposits, are less obvious on Ganymede than Europa but do seem to be there. For Ganymede and Calisto, natural radioactivity in the moon's interior would provide enough heat to maintain a layer of liquid water at that depth, if they have the expected cosmic abundance of radioactive elements (it will be a long time before we know if they really do, but it would be a big surprise if they don't).

8. Europa, Ganymede, and Callisto all provide evidence of a thin atmospheric layer known as a 'surface-bound exosphere'.

This one is a bit of an oxymoron. An exosphere is an atmosphere in the sense that it is a shell of gas surrounding a solid planetary body, but it is so thin it is not an atmosphere in the usual sense. For that, it would have to be dense enough to affect the climate on the surface and exhibit weather phenomena like cloud formation and precipitation. On Earth the layer we call the exosphere is at the very fringe of the atmosphere, where it meets space. There, the density is so low that the very few molecules and atoms it contains have more chance of escaping from the planet altogether than of colliding with each other. Pressure and temperature are meaningless terms under those conditions, and the existence of a minute trace of gas around Ganymede and its icy siblings is really not surprising, since the frozen volatiles (mostly water ice) on their surfaces are exposed to solar radiation, micrometeorites, cosmic rays, and other sources of vaporizing energy. By headlining this as a key result from *Galileo* we ended up with newspaper articles all over the world hailing Jupiter's satellites as bodies with atmospheres, which they are not. The only moon in the Solar System with a proper atmosphere is Saturn's Titan, which we visited with *Cassini*, as the next chapter will relate.

9. Jupiter's ring system is formed by dust kicked up as interplanetary meteoroids smash into the planet's four small inner moons. The outermost ring is actually two rings, one embedded within the other.

This again can be a bit deceptive. Jupiter's rings are very tenuous, which is why they were not discovered until 1979, during the flight past Jupiter of the *Voyager I* spacecraft. Now the popular press and even textbooks sometimes represent Jupiter as having something comparable to Saturn's amazing family of rings, which is not the case. And, although the explanation for their origin is probably right, *Galileo* didn't see interplanetary meteorites smashing in to any of the Jovian moons. There was, however, a comet that smashed into Jupiter, *Shoemaker-Levy* in 1995, a rare and spectacular event. With NIMS we observed the resulting fireball, and measured its temperature at around 25,000 °C, near the cloud tops that are normally about 150 degrees below zero.

10. *Galileo* was the first spacecraft to dwell in a giant planet magnetosphere long enough to identify its global structure and to investigate its dynamics.

Jupiter's massive magnetosphere is invisible to the eye, of course, since the clouds of particles that ebb and flow within it cannot act as a 'tracer' the way that condensate clouds do in the atmosphere. But someone estimated that if we could see it from the Earth it would fill as much of the sky as the Moon does. Within this vast space, magnetospheric variations and charged particle fluxes constitute what has become known as 'space weather'. For many of us it is not as interesting as the real weather in Jupiter's atmosphere, which we were studying intently with NIMS at the same time, revealing a host of fascinating things. If NASA was going to list generalities in its top ten discoveries, then the first detailed studies of Jupiter's meteorology certainly should have been included. Perhaps because they were the first to ply their trade in space, magnetospheric scientists are many in number and maintain a traditional stranglehold on mission politics that often seems to be lacking in the more easy-going atmospheric and geological science community.

That concludes NASA's list. It will be clear that not all of these top ten discoveries are the things that I would have chosen, although they are all good and exciting research and certainly qualify as breakthroughs of one kind or another. What I think should have been included near the top of the list (it is there, but buried deep, in discovery No. 1; it deserves much more emphasis) is the finding that Jupiter's atmosphere at the probe entry site was very dry. If this were typical of the whole planet, Jupiter would have much too little oxygen, which should be present mainly as H_2O, to be representative of the Solar System as a whole. Where could it have gone? Trapped deep in the interior somehow, perhaps.

With NIMS and the camera we found the likely answer: meteorology. The entry site was not typical of the whole atmosphere; it was a 'hot spot', one of many regions where Jovian air (mostly hydrogen and helium) is descending, depleted of clouds and condensates like water. The adjacent rising regions bring air from the warm, moist depths ('moist' meaning ammonia and a lot of other things as well as water) to the colder upper atmosphere, where condensation takes place producing the cloud layers. On the planet-wide scale, this effect, combined with Jupiter's rapid rotation, is what produces the cloudy zones that give the disc of Jupiter its banded appearance. The water/oxygen problem is so important that it spawned a whole new mission, *Juno*, which carries a microwave spectrometer to probe deep into the atmosphere and measure water amounts far below where the probe, or NIMS, could sound, and hopefully below the weather patterns as well. *Juno* launched on 5 August 2011 on a five-year journey to Jupiter, so as I write it is nearly there.

The new understanding of the Great Red Spot and the other giant vortices deserved a rating, too. It's true that we still don't fully understand what it is or how it behaves (but we could say the same about the magnetosphere). Under close scrutiny, the spot was seen to be made up of spiral arms of cloud, mostly ammonia but with that elusive red tint, with the gaps showing no cloud underneath. In other words, it is mushroom shaped and not a solid column of cloud. The swirling cloud cap is higher at one side than the other by ten kilometres or so (not a large tilt angle, since the spot is 35,000 kilometres across) and high winds flow around the edge. The white ovals, brown barges, and other large, compact cloud features all have properties that *Galileo* was able to catalogue from orbit, greatly, but not completely, improving our understanding of how they form and behave.

The biggest missing factor is that the observations were all from orbit, i.e. from above; we still don't what happens underneath these long-lived phenomena nor how deep they extend, and until we do we won't understand them completely. 'Understanding' is relative, of course. We still don't completely understand hurricanes and tornadoes—distant cousins of the Great Red Spot and other Jovian giant vortices—on the Earth, even though we have studied them closely for a long time.

One of my personal nominations for the best achievements by *Galileo* didn't even occur at Jupiter. The fly-by of Venus, although not originally intended to do science at all, produced the first images of the deep cloud structure on Earth's hothouse of a neighbour. These turned out to be tremendously interesting and changed our understanding of what's happening on Venus and why its climate has diverged from our own. No comparable discovery occurred at Jupiter, because it is a fundamentally different kind of planet.

The technical achievements of *Galileo* have to be trumpeted along with the science. It was the first mission to measure Jupiter's atmosphere directly with a descent probe, the first to conduct observations

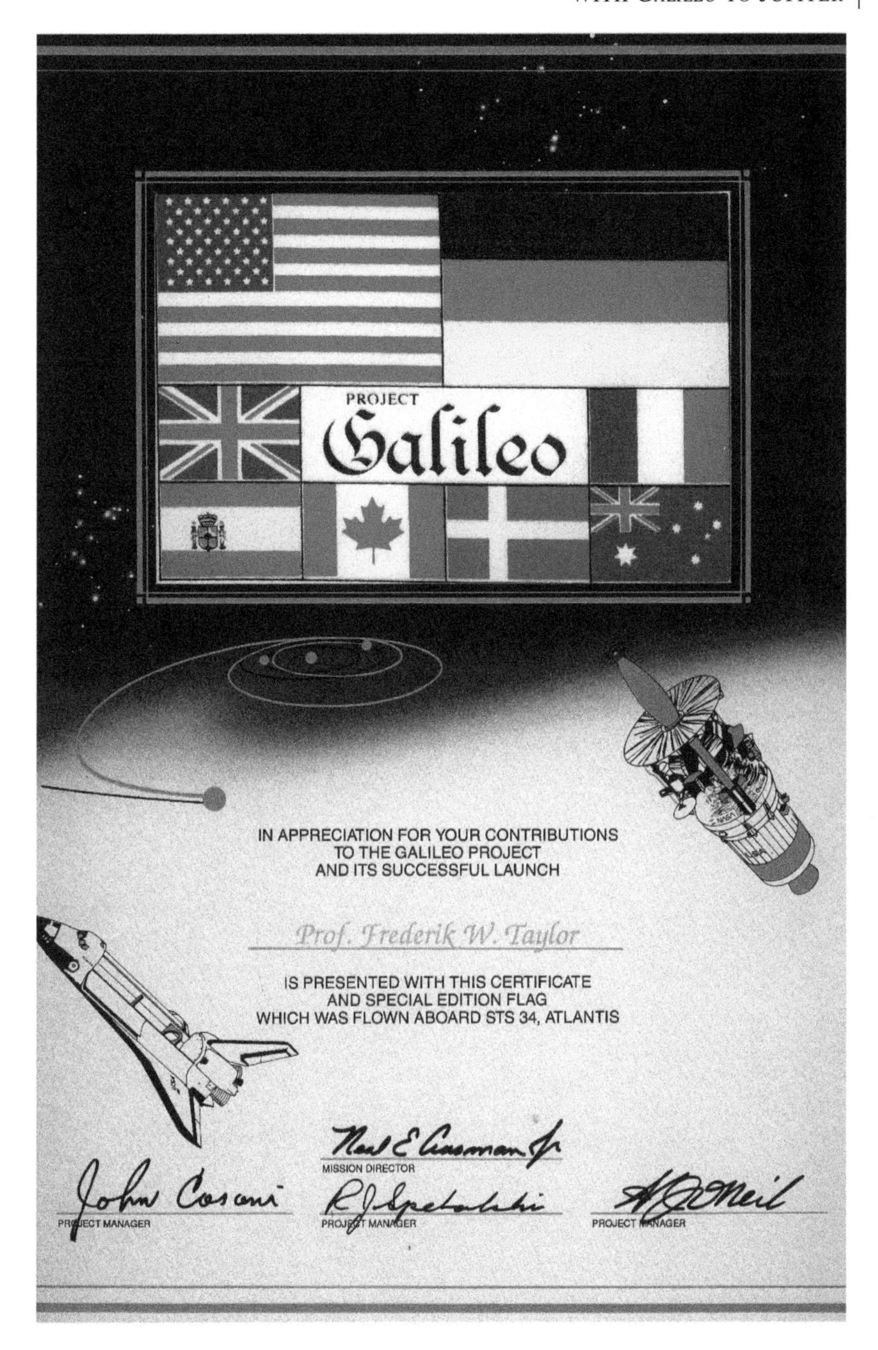

NASA was generous with certificates of achievement for participants in its missions, and I have no less than five related to *Galileo* – for instrument delivery, successful launch (above), the Gaspra encounter, the Ida encounter and Dactyl discovery, and the successful completion of the science mission at Jupiter.

of the Jovian system from orbit and to skim close by the moons, sometimes just 100 kilometres—65 miles—away, an astounding feat of navigation by the controllers at JPL. *Galileo* was also the first spacecraft to fly close by an asteroid and the first to discover a moon of an asteroid.

Requiem

For *Galileo's* final kamikaze mission, the entry point into the atmosphere was just south of Jupiter's equator, at an angle about 22° above the local horizon. The speed at which it entered, over 100,000 miles per hour, would cover the distance from Los Angeles to New York in 82 seconds. This was much faster than the probe entry, eight years earlier, and this time with no heat shield for aerodynamic braking. The orbiter structure had no in-built strength to prevent its rapid break-up, long before it achieved the depths the probe had reached. The last signal from *Galileo* was received at 12.43 pm California time, nearly an hour after its actual demise despite the radio waves travelling at light speed, a final reminder of how far away it was.

An event was held at JPL at the same time as the death and cremation of the spacecraft took place, but I wasn't there. Oxford to Pasadena is a long way to go for a funeral, even for that of an old friend.

6
Back to Oxford

At the Jet Propulsion Laboratory in 1979, I was working on three fronts. We had all the data we were going to get from Venus, but the results were good. The data analysis phase, including publication and presentation at conferences, was in full swing, and set to go on for many more years. We were designing and building an imaging spectrometer to fly to and explore Jupiter. And looking longer term, we were thinking about how to send an instrument to investigate the climate on Mars, a saga that turned out to be fraught with unforeseen obstacles, but ultimately would be successful.

Doris and I were really happy with our home in a canyon on the north-eastern fringe of Altadena, the higher part of Pasadena, and a stable of cars that by now included not just my Aston Martin DB5 and Doris's Lotus Elite, but also (among others, at various times) another Aston, a DB2 convertible, and two Jaguars, a classic Mk II S-type and a new XJ6. JPL had just promoted me, raised my salary, and given me one of the best offices in the science building, with a view of the reflection pool with its waterfalls and fountains, against the backdrop of the San Gabriel Mountains. At quiet times, deer would come and graze among the trees outside my window. I had my own personal parking space with my name on it, a privilege beyond measure. Nirvana!

We had a good social life, too, with a wide circle of friends and activities. Recreation included scuba diving; a group of us would rent a boat and travel out to Catalina Island to dive from the boat into the kelp forests that surround the island. The sunlit waters teemed with fish, from giant goldfish to small sharks that looked exactly like their larger and more deadly kin. The mini-sharks were harmless and friendly, but if they appeared a few inches in front of your face mask they could cause minor heart attacks before you realized they were not the larger version at some greater distance.

The Aston Martin Owners' Club was great fun. We would go on long drives at irresponsible speeds to exotic destinations, like the Historic Car races at Laguna Seca near San Francisco, and the Classic Car Concours at Pebble Beach. We held joint events with the Rolls-Royce Club, whose members were generally elderly and very rich. They liked the youthfulness of the Aston crowd and took us to amazing country clubs near Palm Springs and a vast warehouse near Laguna Beach full of imported beer, wine, and spirits, with enough spare room inside to take all of the classic cars alongside the built-in hospitality suite.

We had started going to church again. Neither Doris nor I had enjoyed the regular attendances we made as children under parental direction and had pretty much put religion behind us when we left home for university. But Neighborhood Church in Pasadena belonged to the Unitarian Universalist Association, an up-to-date version of an old denomination that had started in medieval East Anglia and flourished in pre-revolutionary Massachusetts. Universalists eschewed dogma and mysticism but recognized all religions as inspirational and introspective when developed in a friendly social environment. The minister, the Reverend Brandoch Lovely, was a Harvard-educated intellectual, movie fan, and master poker player who soon became one of my best friends. After a few games of cribbage we would talk for hours about science and religion and how they can complement each other without conflicting, if we first clear away the haze of medieval mythology that more conventional churches still use to keep their congregations in thrall. We taught each other a lot, about where our current understanding of the laws of physics runs out of steam, in understanding the universe at one end of the scale and human behaviour at the other, and basic things like why it was better to trust people unless there was a clear reason not to, and to trust friends absolutely by definition. Like all profound principles it sounds simple but pays dividends.

A call from the Old Country

While we flourished in Pasadena, colleagues from Oxford would come to visit, to discuss our joint Venus project and other interests, and sometimes just to keep in touch. On one of his visits, my old mentor John Houghton and I went for a long walk in the foothills. I gave him the standard advice to always look down at the trail while moving so as not to step on any rattlesnakes; when you want to admire the view, you stop and stand still first. John confided in me that he had plans to leave Oxford to become director of the Appleton Laboratory, a government facility located outside London near Heathrow Airport. The laboratory was named after Sir Edward Appleton, the discoverer of the ionosphere, a layer of charged particles in the upper atmosphere that was vital for long-distance radio communications in the days before relay satellites. The laboratory was founded in 1924 as the Radio Research Station to work on the development of radar and the study of the ionosphere. John wanted to modernize the Appleton and build a role for it in space research, which under his leadership it eventually achieved, and it continues to do so to this day as part of the Rutherford Appleton Laboratory at Harwell, 20 miles south of Oxford.

Someone would have to take over John's position at Oxford, and since I had the right background, would I think about applying and 'keep us in space'? It was an exciting prospect for me, but on the other hand after ten years we were nicely settled in California. My early ideal of returning to England to a professorship at Oxford had faded in the sunshine, and my research, then focussed on Venus, Jupiter, and Mars, would suffer from the longer lines of communication needed to keep in touch with JPL and NASA. On the other hand, I would be in charge of my own stand-alone research team rather than part of a huge laboratory, and the lifestyle of an Oxford don takes some beating. We thought about it very hard for a considerable time.

The deal on offer in the UK was not without problems. John was not going to relinquish his university position, a statutory Readership, immediately, but instead would go on extended leave. My title was to be Deputy Head of Department and acting head for five years, the post funded through the newly merged Rutherford Appleton Laboratory. In July 1979 I went to RAL for interview and was told I was in the top 1% of applicants for a Principal Scientific Officer position, the main career grade in the British scientific civil service. They would offer a salary that was five points above the age-related scale point, and I would be free to apply for promotion to a higher grade after 12 months. The university matched this with a Readership grade, again with a five-point boost ('absolutely unprecedented' according to the University Chest, as the finance office was still quaintly known then). There was also the possibility of a Senior Research Fellowship at my old college, Jesus.

All of this was still much less remunerative than I was used to in the USA, and I did not relish the 'deputy' part; for all my friendship and respect for JTH I knew his genius was mercurial and unpredictable, and working for him would be daunting and possibly dangerous for my own ambitions and interests, which had turned towards the planets during my ten years at JPL. I could handle the underdog role for a year or two, but for the longer term I had to gamble that he wouldn't come back.

I did, and he didn't, but there was a legacy that, apart from knowing anything could happen, I had not anticipated. As JTH's fertile mind looked for new worlds to conquer, he lighted first on oceanography, which he perceptively saw as the missing link in understanding climate change, realizing that climate was becoming the next big thing in global debates. The immediate consequence of this was that a rare new academic post, that was assigned to the department before I arrived, was given to an oceanographer, trained as an applied mathematician, just before I got there. The appointee was excellent, but that wasn't the point; the department was very small—just four senior academics, plus support staff, altogether, not nearly enough to meet our ambitions in space research, especially as I intended to build up the planetary part along with a larger involvement in Earth observation. Getting into global oceanography was exciting, but adding a massive new field to the one on which we had built our reputation and which was full of opportunity for the future was a bridge too far. But fools rush in, and I was driven by the young person's relish for risks and challenges.

Coming home

And so it was that, the day after my 35th birthday, Doris and I set sail for England on the Cunard liner *Queen Elizabeth II*. We had driven across the United States in our Lancia Beta, which of all the cars we owned then was the one that seemed most suited to our daily needs in Oxford. Now, of course, I wish I had brought the DB5; I sold it in Pasadena for more than I had paid for it originally, but today that profit would be at least a hundred times greater.

The QE2 had a lower deck dedicated to carrying vehicles across the Atlantic at a surprisingly reasonable price, and by bringing the Lancia we would have not only the car itself but also a load of our possessions to help us get set up back in England. Our travel agent at Thomas Cook had advised us to spring for the cheapest cabin in first class, which cost about the same as the upper end of the second class range. It was small, therefore, but as she rightly predicted we would spend hardly any time there except to sleep, and during the day we would have access to the first class restaurant and facilities. Thus it was that we enjoyed wonderful meals and entertainment, including gambling beside Rod Stewart at the roulette table in the casino.

In spite of a nasty bit of extortion by the belligerent dockers in New York, who demanded a substantial bribe in order to protect our car from damage, and a storm in the mid-Atlantic which sent plates and cutlery flying through the air across the dining room, the voyage was delightful. For some reason I was one of very few of the hundreds of passengers on board who was not seasick during the storm. I had been expecting the worst after suffering badly on our first trip, on calm seas, in a small boat to go diving off Catalina. Now, however, I had most of the big ship to myself, and as it pitched around on heavy seas I went exploring the almost deserted lower decks.

The door to the below-deck swimming pool was closed but unlocked, so I looked in and saw that the water was mostly down at the far end, extending well out of the pool itself and right up the wall at the opposite end to the door through which I was peering. In less than a minute, the pitch of the ship reversed and a wall of water started to move towards me. I slammed the door, noticing that it had a rubber seal around it, and swiftly retreated. Walking involved banging against first one wall of the corridor and then the other; when I hit a cabin door, sometimes there was a moan from within as the suffering occupants thought I was knocking.

Still not sick, I ventured up on deck to take the air and see what the ocean looked like under these conditions. Hanging on to railings,

On the rear deck of the *Queen Elizabeth 2,* leaving New York harbour, 25 September 1979 to return to become the Head of my old department at Oxford University after a decade in California with NASA.

drenched in spray, I made it to a vantage point where I had a panoramic view, and was amazed to see that the horizon was well above the deck in every direction I looked. This huge ship was at the bottom of a trough in swell that must have had a wavelength of more than a mile. Oceanography, indeed!

The Christmas-card view from our new home on Headington Hill, snapped from the bedroom window one wintry day, looks out over the Oxford colleges and the laboratories in which we would develop and built the experiments to study the Earth and other planets from space.

On arrival at Southampton we set off for Oxford in the Lancia, which had an ugly scrape on the front wing, put there apparently because I talked back to the vile dockers in New York. On the road, one of the first things we saw was a phalanx of joggers led by a familiar figure: the disc jockey Jimmy Saville, on a charity run. At the time he was a popular media personality but we were never charmed by his psychotic demeanour and were unsurprised when, decades later, after his death, his true character was uncovered.

We were heading for Waterperry, a small village ten miles from Oxford where we had the use of a cottage while we searched for a house of our own. Had we stayed a bit longer we would have had as a neighbour a much nicer celebrity, Rowan Atkinson, who lived in a big house in the village, behind security gates that would open from time to time to allow a glimpse of him roaring out in his Aston Martin.

Before long, I had another Aston myself. The Lancia was a reliable servant but dull. Its inherent good looks were spoiled by massive ugly bumpers, the result of new safety legislation that had come into force in California the year I bought it. That particular model was not imported into the UK, so parts were difficult to come by, and of course it was left-hand drive. The problem solved itself in traumatic fashion early one morning in June 1981 when I was driving to a meeting on near-deserted roads just as dawn was breaking. Somewhere on an ordinary country road in north Oxfordshire I was following an unloaded flatbed truck and looking for a chance to overtake when we were both brought to a halt by temporary traffic lights protecting some minor road works.

The lights stayed red for an age and I had a long way still to go. Glancing in my rear view mirror, I saw an unmarked white van approaching. As he got closer, I realized that he was not going to stop. I learned later that the driver had been travelling all night, carrying a heavy load of books, and had fallen asleep at the wheel. Just before the impact I wrenched the wheel to the left and was thrust violently into the ditch, missing the tailboard of the lorry in front by a few inches. If had been stopped any closer to him, the platform on the lorry would have sliced the top off the Lancia and my head with it.

My seat was ripped out of the floor and I was lying on my back, still strapped in, when a most peculiar thing happened. The door was opened by a policeman, who helped me out. The remarkable thing was that he was a big, heavy, black Federal officer in the beige uniform with the big gold badge and the gunbelt that was familiar to me from my ten years in the USA. But this was England. It was only nine months since I had moved back from California, I was shocked and confused after the crash, and I could see his big Chevy with the huge rack of flashing lights on the roof in the background. I was obviously hallucinating.

But in fact, it turned out that the crash had taken place right outside the US Air Force base at Upper Heyford and their security system had picked it up, sending the patrol out in just a couple of minutes. The British police arrived later to take the van driver away and call a garage to retrieve the wrecked vehicles. The Lancia was a write-off, with the boot section a crumpled mess and the front damaged by the descent into the ditch. Amazingly, the middle section was intact and all of the doors still opened and closed, with only my horizontal seat to show that anything had happened. This was a consequence of the California safety laws that I had so despised when they were introduced. The passenger compartment was a reinforced steel cage, and the huge bumpers had absorbed some of the impact. I would have died without them. A few months later, I received a letter from the courts to say that the white van driver had been fined £75 for dangerous driving.

The only good thing to come out of this episode—needless to say, I missed the meeting—was that I needed a new car. I gleefully took myself off to Robin Hamilton, an Aston Martin specialist, and selected a 1974 AM V8. This served me well for many years until eventually I sold it, once again for a profit. It had to go because I found it impossible to always drive it as slowly as the law required, and was increasingly falling foul of the dreadful radar-operated speed cameras that were beginning to appear everywhere. After trying several smaller sports cars, and still in danger of having my driving licence removed, I ultimately settled on a 1984 Bentley Eight, a splendid if unwieldy carriage that also had the virtue of being fun to drive slowly.

Back to Business

Almost immediately after getting resettled in Oxford, I started making trips back to the United States. Several of my JPL projects were still active, and my new group at Oxford had joint activities and ambitions with NASA as well. I took over as Principal Investigator on the most current of those, an experiment called the *Stratospheric and Mesospheric Sounder* on the satellite *Nimbus 7*, which was by then operating in space and just hitting the peak of its data analysis activities. I was a frequent flyer with Trans World Airlines, whose policy at the time was for Gold Card holders like myself to receive an automatic upgrade to the next higher class any time there was a seat available. There always was, and I soon got accustomed to the Business Class cabin where they poured unlimited champagne from proper bottles, all flight long if you wanted it, and asked you how you wanted your steak cooked. I was sorry when TWA went bankrupt, and so did Pan American, the airline that I used for very long flights to places like India. With their demise another golden age ended, and the much more ascetic regime we suffer under today's airlines was ushered in.

Britain seemed quite austere in many ways after spending a decade in California. It wasn't just the relative poverty and low salaries; there was a certain meanness of spirit that permeated everything. I was often reminded of a quote from Quentin Crisp's book, *The Naked Civil*

The wreck of my Lancia Beta, whose US safety specifications saved my life when a white van piled into me from behind while I was on the way to catch a ferry to Holland. I was headed for the European Space Agency laboratory near Amsterdam for a meeting to plan a mission to Mars called *Kepler*. Like my meeting, the Kepler mission itself never took place.

Servant (1968), in which he says 'The difference between England and America is that in America they want you to succeed, because they feel you may drag them forward with you. In England, they want you to fail, because they fear you may leave them behind.' On my first flight back in to Heathrow from Washington I got into a lift to go down to the car park

After the loss of the Lancia I invested in an Aston Martin V8, seen here on a trip to Wales. I took it to Aberystwyth, and to meetings at many other locations around the country, as a member of the Science Research Council's Solar System Committee, which dispensed grants for planetary research in the UK.

and was confronted by a row of blank black buttons. The floor numbers had been painted on the buttons, and years of people pressing them had worn the numbers off. In America, they etch the numbers on the plate alongside the buttons so this doesn't happen. I'm back, I thought.

And yet somehow I wasn't sorry. Underneath the slightly threadbare culture that 1980s Britain presented was an integrity that was rich in meaning and purpose. In America, what you see is what you get. But life in Britain, especially in a place like Oxford, has many invisible dimensions that are constantly, and often unexpectedly, opening up. We don't really believe H.G. Wells's story in which a machine of brass and amber can propel us through the time barrier, but it's a parable for something that really happens. It's a little like a J.K. Rowling novel: you can be somewhere quite ordinary like a railway platform and a hidden door will open to a platform that is not supposed to exist where a train will take you to somewhere fantastic. Philip Pullman's subtle knife let its owner into places that are all around us but which we cannot see. Enid Blyton's wishing chair grew wings and carried children across the mountains to enchanted lands that were not on any map but as children we knew they existed. England to me is like all of that, if in a slightly more down-to-earth way.

We soon moved out of the rented house in Waterperry to a home of our own. The daily commute, and the parking problems in Oxford, even though neither was anything like as bad in 1980 as they are now, soon convinced us to try to get somewhere actually in Oxford. We had saved enough from the sale of our house and cars in California to be able to afford to live on Headington Hill, a quiet and leafy conservation area within sight of the dreaming spires in Oxford city centre. It was a pleasant walk to my office in the Physics Department, down the hill on a largely traffic-free street, then along the river Cherwell and into the University Parks, which have a side entrance that leads straight in to the University science area.

Head of Department

I was very young to be a head of department at Oxford; my peers tended to be old men living mainly on their past glories. I had no training, experience, or clear idea of how to be an effective leader of a university department or indeed anything else. At first, I had no mentor or confidant close to hand. Consequently, I made a number of avoidable mistakes. Most of these stemmed from the automatic assumption that any kind of growth is synonymous with success, which in later life I would learn is not the case, although most managers, entrepreneurs, and politicians, past and present, invariably seem to think it is. I now know, and wish they did, that what actually matters is quality, and its by-product satisfaction. The fundamental problem is that growth is measurable and quality is not. Eventually I worked this out, helped by continued late-night musings with the Reverend on my trips to California, and eventually also in Oxford in conversations with elderly dons over a cup of tea in a dimly lit oak panelled room on a rainy afternoon or with a post-dinner glass of port.

So, as I took on the competition back then, I seized every opportunity for growth, even in oceanography, without thinking too much about whether it was really and truly the right thing to do. This mind set was encouraged by the system of public funding for university research, which tended to be erratic and unreliable (and, although I didn't know it then, was about to get much worse). If you did not have a lot of irons in the fire and one area or project dried up, as it was liable to do sometimes without notice, you were left without the means to support key members of your team and the entire game was over as far as 'big science' was concerned. On the other hand, the middle ground was elusive and if you were successful then you went too far in the other direction and ended up overcommitted and stretched to the limit. I suppose I should grateful it was mostly the latter in my case.

After a year or two back in the UK I was a member of several organizations. One of them was the Solar System Committee of the Science Research Council, which was responsible for awarding the grants that my department relied on for its research. Of course, I had to leave the room when our own applications were discussed, and bite my tongue when I came back if I didn't like the outcome, as was the case more often than not. I was a member of a panel of the European Space Agency that made key decisions about new space missions, and of the European Space Science Committee that endorsed them. In the USA, the National Academy of Sciences had an influential body called the Committee for Planetary Exploration, known as COMPLEX, which advised NASA on its science directions, and the agency took their recommendations very seriously. The members of COMPLEX were mostly American, of course, but they reserved a place for a European liaison member and I took that over when the French super-scientist Jacques Blamont finished his term. One of the things I liked about COMPLEX was that it held a 'retreat' once a year in Santa Fe, New Mexico. Noted for its artistic community and bohemian outlook, this sleepy desert town is a fine place for deep thinking and writing about long-term scientific objectives and goals.

While working in England, I kept in touch with JPL. A letter in April 1982 from my former colleague Reinhard Beer in response to one from me suggesting a new collaboration says:'The plan for the European Space Platform sounds interesting and . . . may be our only chance for the future. As you know, NASA has cut its planetary research so heavily that it now looks certain that only [one of the group] will get any support at all. I have been cut out of [NASA's] program totally. . . . Atmospheric Experiment Development looks as though it too is dead, Mars Climatological Orbiter or not. [Other members of the group] may have to leave because we will have no Pioneer Venus support next year, and so on. . . . .. Your magic crystal ball got you out of our funding mess just in time, good luck! [I am] looking for military funding . . . '. Five days earlier Lee Elson had written 'the JPL VORTEX team is fading rather rapidly. Our funding next year totals $15,000 . . . at a time when real progress is being made towards answering some of the fundamental questions about Venus . . . '.

In stark contrast, I was having some success at building a programme at Oxford and persuading the authorities to fund it, despite the limited resources at their disposal compared to those the Americans enjoyed, even in the slump my JPL colleagues were experiencing. By September 1987, I was submitting a research grant proposal to the Science and Engineering Research Council for a four year grant 'to support the Department's space experiments programme, which includes the ISAMS, ATSR, NIMS, EOR and PMIRR space instruments, plus related laboratory work and data analysis and interpretation'. The total applied for was £7,846,110 and, although the award was somewhat less after it had been through the committees and their paring-down process, we also had other grants, and we managed contracts to build some of the flight hardware in industry. In a review of the Department in 1990 I was able to list 17 grants worth a total of £5,956,122. I calculated that we were bringing in more money than the other four Oxford physics departments put together, and more than 10% of the income of the entire University.

If I expected rewards or even praise, for that, I was disappointed. In those days Oxford was mostly still the domain of gentlemen scholars to whom 'research' meant spending a little time in the library, and the very mention of money was abhorrent. Our riches were if anything, slightly embarrassing, and a bit of a nuisance with my constant demands for more and better facilities for our space work, and perhaps a few more permanent posts with which to attract and retain more of our key players. The Vice-Chancellor's office had a body called the General Board that dealt with such requests and they used a formula in which external income was a factor. Ours sent the result off-scale, so the Board decided to discard the formula in our case and give us some much smaller amount which they deemed appropriate but which had seemingly been pulled out of the air. There was no acceptable way to appeal or take issue with this, but it happened at one point that an atomic physicist, Patrick Sandars, became the Chairman of the General Board and, since I knew him, I felt able to ask privately why they declined to apply the formula to Atmospheric Physics. 'It gives the wrong answer', he said, without a trace of irony.

Patrick's predecessor as Chairman of the General Board was Clark Brundin, an engineer who happened to be a member of my college. I was sitting next to him at lunch one day and he remarked that Atmospheric Physics was not a statutory department of the University despite what it said on the letterhead that I had inherited from John Houghton.

I was surprised, and said that was news to me, but would it make any practical difference if it were? The answer was 'not really'; Oxford was full of anomalies from a gentler time where things arranged themselves by common consent without worrying too much about rules and regulations. However, I was entitled to an extra supplement to my salary as Head that would only materialize if the department was properly on the statutes. A few weeks later, on 28 June 1985, the *University Gazette* published a notice that 'The Department of Atmospheric Physics shall be assigned to the Reader in Atmospheric Physics, who shall have charge of it and make provision for lighting, warming, water supply and cleansing of the Department.' My salary went up accordingly.

Admin and Teaching

By the late 1980s, I was trying to run a university department, head up a collaborative institute with two powerful government organizations (of which more later), lead Britain's most expensive scientific space project (again, that's another chapter), and maintain a world-class planetary programme, all at the same time. Any one of these on its own required the sort of effort that would kill a person with the constitution of an ox if done properly. All four, together with the usual load of teaching, committees and short- and long-distance travel to meetings and conferences, was frankly impossible.

I empathized with the Nobel Prize-winning zoologist Sir Peter Medawar, who took over a department at Birmingham when he was about the same age (mid 30s) as me. He wrote about it in his memoir, published in 1985, which I read just a few years after I took over at Oxford. He had found he was expected to be '. . . a teacher, administrator, probation officer, psychiatrist, and employment agent . . . also a fund raiser, and this last was my most onerous job . . . '. He missed out politician, innovator, publicist, examiner, world traveller, and quite a few others.

On the teaching front, I introduced a course for the undergraduates called *Climate Physics*. This involved two lectures per week for one term and was designed to show the student how the basic physics they had learned in other parts of the course—quantum mechanics, thermodynamics, optics and electronics—could be used to make simple models of the atmosphere, ocean, surface, and cryosphere that were powerful enough to explain the main features of the climate. The average surface temperature, cloud cover, ozone layer, role of the ocean, and other basic quantities could all be calculated in an approximate but realistic way, and simple experiments in climate change—what happens as the carbon dioxide in the atmosphere increases, for example—could all be calculated without a computer. Later, I added a course on *Planetary Atmospheres* to the final year syllabus, again using fundamental physics to explain the Solar System and its diversity of atmospheres and climates. Textbooks were written, lectures delivered, exam questions set, marked, and graded. I did two stints as a finals examiner for the whole Physics course. Examining was fun, but very intense and time consuming.

Unification of Physics

When I arrived in 1980 there was a body called the 'Inter-Departmental Committee for Physics', which met twice a term to discuss teaching in physics, since all five physics departments contributed to the organization and teaching of a single undergraduate degree. It was also a chance to discuss any other items of common interest. The departmental heads took turns at chairing this, and it was during my three-year tour of duty starting in 1986 when we decided to merge the five physics departments into a single, large administrative unit under a new position of Chairman of Physics. It was my job as current chairman of the interdepartmental committee to steer these changes through, and I was not entirely keen at first. Our space experiments brought in a lot of money, but in terms of academic staff numbers we were small, and I was worried about being overpowered or swallowed up by the larger departments.

However, just at that time, the national government chose to launch one of its typically numb initiatives, this time having decided to merge smaller university departments into larger ones. This was to be achieved by forcing academics and their staff and students to move across the country if necessary. A task force was set up to implement the policy; they announced that they would do geology first, and physics next, then the other subject areas one at a time. In the document they sent out, it was stated that there were a number of small, specialized areas *like atmospheric physics* (my italics) that would probably benefit from a larger economy of scale. Mine was the only department in the UK with atmospheric physics in its name! What were they thinking and why, and what would they do to us? Before I could find out, their task force tore into geology, scattering bits of hapless departments everywhere. The Geology Department in Oxford, as one of the largest, was on the receiving end of some of these fragments and was one of those that grew. That would not be my fate, however, when they turned the spotlight on Physics. Suddenly, voluntary unification of my department with the other physicists at Oxford did not look so unattractive.

The first Chairman of Physics, Chris Llewellyn Smith, was a decent chap and a good manager as well as a brilliant theoretical physicist, and the future looked promising. Now, as the head of one of six physics subdepartments, together forming the largest physics department in the country, I was part of a large family of outstanding scientists working on everything from subatomic particles at the small end to galaxies and black holes at the large, with the Earth and planets comfortably in the middle. Because of the human scale of the things we dealt with, atmospheric was actually the oldest discipline of any that now made up the integrated Physics Department, as well as the most modern in that we were engaged in space research while the astrophysicists were still using telescopes. In the Clarendon archives I uncovered a syllabus from the early 1800s, which showed that 'natural philosophy' students of that time mostly studied things that were now in the bailiwick of the Atmospheric Physics subdepartment. The one-page syllabus had a footnote that said that anyone attending a lecture course who was not a member of an Oxford college could do so on payment to the lecturer of a fee of two guineas, or one guinea 'for any fubfequent coufe'. This very large sum, for the time, was presumably intended more to keep out the riff-raff than to make a profit. There was also a warning that, just because the lecturer had died, there was no avoiding the fee.

'Oxford five link to form Physics giant', said the headline in the *Times Higher Education News*. I was the only head of one of the new sub-departments who was not a professor. This was duly rectified; the Times article said: 'A professorship of atmospheric physics will be established in recognition of the distinguished work of the atmospheric, oceanic and planetary physics unit. The 1989 University Funding Council review of meteorology placed Oxford in category A of universities active in the subject, and drew attention to Oxford's large concentration of atmospheric scientists. The new professorship will go to Dr FW Taylor, presently reader in atmospheric physics at Oxford.' There were no interviews this time. An integrated physics department would also mean that I was relieved of a large upward administrative burden, and was therefore a good thing, although of course it could be a double-edged sword, as I was to discover.

I took the chance to update the name of the department (now officially a subdepartment) to Atmospheric, Oceanic and Planetary Physics. In the early 1960s, before I came as a research student, John Houghton had changed Dobson's name for his group from Meteorology to Atmospheric Physics, in the interests of modernization. Now, with our scope increased still further, I thought 'Terrestrial and Planetary Physics' would be good, but fell foul of the Professor of Geology who insisted that 'terrestrial' meant, or at any rate included, 'the land'. I could see his point so I gave way, only to be affronted not long afterwards when he changed the name of his department from Geology to Earth Sciences (and one of his groups, from 'Seismology' to 'Geophysics'). Earth Science sounds like it includes the atmosphere, and Geophysics sounds like part of Physics, so I was incandescent, much to the amusement of Patrick Sandars, who was now Head of the Atomic and Laser Physics subdepartment. He referred to it as a 'perfect storm in a tea cup', which I suppose is what it was, and I soon calmed down. As Lord Byron said in 1810, the relish for the ridiculous is what makes life tolerable.

EXPERIMENTAL PHILOSOPHY.

INTRODUCTION—The laws by which inquiries in Natural Philofophy fhould be governed—matter and its properties—the laws of motion—the attraction of cohefion—gravitation.

MECHANICS.

The lever—center of gravity—the balance—pully—wheel and axle—wedge— inclined plane—fcrew—the means of altering the direction of motion—compound machines—friction—ftrength of materials.

Uniformly accelerated and retarded motion—projectiles—central forces—the Solar fyftem.

Bodies moving down inclined planes—pendulums—weights and meafures.

Collifion.

Springs—watches.

HYDROSTATICS, PNEUMATICS, HYDRAULICS.

Properties of fluids in general—thofe of incompreffible fluids—fpecific gravities—properties of elaftic fluids—height and preffure of the atmofphere—found—fteam—the motion of water through pipes—fiphons—water fpouting from an orifice—time in which veffels will be emptied—refiftance of fluids.

METEOROLOGY.

Barometer—Thermometer—Hygrometer.

OPTICS.

Dioptrics—refraction—foci of lenfes—ftructure of the eye—microfcopes—telefcopes.

Catoptrics—reflection—foci of reflecting furfaces—aberration produced by fpherical furfaces—reflecting telefcopes.

Perfpective—camera obfcura—camera lucida—anamorphofes.

Colours—rainbow—difperfion of light—achromatic lenfes.

ELECTRICITY.

MAGNETISM.

Terms of attending the Lectures—For the Firft and Second Courfes, each Two Guineas, and for any fubfequent Courfe, One Guinea.

*** All Courfes attended in the time of the late Reader in Experimental Philofophy to be reckoned.

S. Collingwood, Printer, Oxford.

The Oxford 'Natural Philosophy' (i.e. Physics) syllabus from around 200 years ago. Topics on a human scale, like meteorology, optics and planetary science, dominated then but later lost out to particle theory, cosmology and other 'modern' physics. With the growing importance of climate science, space technology, and computer modelling, all in my domain, I sought to use whatever influence I had as a professor to reverse the trend a little.

We're all Europeans now

Unification was the name of the game not only in Oxford Physics but in the whole of Europe as first the Common Market and then the European Union gathered momentum. A side benefit of this was the flourishing of the European Space Agency (ESA), a 20-nation alliance with Germany, France, Italy, and Britain as the major partners. Since the ossified mind set that prevailed among many British politicians at the time regarded any kind of activity in space as frivolous and extravagant, a low priority at best, this breakthrough was timely and brilliant. The naysayers made sure the UK was not committed to any kind of *manned* activity in space, but that was OK as far as I was concerned. It was not that I did not think that mankind had a future in space travel, but rather that my years in the States had made me sensitive to the way the manned programme could suck up all of the money that would have funded a much more vigorous unmanned programme, including planetary exploration. In any case, Europe had to walk before it could run.

ESA sits alongside, but is not governed by, the European Union, a much larger and much less efficient entity. But the EU has an awful lot of money, giving out billions in a singe tranche as grants to researchers, and a little of this (relatively speaking) ends up supporting space research, including participation in the programmes run by ESA. I could, therefore, as a European once more, benefit from that too, once the necessary networks were in place.

Within this support framework I had, over the years, a number of delightful collaborations. These ranged from Helsinki in the north of Europe to Heraklion in the south, taking in Berlin, Paris, Granada, and Rome, plus a large number of smaller interactions, all of them productive in ways that showed the leverage that comes from an expanded outreach. My long and important partnerships with Athena Coustenis and others at the Observatoire de Paris on the fabulous *Cassini* mission to Saturn; Ulf von Zahn at the University of Bonn on the brilliant but finally, and undeservedly, unsuccessful *Kepler* mission to Mars; and Manuel Lopez-Puertas in Granada on studies of atmospheric radiation, were all part of the modern European spirit of cooperation. And, of course, it did not have to stop there.

Physics and Politics 'down under'

Towards the end of 1994 I received an invitation from the Australian Research Council to sit on a review board that was to investigate and report on the state of physics in Australian universities. The chairman was John Carver of the Australian National University in Canberra, a physicist of such renown that the main Physics building at his university is now named after him. I had met John at a conference on one of his visits to Europe, where we had discussed his work on the early climate of the

Earth, when the composition of the atmosphere was quite different to that today. The rules set up for the review specified some foreign membership and these were satisfied with the appointment of one member from the USA and one from Europe, the latter being myself. It would be necessary to spend six weeks touring all around Australia, spending a few days in each of around a dozen physics departments.

Before I set off on the long journey I nursed a naïve belief that Australian politicians are more sensible than those in the UK; just as before my first trip to Canada I had pictured that country as a haven of peace and prosperity in a troubled world. Reality is a harsh mistress and, while no doubt many of the people who govern both of those great countries are first-rate, some of what they wrought seemed doctrinaire and ill-considered to an even greater degree than I was used to in England. I was able to experience the actual process briefly when I had some time off in Canberra and used it to walk up the hill to the Australian parliament. This is underground and you can actually walk on top of it, but more surprising than that was that, in 1994 at least, I could walk in and go straight to a gallery overlooking the chamber in which a debate was taking place without encountering any of the security barriers that gird the equivalent at Westminster. The topic was something to do with farming land disputes and I didn't listen for long.

While I was travelling around Australia, discovering what a delightful country it is, the newspapers headlined one of the occasional attempts by republicans in the government to ditch the links with Britain. This puzzled me: why give up a valuable friendship and collaboration, when

While in Australia for a review of the nation's Physics departments, I visited my cousin, Professor Jim Taylor, at the University of Southern Queensland. The newspapers were interested in my views on climate change and on (not) severing Australia's British connections.

you already have, or can have, all the freedom and self-determination you want? I wrote a letter to *The Australian* before I left saying so. The published version of the letter did not show any address, but still I got a few letters in response from the readership, all of them supportive, that found me after I got back to Oxford.

As our panel travelled around the huge, empty country, in a series of hops by plane, it became clear that Australian physics was in fact well on the way to becoming a wasteland. There were plenty of good people and projects, but, except in the strongest departments like Canberra and Sydney, there were also empty workshops and disillusioned and depleted staff. There were various reasons for this, and of course I well knew that expensive research often has difficulties in the best of all worlds, but one problem stood out like a sore thumb. A year or two earlier, the government in its wisdom had decided to make a sea change in the way universities were supported. This had two main parts: firstly, they would sharply reduce the funds that went directly to universities, and put the money into research council grants where the academics could compete to get it back. This would ensure the money ended up supporting the best departments rather than being dispensed by formula. Secondly, they would stop discriminating between universities and technical colleges and the like; they could all be universities and compete for funding on level terms. This would discourage elitism.

Competition and equality sound all right: what better way to get the best research and teaching, and value for the taxpayer, eh? But there are all sorts of unintended consequences, one of which is that grants are awarded piecemeal by a system that is capricious in the extreme. Consequently, the funding of an individual or group, and the whole department, seesaws wildly according to whether you have been lucky or unlucky recently. Even if the department is large enough to average out to something relatively stable, there are rules about transferring grants awarded for one purpose to another. Sooner or later, you have to lay off your best people and downsize your workshops, and the arrival of a big grant six month after you have bitten that bullet is no help at all, you need those people but they are gone.

The merging of universities with polytechnics created a university base that was suddenly more than twice its former size, with no gradual evolution and no additional funding. (It would also, in time, create a shortage of people with technical training, but it was not our job to consider that.) The net outcome was a savagely competitive fight for funding in which scholarship was lost in the dust of the stampede to survive. Far from increasing efficiency, an expensive management culture was replacing the academic self-government that the Aussies had inherited from their European forebears. I was not in Australia long enough to see this evolve there, but I little realized then that the same approach was to be adopted in the UK a few years later and I would certainly feel the consequences there at close range.

I enjoyed the Australian visit, the people were super, both in and out of the universities, and despite the ill-considered tinkering there was still much good work going on. I had my first and only kangaroo steak, which (as far as I could remember) was similar to the horse steak I ate at Harvard and the venison we grow at Oxford, a stronger-tasting version of beef and a bit tougher to chew. I also discovered that Foster's lager, like many French wines, is better in its homeland. The kind they used to market as 'nectar' in England tasted like dingo urine diluted with rainwater. There was no trace of real ale anywhere I went in Australia, and I'm told there still isn't.

I visited both of my male cousins, who had emigrated there years ago: Jim Taylor was in charge of a large and very impressive centre for distance learning at the University of Southern Queensland in Toowoomba. His older brother Reg, like Jim a long-term resident of Amble, Northumberland before emigrating, had retired as a marketing manager for Cadbury Schweppes. From him I learned that Cadbury had been in Australia for nearly a century and made their celebrated chocolate in a factory with a model village attached, based on a design similar to the famous original in Bournville. Instead of Birmingham, however, this one is a suburb of Hobarth, Tasmania.

I made suggestions that went into the report of our review board that I hoped would help with the problem areas across the system as a whole. One of them was that they consider introducing renewable programme grants, like those that worked well in the UK to give the larger departments some stability. After I was back in England, I heard that this recommendation had been dismissed on the grounds that it was contrary to the spirit of competition that was to be encouraged. The big departments would have to live on scraps and, if they couldn't do that, maybe market forces were indicating that they shouldn't be so big or so durable. Not only were my attempts at the amelioration of damaging changes in Australia nugatory, those same changes would soon be introduced in England and I would be trying to tackle some of the same problems myself.

In Marble Halls

In 1996 I received a letter from Ilias Vardavas to say he was moving to the University of Crete where he hoped to set up an atmospheric physics group. As I was head of an established group in the same area, would I help him with advice and possible collaboration? I flew out to Heraklion for discussions and to see his set-up for myself.

Ilias turned out to be more Australian than Greek. Born in Rhodes, his family had emigrated when he was young and his scientific career had been built in Australian laboratories where the senior people he worked with included John Carver, with whom I had recently toured the major Australian universities. Ilias had responded to a call from the University of Crete to find expatriate Greek physicists willing to return and staff their rapidly expanding department. The Cretans were investing in their university, gradually moving it out of cramped and dusty accommodation in the old city to a new site on the surrounding hills, with spectacular views over the Mediterranean Sea. Physics had been one of the first departments to move, and now occupied a building with marble halls and sweeping balconies, allowing the occupants to work, eat, and think in the open air as well as in their offices.

The opulence of the surroundings was slightly offset by the fact that there was not a great deal of money for research from sources within Greece ('It all goes to Athens, anyway', said the Head of Physics in Crete grimly when I spoke to him on one trip). There were also crippling internal political vendettas that, as a visitor, I never understood, but they clearly darkened the atmosphere under the brilliant Mediterranean sunshine and made Oxford seem like a monastery by comparison, which in some ways it still was. Incongruously, Crete reminded me of my first visit in the 1970s to the US National Center for Atmospheric Research in Boulder, Colorado. This also occupies an architectural masterpiece of a building on a glorious site high in the hills overlooking spectacular views. In the case of NCAR, these are the foothills of the Rocky Mountains, surrounded by trees, and the views are of the Great Plains stretching into the distance beneath. Like many another first-time visitor, I exclaimed to my hosts that they had a wonderful working environment. 'Yes, *physically*' said one of them darkly.

For the atmospheric physics group in Crete, the most promising source of funding was the European Union. Ilias was very good at assembling teams and competing for EU funds and we had many years of interesting meetings, focussed on the theory of atmospheric radiation (what happens to the energy from sunlight, basically, in its exchanges with clouds and the surface) and its applications to problems like global and regional habitability and agriculture. On one occasion while I was there, Ilias was lecturing to the undergraduates (in Greek, of course) and I found I could read and understand his lecture notes despite the language problem, because the basic material was so familiar to me.

The notes were much better than anything I had produced in Oxford and in the ensuing discussion we resolved to write a textbook that would be useful to students, including graduates and young researchers, in both of our universities. Over the next few years we laboured on this and the result was *Radiation and Climate*, published by Oxford University Press in 2007. The only problem with this was that as a hardback tome it was too expensive for the students at whom it was aimed and sold mainly just to libraries. We rectified this four years later with a second edition that included a cheaper paperback version, and I look for it to become established as a classic in the field.

Finlandia

In many cases collaborations led to direct participation in the academic life of a country unified within the European community, but still steeped in its own traditions. When I was invited to preside over the examination of a doctoral thesis in Helsinki, for example, I found that it involved very formal dress—white tie and tails—for the viva, which is held publicly. It was followed by a lively party with games and masquerades that went on well into the small hours of the night; more like a wedding than an examination. As the official 'Opponent', I was alone on a large stage with the candidate for exactly two hours, watched by a large audience of the colleagues, relatives, and friends of the would-be Doctor of Philosophy. The format was that he would speak about his work for 20 minutes, and then I could ask him searching questions about it for most of the rest of the allotted time. Then in the final five minutes, I was to summarize and deliver my verdict.

On the way in to the great hall that held the event, I could not fail to notice two large tables in the lobby laid out with glasses and plates for food, ready for the post-exam celebration. The ladies responsible for this must have had their ears to the closed door, because when I stood up as instructed at five minutes to four with the words 'I will now summarize and conclude' there was a loud popping of champagne corks, outside the door but easily audible throughout the hall. 'What if I had failed him?' I later asked the professor in charge on behalf of the university, known as the Custos. 'We have already examined him internally', he said in the dry Finnish way, 'we don't expect him to fail.' Then he added, 'We drink the champagne anyway'.

7
Forecasting Weather and Climate

By the 1980s it had become fashionable to set up Institutes in universities. These are monolithic organizations that cut across the normal departmental boundaries in order to achieve the optimum management structure required to address some specific major problem. Collaborators in other universities may be involved, and sometimes people and funds from research groups in laboratories run by industry or government. I knew something about Fred Hoyle's activity at Cambridge, when he set up the Institute for Astronomy there in the 1960s. Much of the correspondence that documents his progress has survived, in which he wrote pithy remarks such as 'The aim of the Institute is to deal in problems of some magnitude and grandeur . . . [which] cannot be done in an atmosphere of petty intrigue.' I wish I had known him better and could have benefited from his lessons learned, before my own experience with this kind of venture.

It started when, in 1983, John Houghton left the Appleton Laboratory and moved to Bracknell as Director General of the Meteorological Office, usually called the Met Office, where he was to remain in charge until his retirement in 1991. This freed up the Readership at Oxford, as I had anticipated, although protocol demanded that I had to go through the full gamut of applications, selection board, and interviews. It could have gone to someone else, of course, but it didn't and in due course a formal letter of appointment came. I was at last in a position to think about relinquishing my legal resident status in the USA, which I had hung onto in case I needed to ask JPL if they would take me back.

By Hooke or by crook

One day not long after taking over at Bracknell, John called by his old Oxford office, now mine, with one of his visionary ideas. As I have said, there was a fashion at the time for setting up 'centres', 'units', or 'institutes', loose conglomerations of separate organizations created under a single administrative structure to address a specific major topic or issue. The resultant focus was expected to attract attention and especially funding—important at a time when support for research groups, even in universities, was deep into the transition to piecemeal funding and detached professional management, American style.

Ahead of the game as usual in his thinking, John came up with the idea of a joint institute between his old and his new domains. It would be named in honour of Robert Hooke, the innovative instrument maker and contemporary of Newton, Boyle, and Wren. Hooke had lived in Oxford in the years around 1650, working on optical, astronomical, and atmospheric problems. These included, it is said, heavier-than-air flight, although this turned out not to be possible despite the application of Hooke's genius, because it needed a motor of some kind with a higher power-to-weight ratio than anything that could be conceived at the time.

When John Houghton formally resigned from Oxford University, in order to move to the Meteorological Office, we held a farewell dinner at which as his successor I presented him with an Orrery, an inscribed mechanical model of the Solar System. Marie Corney is in the background.

To create the Institute, the Met Office proposed that one of its research groups would move from Bracknell to Oxford and work alongside the Department of Atmospheric Physics, and its work would 'concentrate on topics where substantial overlap exists with the University Department'. The theme was to be Satellite Meteorology, the art and practice of providing the data needed for weather forecasting using the latest space-based equipment and techniques. This would be broadly interpreted to include the development of new instrumentation for future satellites, and research into the meteorology of the stratosphere, which was thought to have an unknown but possibly important role in controlling the weather at the surface. Since our Oxford department was one of the leading research centres in the country for developing methods for remote sensing from satellites, for building instruments, and for understanding the stratosphere, while the Meteorological Office was responsible for operational weather forecasting; this was a match made in heaven. I was enthusiastic; it surely was a mutually beneficial way to grow our research work, but more importantly it would give us a new and exciting outlet from academia into the applications area.

The Hooke Institute for Collaborative Atmospheric Research (1984–1992)

After some difficulty in acquiring the additional office space for the people moving from Bracknell, the first phase of the Institute was up and running in July 1984, with myself as its titular head. The scientific interface developed as expected, but it soon became clear that our Met Office unit served under not only their local group leader, John Eyre, but also a hierarchy of remote managers who maintained a watching brief from Bracknell, 50 miles away, and showered us with letters and memos when they saw anything they did not like. This stiff formality presumably resulted from the Met Office's historic position as a part of the Ministry of Defence. Unfortunately, some of the people in the chain of command resented staff and funds being diverted from Bracknell to Oxford and did not care for the whole idea of the joint institute. They couldn't say so, of course, since their boss had set it up, but they subtly imposed an iron civil service regime on their unit designed to make sure no fancy ideas of academic freedom crept in. This naturally had knock-on effects on my department as well, as we began to work in tandem.

I had naïvely imagined that the small Met Office group would be pretty much left to its own devices to get on with our agreed programme under the direction of John Eyre and myself, and the work of the Institute as a whole reviewed from time to time by the steering group that had been set up for this purpose. The Met Office Director of Research, who barely concealed his distaste for the whole idea of the Hooke Institute, made it clear that the Met Office's contribution would be whatever they decided, and it would be directed from Bracknell. If they wanted to slip in extra tasks that had nothing to do with our agreed programme but happened to be convenient to the Met Office as a whole, then they would do so. I was repeatedly reminded that the Met Office people and their programmes of work were answerable only to their line managers and no one else, in particular not to me. My reply was along the lines that this was undoubtedly true, but we had signed an agreement that specified in outline what would be done in the Institute and as its head I was responsible for the success of that. If Met Office people did not work on the agreed programme then that was fine, but if it persisted they would have to leave Oxford and do the work somewhere else, like back at Bracknell. This went down very badly and led to a darkly opaque warning that I was upsetting people, and the first of many threats to withdraw.

The Met Office unearthed some arcane documentation that decreed that academic staff at Oxford are constitutionally responsible to 'The Chancellor, Masters and Scholars of the University'. Their interpretation of this was that there was no obligation on them or anyone to trouble me before arranging particular changes or activities in 'the Hooke' no matter what they were or who they involved. This was typical of the treatment I received from the management at Bracknell, even those people with whom I enjoyed a warm relationship outside the confines of formal meetings and correspondence, which was most of them. Reliably and generously funded by the government, Met Office staff enjoyed large salaries, undemanding hours, and plush facilities, that put the hand-to-mouth existence of most university operations to shame and this, I supposed, was what imbued them with a sense of unquestioned dominance over any collaboration with a mere centre of higher education, even one as ancient as Oxford.

I, on the other hand, was acutely aware that I had chosen university life for the freedom it gave me to pursue my interests, and I wasn't about to take orders from remote administrators stultified by the civil service mentality and conditioned mainly to churn out an excellent but basically standard product. I was the head of the Institute, which was a research group in my university, and I was determined that it was going to have an academic approach to its research and stick to the theme that we had agreed. Just creating the Hooke venture involved a lot of work and commitment on my part and put my reputation at the University on the line.

In spite of these tensions with Bracknell the activity at Oxford rubbed along very well, superficially at least, although the iron grip that was kept on the visitors meant that very little joint work could be done, and nothing truly innovative could be got off the ground. Seen from the outside, however, it looked pioneering and successful, and for a time we were the envy of our peers at Oxford and elsewhere in academia. It was not long before we were looking for ways to make the venture even more efficacious, and of course larger. Automatically going for growth is a common but fatal mistake made by many a leader in all walks of life, but by the time I realized that it was too late.

Half of me is ocean, half of me is sky

Funding to universities for research in the area defined by the Hooke Institute charter came under NERC, the Natural Environment Research Council. It was clear from the outset that their grant support would be vital to the success of the Institute's programme. As well as providing grants to universities, NERC had several institutes of its own, but none that dealt with satellite meteorology, or even with atmospheric science. This was clearly an oversight at best, and possibly an actual deficiency. Perhaps, I mused, as well as the existing system of grant support for the university component, NERC would be interested in contributing a directly funded unit to make a tripartite institute in that area? It turned out they were, but on their own terms.

The area of joint interest to which our exploratory discussions with NERC led us was not satellite related at all, but had to do with understanding the ways in which the oceans and the atmosphere interact and the development of general circulation models in which the ocean was an active part. Most of the models in existence at the time treated the ocean as a lower boundary to the atmosphere, but of course it was an enormous reservoir of heat, moisture, and momentum so that a proper treatment of the exchange of these key parameters between the ocean surface and the overlying air mass was vital for longer-term weather forecasting and predicting possible climate change. This was the area in which our newest senior academic, David Anderson, worked, and so our university department already had an interest and at least a small contribution to make. I wrote an upbeat article for the journal *NERC News* in September 1985, which concluded: 'If their [the staff] enthusiasm is any measure, the future success of the Hooke Institute in the scientific areas it embraces is assured.'

I still thought satellite meteorology made more sense than oceanography as a theme for the Institute at Oxford. There were already several leading academic groups working on ocean–atmosphere coupling, including one in the Department of Applied Mathematics and Theoretical Physics at Cambridge. However, for reasons of internal politics I never fully understood, despite its internationally acknowledged excellence, the Cambridge group was apparently considered surplus to requirements by their university. It was supported by soft money, mostly from NERC, and consisted of about a dozen researchers, led by three extremely bright scientists, Adrian Gill, Peter Killworth, and Mike Davey. Our own David Anderson had been part of this group before he joined us in Oxford as a senior academic member of my department and knew Gill and co extremely well.

The Met Office was, of course, a world leader in the development and use of computer models of the atmosphere for the purpose of weather forecasting. Under John Houghton it had plans to extend its forecasting skills well into the future to predict the looming threat from climate change. It was to set up in 1990 on its premises at Bracknell the

Hadley Centre for Climate Research and Prediction, which was dedicated to making progress in this area. Back in 1984 when the Hooke Institute was still finding its way, however, a different way forward was being planned, and wheels were turning of which I was initially unaware. Secret talks between the Met Office, NERC, and Cambridge had led to a deal to move the whole group, including students, to Oxford to join the Hooke Institute and work on problems of long-term forecasting, especially those involving ocean modelling, in my department.

It was presented to me as a done deal in a telephone call to my home late one Saturday evening while I was entertaining guests to dinner and was not exactly in any state for a critical discussion of strategy at the highest level. Since we were all on such good terms personally, I could only assume that I had been kept out of the negotiation for fear I would baulk at the hard work required to assimilate the Cambridge group and set them up with accommodation and support facilities. There was some truth in that. Since I was the local manager and notionally the head of the Hooke Institute, it would fall to me to be in charge of absorbing the incoming team and making them happy. The individuals involved made it plain that they really didn't want to leave Cambridge, but had been forced to move or face destitution, so the task of brightening their collective existence was far from easy. And of course the gimlet eyes of my senior colleagues at Oxford—the Chancellor and Masters if not the Scholars—were on me: the academic performance of this mass intrusion of mine, as they saw it, had better be world class. At Oxford, nothing less is acceptable.

Timeo Danaos et dona ferentes

In spite of the many practical difficulties, I swallowed any doubts I might have had that making this unexpected windfall work somehow was a game that was worth the candle. Concerns about the climate were looming globally, and debates among scientists everywhere about how serious it was were beginning to spill over into the popular media. Even the politicians were beginning to show an interest. Viewed scientifically, climate change was a beautiful problem that could, indeed had to be, addressed by the application of basic physics. The key disciplines were radiation and fluid dynamics, and potentially the equations could be solved with the help of the latest and most powerful computers. In terms of applications, and what we nowadays call 'impact', nothing less than the future safety and wellbeing of world society was at stake.

Such research was, of course, right up our street, with related theoretical and experimental work in the area already underway in our department. It seemed there was nothing not to like about building up our work in an area of growing importance, and the involvement of the outside bodies would give us easier access to the big models and powerful computers that we would need. A diagram that I doodled on a pad while on a long flight (there were no laptops then) shows my dream of a department that had four interlocking areas of research: space experiments, planetary exploration, satellite meteorology, and climate modelling related to TOGA. TOGA stands for Tropical Ocean, Global Atmosphere, which was the title of a big international programme to coordinate research on coupled ocean–atmosphere modelling. Our newly minted effort at Oxford became a big part of this, mainly through Adrian Gill who was the chairman of the TOGA Programme Committee, and we were soon using TOGA as shorthand for the group and its work.

We started work on revising the agreements and contracts to turn the Hooke Institute into a full tripartite operation with two, rather than one, major themes. NERC duly signed up to add 'coupled ocean–atmosphere modelling' to the Institute's charter. However, the research council was about to undergo one of its periodic shake-ups and this time it was reorganized from the top down. In particular, there were big changes in the way it was managed.

This would reflect the culture shift that was underway in the 1980s, which saw the end of the days when the research councils were responsive to the needs of universities, as had generally been the guiding principal at the start of my career. Now, the prevailing ethos was that the councils could not only hive off a large percentage of the funds designated by the government for research and other work in their own centres, but also that they should direct most of the remainder into specific topical areas seen as political priorities. 'Blue skies' research was not encouraged, and remains unfashionable to this day except as a largely unfulfilled ideal amongst the academics themselves.

The new structure brought in at the Natural Environment Research Council in 1986 was based around the creation of three new posts, to oversee each of the Council's three main areas of responsibility. The oceanographer John Woods returned from a position in Germany to become Director for Marine Sciences. Rather oddly, but accurately reflecting NERC's priorities at the time, the definition of oceanography in his brief included atmospheric research. The plan sought to attract high-powered leadership by offering each of the new 'super-directors' a budget and sponsorship for a part-time university position to go along with their administrative duties. On his appointment to the ocean–atmosphere role, John Woods decided to base himself at Oxford and came one day to ask me if I would host him in my small Atmospheric, Oceanic, and Planetary Physics empire.

This seemed like a good idea, virtually guaranteeing political support for our goal of pushing climate forecasting into a new era. It did not go down well in the wider UK academic community, however, where research groups were desperately scrabbling for funds and saw Oxford getting an unfair inside track. The *New Scientist* thundered that it was not 'justifiable that one of the new directors has, in his first year in office, moved his own research team to Oxford. . . . With many academics now finding it hard to gain support, even for their best research projects, their anger is understandable'.

Whoever those 'many academics' were, they need not have been so jealous. John's first act on arriving was not to enhance the support for myself and the programme of our institute that I naïvely expected, but the draconian cancelling of all of our existing NERC grants. This was allowed by NERC's rules, but almost never implemented except in extreme cases like the death or desertion of the grantholder, and it affected all aspects of our research, not just oceanography. He explained his actions by saying his arrival meant a fresh start for us.

Growing like Topsy

Part of the fresh start was to be a big new oceanography research project called the Fine Resolution Antarctic Model, FRAM. (This was a double or even a triple entendre; as well as being the acronym formed from the initials of the name of the new project, *Fram* was the name of an historic ship used by oceanographers in Victorian times for now-legendary journeys to the polar regions of the Earth. The name means 'Forward' in Swedish.) As one of John Wood's and therefore now NERC's pet schemes, FRAM had been identified for direct funding as a flagship project, and it needed a base somewhere in a congenial environment where a good team could be attracted. The ultimate goal was to make an enormously long run of a computer model with much higher resolution that ever before, in order to understand the importance of small-scale motions such as the eddies that occur on all scales in the oceans. It obviously had some scientific relevance to our coupled ocean–atmosphere modelling project, but the link was tenuous and really TOGA and FRAM were completely separate intellectually, as well as distinct in term of staff and budget.

FRAM was not bad science, of course, and the path of least resistance would have been just to add a fifth major string to the department/institute bow and get on with it. But we had pushed our neighbours in the Oxford science area to the limit, and there was nowhere for Dr Woods and his NERC contingent to reside that was near the rest of us. They reluctantly ended up in a big old house at 62 Banbury Road, more than half a mile away from the Physics Department. Even acquiring the use of this unloved edifice was an uphill struggle that took up a lot of my time and goodwill that I could have used better elsewhere, but NERC wanted more satisfaction. I was brusquely informed that it was my job to find them better accommodation, and the threat that NERC would pull out completely and leave the whole institute high and dry was regularly and ruthlessly brandished.

We were already trying to house the TOGA team properly. FRAM, it seemed, would be on the same scale or larger. But to make the problem even more complicated, there were other opportunities knocking at the door as well, and further problems around the corner.

Come one, come all

The novel idea of the Hooke Institute, as a place where powerful people in government organizations could buy their way in to an ancient university and set themselves up for 'a life of the mind' in the groves of academe, caught on like wildfire.

The Meteorological Office in its by now familiar unilateral mode decided to close the Geophysical Fluid Dynamics Laboratory at its headquarters in Bracknell and transfer its principal intellectual, Raymond Hide, and his group with all their equipment, to Oxford. The equipment was massive, several large and very heavy rigs that contained big tanks of fluid that could be differentially heated and rotated at varying speeds and fed with tracer dyes to follow the resulting motions, which were often spectacular. The aim was to uncover the way in which the circulation organizes itself in planetary atmospheres. When I learned about this I was very keen to have Raymond, and his younger colleague Peter Read, with us in Oxford to add a new dimension to our planetary work. But where would we find space for them, especially with that vast laboratory, which was so big and heavy that it could only go in a large room with a solid or specially reinforced floor?

Several other leading Met Office scientists, when they heard of Raymond's planned move, made informal contact to see if there was any prospect that they could move here as well. While we were pondering this, a famous planetary scientist from another university called on the phone and asked if he could move to Oxford, saying the Royal Society would sponsor him. He in particular was engaged in exciting research that was relevant to my own, but I had to be ruthless and explain that it sounded marvellous, but there were too many practical problems. He did not buy this explanation despite its verisimilitude, and remained seriously miffed for many years to come.

The Head of the Space Science Department at Rutherford Appleton Laboratory, which through its role as a research council sponsored organization had provided technical support to us on several space projects, declared his intention to move himself and his personal entourage, plus no less than four substantial research groups, from RAL to Oxford. This made no sense to me at all; we needed to keep collaborators at his well-heeled national laboratory to make our space projects possible, not move them in to Oxford where by now we were bulging at the seams. My letter with a polite but not very encouraging response was returned with a wildly aggressive message scribbled on it. Later, they started to move anyway, having decided it was sufficient to get the Met Office to say it was OK, but by then it was too late and the move was never completed and soon reversed.

Next I received a call from the Provost of University College London, Sir James Lighthill, who would later die in the water when the mitral valve in his heart ruptured while attempting a nine-mile swim around the island of Sark at the age of 74. UCL had decided they wanted to move their entire small Atmospheric Physics group from London to Oxford, to merge with us. This was by no means an unattractive proposition: the head of the group, Dr David 'Dai' Rees, was an outstanding experimentalist interested in space instruments, and certainly rather isolated from the mainstream of his department in Bloomsbury. The practical difficulties involved in achieving the move of a tenured academic would be enormous, but Sir James was sure if anyone could make it work, I could. Would I get on with it please?

I obviously had developed some kind of reputation as a prime mover (or, perhaps, as a patsy) that I am not sure was entirely enviable. Time and again my inability to gratify everyone's plans for my department would bring opprobrium down on my head, and those that I could and did help found gratitude in short supply. The level and variety of scientific activity was exhilarating, but the logistics problems associated with all of these initiatives were overwhelming. It was not just that physical space is always at a premium in Oxford, where the departments, Physics included, were clinging to their sites in the historic centre and resisting Cambridge-style migrations to the suburbs and beyond. This kept us close to the historic colleges, but resulted in fierce competition for every square inch of space. In addition, politically the sort of empire-building represented by all of this continued to be increasingly firmly resisted by my senior colleagues in the neighbouring departments. They noted with some justice that non-university scientists now not only outnumbered the hosts in my small department, they were displacing or constraining the accommodation for the real academics like themselves who worked in the surrounding buildings. Furthermore, it was frequently and legitimately lamented that, except for supervising graduate students who worked on their projects, the visiting pseudo-academics did little or no teaching.

The Old Observatory saga

Just as it seemed that everything must implode and fall apart, a rare opportunity came along. The Nuclear Physics department had been shrinking as support for that area of research began to decline nationally. A huge amount of space suddenly became free when they closed their Van de Graaff generator, an early kind of particle accelerator that had become obsolete. This space was not a lot of use to us, as it was in a building that was too far away. However, the Astrophysics department, who occupied the historic University Observatory that was quite close to us, made a bid to move lock, stock, and barrel to the much more modern (and, incidentally, extremely ugly) Nuclear Physics building.

Stretching goodwill to the limits, I petitioned the University to get some of the office space being vacated in the Observatory. Since I was already wrestling with an accommodation problem, I was quicker off the mark than the other departments. In a letter to John Houghton in August 1987 I noted that 'the rest of Physics has quite awful space problems and these were (fortunately for us) not fully appreciated by the new Physics Management Committee when they gave the green light to our takeover of Astrophysics', and they 'would be enormously relieved were it to fall through'.

It worked in our favour that renting the space to government-sponsored organizations like the Meteorological Office and the Natural Environment Research Council brought in some very attractive extra income. I expected to get most of that myself for the subdepartment, but the University decided they needed to keep what should have been my share because of the cost of moving everybody around. In vain I pointed out that they were planning to move anyway. Now I had not only a huge extra management job but inadequate resources as well. Incredibly, I was happy, most of the time at least, but something had to give.

My institute partners were appointing new support personnel—programmers and so on—for the expanded operation and I demanded that they pay for an assistant for myself so I could carry on supporting them and still have some time for my own research, most of which was not actually inside the Institute. With some reluctance they agreed, and I was able to hire the estimable Marie Corney, who had been working for us as a data manager but also had experience of banking and other high-level administrative work before she came to the University. This worked very well until the powers-that-be at Bracknell noticed that Marie was on their payroll but not working for them and tried to redirect her efforts. I responded that Marie's post was a *sine qua non* for my own efforts on their behalf to deal with all of the administration; I was overloaded and could not continue without practical support. I remembered picking up a self-help book on a trip and reading 'if the best thing that happens to you in a day is that a meeting gets cancelled, you have not organized your life properly'. That struck a chord.

Pressing on, I succeeded in getting a Visiting Professorship for Raymond Hide, and the Met Office pledged a large sum of money so that we could convert a basement in the Old Observatory into a laboratory for his research on rotating fluids. This involved removing a huge solar telescope and an old and also very large stellar furnace. The latter had been used by an earlier professor of astrophysics to simulate conditions in the outer layers of stars; it was electrically heated to the requisite extremely high temperature and could only be turned on in the middle of the night. Otherwise, it was said, the lights would go off all over Oxford as the furnace hogged all of the available current.

John Woods started to build up the FRAM high-resolution ocean modelling project that was his particular interest, and wanted to move the members of the core team of the project from their own institutes

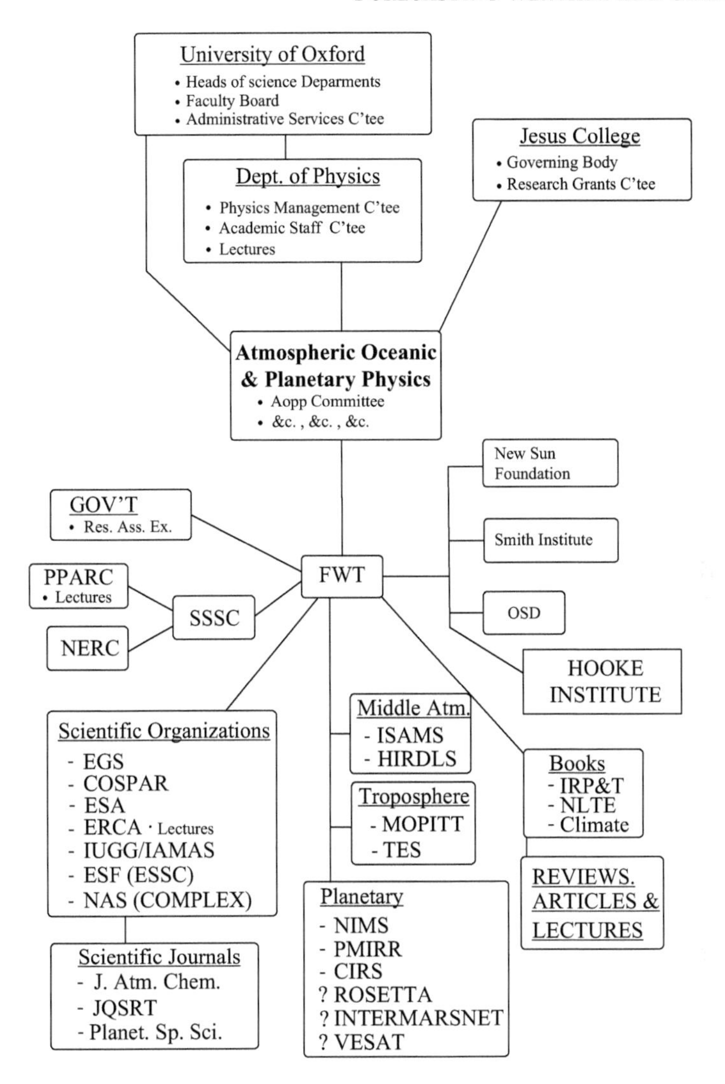

A space scientist must live and breathe organization charts. Here is a well-worn and altered version that I did for myself to try to organize my main commitments, while trying to work out why I had no spare time. The meanings of the various acronyms mostly can be found in the glossary at the end of the book.

to his new domain in Oxford. The NERC people already in Oxford started to move to the Observatory from Banbury Road as their numbers increased. Whatever its scientific virtues, I did not think building up FRAM helped the Institute, because it involved no collaboration with the University and not even, as far as I could see, with the Met Office. However, John Woods was devoted to FRAM and adamant that he wanted his own research programme. Any idea that I had had about his presence in Oxford strengthening the joint work on TOGA, already thriving with Adrian Gill now employed by the Met Office and seconded to Oxford to work with David Anderson, was wishful thinking and well off the mark.

There was little to be done about it since as a director of our main funding agency John wielded great power. The best thing was to learn to love FRAM as much as he did, and it was actually not too hard. The Antarctic is unique among the world's oceans in that it extends right around the globe. As a result, very strong currents flow and many remarkable instabilities form, somewhat like they do on Jupiter. It would be very interesting to see if a computer model, programmed with our understanding of the physics of fluid flows, could reproduce and explain what was going on there. NERC had acquired a Cray supercomputer, one of the few big number-crunchers in the world capable of taking on such a task.

It was not a bad idea for John to have his own small group working on such a programme, keeping him connected with the academic community for whose funding he was now responsible, and giving him a research project to work on during gaps in his administrative load, although the lack of any collaborative element within the Institute remained a concern. The trouble was, he had a much bigger venture in mind: groups were to be imported from various parts of the FRAM family to make a 'core group' at Oxford, from where the whole thing would be run. Not only that, additional projects including 'Buoyancy forced motions in the ocean' and 'The seasonal boundary layer of the ocean' began appearing from nowhere. The planned size of John's research group had risen from two to 20 people, and there were plenty of signs that this would soon grow larger still.

We were full up again, and somewhere a line had to be drawn. First I had to declare a hold on the expansion of the FRAM modelling group. I also had to tell the other high-powered people who wanted to move to Oxford to join our happy band of brothers that there was no chance for the foreseeable future, and sorry but they should stay where they were. It was just as well that the negotiations over moving the Atmospheric Physics group from London fell through when UCL, who had instigated the move but were now under new management, backed off to a plan that would transfer the people but none of their funds or other supporting resources. I would have had to find not only a lot more space but also two senior university posts and a lot of money and workshop time, and by no amount of effort would this be remotely possible. Dai Rees, fed up with being messed around, called off his move with apparent good grace.

Trouble in Paradise

The acquisition of the Observatory with all of the extra space it granted was a big step forward and ought to have solved many of our problems.

In fact, things soon started to get much worse. An organization like the Hooke Institute is, by its very nature, inherently unstable, because it is subject to powerful forces while being only loosely held together. In a way, the business of creating something and managing to exist at all in our original cramped environment had helped to keep us united in a common cause; but once the first phase was over the independent elements could and did assert themselves.

It soon became apparent, although it was never publicly revealed or discussed with me, that the Meteorological Office was syphoning off some of the expertise in the Hooke Institute for its clandestine work with the Ministry of Defence. Whether or not this sort of thing was OK under our agreement, I was uncomfortable with them introducing a military element into the work of the University. I knew it was something students could be counted on to vigorously protest, and I had always avoided deals with the military where the department was concerned. Anyway, what was happening now was on too large a scale, and was affecting the work of the Institute on its stated goals, and I certainly felt I should have been at least told what was going on rather than having it actively kept from me.

It all came to the surface when Adrian Gill angrily sent me a copy of a secretive memo from the Director of Research at Bracknell to Adrian's Met Office line manager, instructing him to 'arrange for Gill to develop an ocean forecasting model to cover the Iceland–Greenland gap area', adding that 'I envisage that the main development will take place within the Hooke Institute. . . . the expectation must be that the task will be undertaken largely within resources available or planned at Oxford.' This was a very large task and nothing to do with TOGA. Crucial resources were being syphoned out of the Institute's programme for inappropriate purposes, and it was being kept secret from me, although I was responsible for the Institute's reputation and performance.

Until he received his instructions from on high, Adrian didn't know about it either. His note scribbled on the copy he sent me of the Met Office memo read 'Fred: This is news to me. I don't like it. Adrian.' The resulting slow-burning furore was threatening a crisis when it came to a sudden and unexpected ending with Adrian's untimely death in April 1986. I had noticed when chatting to him that he sometimes put his hand on his brow and shook his head slightly, but I thought nothing of it until a brain tumour carried him off in only a couple of weeks after it was medically diagnosed. This was a shattering blow to his colleagues, including me, as well as to the Institute, its organization, and its programme.

A few months later, the oceanography part of the Hooke Institute had fragmented into three, perhaps four, separate components with surprisingly little interaction or evidence of mutual interest between them. The undercover work continued and spread to NERC, whose unit at Oxford was headed up by Peter Killworth, one of the original Cambridge oceanographers. Peter had been taken onto the staff at NERC's

Institute for Oceanographic Sciences at Southampton and seconded to Oxford, nominally to work on FRAM. One of the computer programmers in the Institute decided to quit and came to me in person with his letter of resignation. I asked him what was wrong and was shocked to learn that, far from supporting David Anderson's research as intended, he had been pressed into working on a secret modelling project for the Ministry of Defence. I wrote to Peter that this was 'unconnected with the agreed programme of the Hooke Institute and, worse, unconnected with the stated programme of the grant from which he is paid. I am the Principal Investigator for that grant and I am horrified to find that resources awarded in my name have been misappropriated in this way.' I had seen what happened to academics who were discovered using the funds in their research funds for unapproved purposes, and I did not want to join them in purgatory.

Research students are an absolutely crucial part of any academic endeavour. One thing on which we could all agree was that we wanted lots of them in the Hooke Institute. Not only would we train bright young people and encourage them to become the future of an important and growing field, but also their ambition and enthusiasm was a key part of the effort available to apply to any research project. Perhaps even more important were the post-docs; these were young scientists with freshly-minted doctoral degrees from our own and other universities who worked on soft money from grants for a few years as they sought permanent positions elsewhere. Again, their work ethic and youthful enthusiasm was a key part of every successful research group.

Appointing research students and assigning them to projects and supervisors was my responsibility in our subject area. The research supervisor did not have to have be an academic provided that arrangements were made for some kind of close academic involvement and the work the student did was appropriate material for a successful thesis. The high-powered scientists in the Institute were all keen to have several students under their supervision, and I said that was not only possible but almost essential if the Hooke setup was to retain the wider support of the University. In one memorable case, I was myself the co-supervisor for a very clever young woman who worked on ocean circulation problems for her doctorate. I enjoyed the exposure to the details and helping to steer the work, although of course it was her joint supervisor in the Institute, in this case Peter Killworth, who dealt with the finer points.

Students can only be recruited if the university has a special kind of grant called a studentship to pay their fees and living expenses. There was great competition for those, and we achieved a typical intake of only about five students per year in total, for all areas including planets and atmospheres. Now that we had all these new research programmes and high-powered, willing supervisors on board that would no longer be enough. My simple solution was for the Met Office and NERC to increase the amount in the contracts they awarded to the University, initially to pay for accommodation and support staff, to include money for as many students as they wanted. I was surprised when this approach, which I thought reasonable, precipitated a fierce row, but eventually it became clearer.

The underlying problem was that the view of the Hooke Institute from the parent organizations in Bracknell and Southampton was that their money was being syphoned off to Oxford to support what was essentially a vanity project in an ivory tower on behalf of their leaders, who really ought to be back at base solving problems for their loyal staff, rather than swanning around Oxford and enjoying academic life there. Something like that, anyway. Those of my academic colleagues in universities around the country who dared to speak up, not being sure where I stood in this conspiracy, except that I was on the receiving end of an enviable lot of funding, each expressed similar feelings in terms appropriate to their different vantage points.

John Houghton reacted by publicly withdrawing himself from the Institute and naming his Director of Research, an irascible scot named Andrew Gilchrist, as the Met Office champion. John Woods went even more public, with a letter published in *New Scientist* in which he vehemently denied feathering his own nest at Oxford. Both knew it would be dangerous to redirect any more public money, and wanted the University to deliver the students. In fact, Woods wanted full rights for himself and the other NERC people to apply for not only studentships but full grants and contracts to cover further extensions of their research, all from traditional university sources, including the fraction allocated to university funding by NERC itself. I pointed out that those funds were very finite and that in effect what he was proposing would put the external Hooke scientists in competition with their university colleagues; we were unlikely to obtain more studentships overall because the supply was so limited and so hard fought over by all of the universities in the land. I wanted to bring in fresh funds for students from the much larger part of NERC's budget that was earmarked for its institutes.

As tensions increased, there was yet another blow. The Institute was, after all, supposed to be fundamentally about improving weather forecasting using satellites, and this had been almost forgotten in the rush for growth in oceanography in what had become a kaleidoscope of initiatives. John Eyre became disillusioned and left for a sabbatical in the USA, leaving the satellite meteorology project rudderless and me with yet another problem. It seemed likely that the Institute might be heading for a future in which all of its work was in geophysical fluid dynamics, especially oceanography, which was all back-to-front from my perspective.

Eventually the move of Raymond Hide's lab from Bracknell was completed, but even that welcome development was rife with problems. The arrangements, including moving Astrophysics out of the Observatory, and then carrying out the building works in what was a very strange old listed building, took an inordinately long time and led to many frustrations for Raymond, which he passed on to me. Raymond, whose character was that of a natural professor, now had the title to show it, plus a college fellowship and expensively refurbished accommodation. Most of this was the result of a lot of effort by myself, including fielding a string of mostly minor complaints. What it boiled down to was that the respect and support services Raymond got from the University were nothing like what he was used to in the Met Office, which was directly funded by the government, lavishly by university standards. We had actually worked wonders by local criteria but I knew that the setup at Oxford never really pleased Raymond. In fact very little really pleased him after the pre-emptory closure of the laboratory he had devoted his career to building up in the Met Office. I suspect they forgot to tell him they were doing it until it was done.

Exit Strategy

One of the Natural Environment Research Council's demands was that the University must accept the 'NERC Council decision to establish the FRAM Core Team at the RHI'. Furthermore, they wanted direct management control over all of the oceanographic work in the Institute, including people employed by the University. Failing this, NERC would withdraw its contractual support for the Institute and begin to pull out. The Met Office simply said that the whole venture was not viable without NERC and if NERC went, they would pull out too.

NERC increased the pressure by refusing to sign a renewal of the contract that paid the University for the rental of space, staff, and other support unless its demands were met. Among those demands were that I should proceed at once to make appointments of staff whose salaries were to be paid out of the contract that they were refusing to sign. When the contract finally appeared, it made reference to the Constitution of the Institute, a copy of which was attached. It turned out that this copy was of a version they had revised unilaterally in ways that had been discussed but rejected at a meeting of the Steering Group at which they were fully represented. In a further element of farce, it also offered the Institute a consumables budget of '£1000 k per head per annum', which would have come to something in the region of a hundred million pounds. In pointing this out to the Secretary of Faculties, acting as contracts officer for Oxford University, I noted dryly and perhaps a little bitterly that this seemed excessive, 'even given the well-known profligacy of the research council centres when funding their own research.'

There was still no research council money for the University's own research, following the loss of our grants prior to 'restructuring', despite the fact that, in my book at least, funding universities was the main reason that the research councils existed. I made various representations about this, and also made specific proposals that would have led to more atmospheric work in the Hooke Institute, including the possibility of strengthening the experimental work in space. These, of course, were the Department's main interests, and consisted with my thesis that world-class science can not be done with computers alone, despite a strong recent trend in that direction. It all fell on deaf ears.

Life was to get even more complicated. With the unification of Physics in 1990, I had an identifiable boss for the first time, in the form of the Chairman of Physics, Chris (later Sir Chris) Llewellyn Smith. The outside bodies saw their chance to get their way by going to the Chairman without reference to me. Chris made no secret of the fact that he really liked the payments that were coming in to his coffers from NERC and the Met Office as rent, support fees, and overheads, and that he would do all he could to keep it flowing. He would meet clandestinely with the conspirators and tell me about it afterwards (once I received, I think mistakenly, a copy of a memo to John Houghton that began 'Further to our recent meeting in Marks and Spencer's . . . '). Amazingly, but mainly by turning the other cheek, I was able to stay on at least reasonably good personal terms with everyone during all of this manoeuvring.

Powered by Chris and his formidable negotiating skills, we found a way forward. This was to be an 'independent' review, under the chairmanship of Sir Richard Norman, a chemist and former Scientific Advisor to the Ministry of Defence, who was now Rector of Exeter College in Oxford. The announcement of the review included the statement that:

> 'in order to represent Atmospheric Physics from an independent standpoint as a member of the review, and to allow himself more time to concentrate on his own research and his department, which was about to be enlarged by the appointment of two new lecturers, Dr Taylor had decided to offer his resignation as the Head of the Hooke Institute.'

It was not a coincidence that the same day: 'the General Board decided to recognise the expansion of Atmospheric Physics and the increased importance of the subject by upgrading the Readership in the subject to a Chair and had decided that Dr Taylor shall be appointed to the Chair.'

As we gathered around the table for the first meeting of the Hooke Review, Sir Richard had obviously been thoroughly briefed, as he duly produced the conclusion that all the outside parties' demands should be met. In particular, the Hooke Institute should be a stand-alone university unit with no formal departmental attachment. An unexpected result of the activity generated by the review was that other Oxford departments (most notably Chemistry, Earth Sciences, formerly Geology, and even Mathematics), scenting money, expressed an interest in joining a Hooke Institute that was to be vastly expanded in scope.

Raymond Hide was to be the new Director, on a three-year appointment. Sir Richard over-egged the pudding by supporting some of my collaborators' more unreasonable demands, such as professorships for the senior civil servants with an office in the Institute, the right to apply for grants in the name of the University, and the overheads from the Met Office and NERC payments to be made available to the Director as part of his budget. There was a vague mention of support for improving the environment for students but no specific teaching commitment for outside staff, not even those wanting academic titles and college fellowships.

I had agreed in advance to accept the outcome of the review, knowing full well what it was going to be, and had already stood down as Head of the Hooke Institute in Raymond's favour and wished him well. My feelings were mixed: I was very glad to be shot of it, but on the other hand it wasn't going to go away. Instead, there would be in effect a separate department of atmospheric and ocean science over which I had no control, but which was located in the university where I was the professor responsible for the subject. Worse than that, in its new form it was founded on little more than hubris and almost bound to fail, especially without a proper academic hand on the tiller. The outside bodies had seriously underestimated and underappreciated the time and effort I had put into protecting and promoting their schemes, even those of which I did not really approve. There was so much covert strife of one kind or another, much of it not involving myself directly, although I knew about most of it, and I was sure it would tear itself apart in no time. And Raymond, whom we all loved, was nevertheless no manager and no match for the forces lined up against him, nor indeed those notionally in support.

At his first Steering Group meeting as Head, 'Dr Hide commented that the spirit of collaboration within the Institute was apparently not being welcomed uniformly in all groups, with the NERC group being less outward-looking than the others.' NERC responded by announcing that they were going to move the Core Team of another big community ocean modelling project to Oxford, similar to but larger than FRAM, and they made it clear that they expected no resistance this time.

I had been asked to write an annex to the Norman Report giving a position on the new arrangements. It was not difficult to make this positive, verging on jaunty. We had succeeded in getting a Statutory Chair in our subject area at last and I was in it, we had two new senior academic posts to fill, and our commitment to the Hooke venture was now limited to two well-defined areas. The TOGA work had recovered from the loss of Adrian Gill and was showing signs of achieving a breakthrough in medium-term weather forecasting by showing how to read the advanced signs in the temperature of the oceans. The satellite meteorology work had left the unit, but not before contributing important work to the transition towards satellite-based weather forecasting from more traditional methods.

On the experimental side, the international weather forecasting community had started using data from a series of instruments known as Stratospheric Sounding Units, developed by the Met Office in the UK and flown on American operational weather satellites from 1978 to 2006. The SSUs used the pressure modulation method that I had developed for my doctoral thesis, and since we had patented it, the Met Office paid the Department a modest but useful royalty. But the science was the best thing; the stratosphere was showing odd long-term changes in temperature that were linked to the early stages of global warming at the surface.

Finally, I gushed, there was a new collaboration in the Hooke which is 'extremely powerful and can be expected to produce results of far-reaching importance'. This involved the move of the Met Office stratospheric data analysis group under the excellent Alan O'Neill to Oxford, specifically to work with us on the data from the Upper Atmospheric Research Satellite, UARS. Since the Department was then getting ready to deliver a large new instrument called ISAMS for this satellite, this was a marriage made in heaven, and I had campaigned for it as part of the Hooke reshaping. The full story of UARS and ISAMS is in the next chapter.

Requiescat in Pace

The revamped Hooke Institute for Geophysical Sciences lasted less than a year, finally closing in 1992, shortly after John Houghton's term as Director General of the Met Office ended and he left, with a well-deserved knighthood, to be replaced by Julian Hunt, an applied mathematician from Cambridge. Julian was to be the last in a long line of academics to hold the most senior position at the Meteorological Office, which had moved its headquarters to Exeter. Nowadays, its leadership is assigned to a senior civil servant.

When Julian took office and surveyed his new domain he took a look at the Hooke Institute and, unsurprisingly, decided to pull out. He sent me a personal note saying it was all about limited resources and he hoped the Met Office would retain strong links with my department at Oxford in other, more normal ways. I wrote back saying I quite understood, we would collaborate, and we did.

Julian later became Lord Hunt under Tony Blair's reforms and was pleased to entertain some of his meteorological colleagues to dinner in the House of Lords, where we wined and dined and late at night left through deserted corridors, rich in history.

The New Sun also sets

Another tragi-comic episode, with echoes of the Hooke saga but probably unrelated to it, materialized in 1997, when I received an unexpected request from the Secretary-General of the World Meteorological

In Geneva, 8th December 1993. Professor Obasi, the Secretary-General of the World Meteorological Organisation, presents me with a certificate and an engraved pen to mark the creation of the New Sun Foundation, which was to help developing nations find high-tech solutions to problems caused by climate change and other factors.

Organisation. The WMO is part of the United Nations powerhouse based in Geneva; in charge at the time was the long-serving Nigerian scientist-diplomat Professor Godwin Obasi. Would I be interested in joining the Board of Directors for a new initiative he was launching, he asked, to be called the New Sun Foundation? The letter and subsequent phone call explained that it was to be a philanthropic venture, helping the developing countries of the world to solve their problems using modern technology and techniques. Assuming that this meant meteorological measurements from space in my case, I expressed interest in helping with this good cause. To learn more about what I could contribute, I went to the inaugural meeting in Geneva, and was lavishly entertained. What followed in the meeting itself puzzled me, however.

The initiatives that were discussed were things like irrigation and bridge-building, with little to do with space, or meteorology for that matter, as far as I could see. Any attempts I made to put satellites on the agenda were futile. The other people at the meeting seemed to be mostly company executives, looking for business. The second meeting left me equally bemused. At the end of the day, before leaving for the airport, I spoke to Professor Obasi and said I was concerned that I was the wrong person for this job. Well, he said, we wanted someone from the UK. But he admitted that I was not their first choice. Who was? I asked. Prince Charles, was the reply.

Back in England, I made enquiries. Apparently the Palace had referred the WMO's initial approach to the Met Office, who had pointed them to me as a suitable candidate. I asked the Chief Scientist at the Met Office whether they supported the New Sun initiative and whether they were interested in contributing to the effort and resources it seemed to require. I can't remember the exact reply, but it was along the lines of you must be joking. Shortly after that the Foundation sank, apparently without trace, except that I got a call more than a decade later from a former WMO officer who said he was looking into the New Sun records and asking me if I could shed any light on its activities and demise. I was able to reply truthfully that I could not.

8
The Ozone Layer and other Crises

Ten years to terror

On 11 September 2001, I was at the Goddard Space Flight Center in Greenbelt, Maryland, just outside Washington, DC. I had been to Goddard many times over the years, but this time was special: a meeting to mark the tenth anniversary of the launch of NASA's *Upper Atmosphere Research Satellite*. I was the Principal Investigator for one of the ten experiments on this very large satellite and, along with the other PIs, I was due to give a talk later in the day with an historical survey of the results we had obtained. However, it did not happen, not that day. This was 9/11.

Our experiment was called ISAMS, which was the short name we used for the I*mproved Stratospheric and Mesospheric Sounder.* The stratosphere and mesosphere are the layers of the atmosphere that extend from about ten to about 80 kilometres above the surface, and collectively they are often called the 'middle atmosphere'. This is where the ozone layer lies, and that was our target. UARS really should have been called MARS, because it studied the middle and not the upper atmosphere, which is even higher, but that name might have led to confusion.

ISAMS was designed to measure the amount of ozone, and that of several of the important trace gases that were known or suspected of reacting with ozone chemically. At the heart of the instrument was a battery of pressure modulator radiometers, advanced and professionally engineered versions of the home-made device I had flown on a balloon as part of my doctoral project more than 20 years earlier. Each one contained a different gas that we wanted to observe in the atmosphere: water vapour, methane, nitrous oxide, and so on. Sadly, we couldn't put ozone in a cell as it is too reactive and would decompose quickly. We did look at ways to keep the ozone topped up with a small electrical discharge, but eventually decided it wouldn't work. Ozone is sufficiently abundant in the stratosphere that we didn't really need the modulator technique to study it; an ordinary solid-state filter would be fine.

We did, however, include a new kind of device that was a revolutionary step not just for ISAMS but for the future of infrared remote sounding from space. This involved using pairs of high-pressure modulators containing helium to drive miniature refrigerators in what thermal engineers know as a Stirling cycle. This was able to cool the infrared detectors to around –200 °C, which increased their sensitivity by a large factor. Altogether, ISAMS was a large and very complicated device, much more ambitious than anything the Oxford group had ever attempted before.

Up UARS

The *Upper Atmosphere Research Satellite* itself was revolutionary, the first of a whole new generation of Earth-orbiting satellites, developed by NASA as a successor to the now obsolete *Nimbus* series. Satellite technology had raced ahead since *Nimbus* was designed in the 1960s, and this new platform was a much larger and more sophisticated affair than anything we had ever seen before. Its goal was to investigate disturbing trends in the amount of atmospheric ozone, the three-atom version of oxygen (O_3, rather than the more familiar O_2 that we breathe), in the stratosphere. It had been known for more than half a century, particularly since the work of Gordon Dobson, my predecessor twice removed as head of the group at Oxford, that ozone forms a layer about ten kilometres thick centred about 25 kilometres above the surface. Ozone is not very stable,

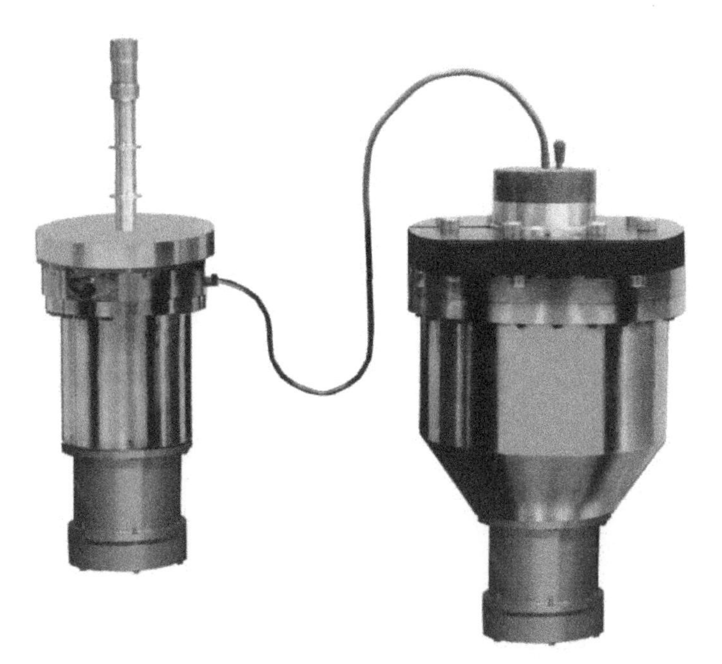

Cutting-edge technology: this miniature cooler was invented in our Oxford department by Guy Peskett and others to give our ISAMS instrument exceptionally high sensitivity to the heat radiation emitted from the atmosphere. For the technically minded, it uses two pressure modulators working together to compress and expand helium gas to work as a Stirling-cycle refrigerator. This is capable of cooling the infrared detectors in the instrument to a temperature as low as that on the surface of Pluto.

so the amounts in the stratospheric layer fluctuate a lot, depending on a complicated balance between production and loss mechanisms that vary with latitude, season, and a variety of other factors.

The ozone is formed by the effect of ultraviolet radiation from the Sun acting on ordinary oxygen, and it acts like a shield to prevent these harmful rays from reaching the surface. If it were not there, life would be badly affected, with radiation damage to skin, eyes, and crops. Ordinary sunburn is a relatively mild indication of the harm too much ultraviolet can do to humans, and animal and even plant life suffers serious damage if the levels get too high.

In the 1980s the world had become aware that man-made pollution was causing the ozone layer to become thinner. The measurements of ozone that were now routinely made around the world, in particular near the South Pole by the British Antarctic Survey, were showing declining amounts, and clever research by atmospheric chemists in the United States and elsewhere had shown that small amounts of certain pollutants were capable of wreaking enormous damage. But exactly what was going on, and what might be needed to fix it, was still very uncertain in the early 1980s. NASA came up with UARS as its main weapon to make measurements that would shed some light on the problem.

The worldwide community of atmospheric and space scientists was invited to make proposals for instruments that could make the necessary measurements, and from the many submissions it chose the best. Not surprisingly, the chosen few, ten in number, were nearly all American, but one of them was British—ISAMS.

Probing the stratosphere

The idea for ISAMS had been developed at Oxford while I was still in California. When I arrived back in 1979 to take over at Atmospheric Physics, a proposal had just been submitted by the department to participate in a study of possible instruments to form the payload on a conceptual UARS mission. Such evaluations of 'straw man' payloads are the usual precursor to large missions, and are used to help develop the spacecraft design specifications, such as pointing stability and power requirements. No commitment is made to fly anything until rigorous and independent assessments show that the concept has been fully worked out and optimized.

The Oxford department already had a similar but simpler instrument, SAMS, operating in orbit on *Nimbus-7*, the seventh satellite in that series. Its predecessor, *Nimbus-6*, had carried into space the first satellite radiometer using pressure modulators based on my D Phil work, described in Chapter 2. SAMS was the first pressure modulator technique to measure composition as well as temperature. Its success led to plans for an even larger, and much improved version. NASA had terminated the *Nimbus* programme with number 7 when it found the resources for the new generation of larger and more powerful satellites, the first of which was the Upper Atmosphere Research Satellite, UARS.

John Houghton was the original Principal Investigator for SAMS and wrote the first proposal for an Improved SAMS, but handed both roles to me when he left the Department for his new job at the Appleton Laboratory. I was busy with several planetary experiments at the time, but with youthful keenness for excess decided I could handle this as well. SAMS was producing a lot of data, and it looked interesting. As the new Principal Investigator, I had to make sure the team published the SAMS results in journals and presented them at conferences, with suitable analysis and interpretation. That is the fun part of a space experiment, and we all set to with enthusiasm.

There was a lot of interest in SAMS data from all over the world, and we had a string of visitors who came to spend time in the department so they could access and use it. The interaction that had the most of what we would now call 'impact' was that by Paul Crutzen from the Max Planck Institute in Mainz, Germany, who came on several visits including a long one in 1982. Paul was an atmospheric chemist interested in the action of oxides of nitrogen on the abundance of ozone in the stratosphere. These relatively scarce gases are emitted by aircraft engines, and the possibility of the new generation high-flying aircraft like *Concorde* destroying the ozone layer was a serious concern. Paul later told us that the work he did in Oxford was an important part of the achievement for which he won the Nobel Prize in 1995.

SAMS was going strong and won awards; ISAMS, on the other hand, was not yet selected for flight by NASA, nor did it have the funding required, which was considerable, something of the order of ten million pounds. This would have to come not from NASA but from our own government in the UK. The political arrangement was that, if they did select us, NASA would cover the costs of putting our experiment on the spacecraft, launching, and operating it, but they would not pay us to build the instrument, as they would have done if we were in the USA. The golden rule was that hardware and people would mix and match as necessary to carry out the project, but no money would cross the Atlantic. We therefore had to fight and win in two very different arenas: one for the flight opportunity and the other for funding; and since neither of these was any use without the other it all had to be done in a coordinated fashion. The campaign would have to start with a more detailed design for ISAMS and a full proposal, including costs and justification for the funds required.

It would be tedious to present all the details of the complex and often painful process of getting selected and funded for such a large experiment, in the teeth of course of rabid competition from one's peers who have plans and ambitions of their own. NASA had elaborate procedures to deal with the mass of proposals they received, which for a large project could literally fill a large room with paper. The Agency has to move with great care and formality, since flight opportunities are scarce and unsuccessful proposers and their groups can face oblivion, so they are often desperate and even litigious. One short vignette will give some idea of what was involved in the early stages. NASA management contacted me at short notice and said they had concerns about the ISAMS proposal and wanted to meet with the PI in their Washington, DC, headquarters. A meeting of 15 minutes' duration would suffice and was scheduled for the following Tuesday morning at 10.15am.

It turned out that what they wanted was a quasi-legal commitment by me to their programme and they wanted it delivered in person; it actually took less than 15 minutes, but I had to make a seven-hour flight to Washington to deliver it. At the time, it was possible to fly supersonic to Washington in only three and a half hours, at a price of course, on *Concorde*. I decided I would pay the price, since this sort of thing was what supersonic flight was in aid of, and hang the expense, but it turned out that the daily *Concorde* flight arrived at Dulles Airport an hour or so too late to get to downtown Washington in time for the meeting, and I would have had to stay overnight anyway. Supersonic travel only really works for businessmen who can schedule their own meetings.

After selection, there were many more meetings but they were usually two or three days long. Quarterly progress reviews, science team meetings, engineering interface meetings, and contingency events all had to be attended by the Principal Investigator in person, accompanied as appropriate by the relevant members of his team. I was a frequent flier on Trans World Airlines, which flew into Baltimore Friendship airport, which was actually more convenient, as well as friendlier, than Washington Dulles, which was many miles outside the city and on the wrong side as far as NASA's Goddard Center was concerned. I had got used to TWA when I worked at JPL, because the Americans insist on the use of their own 'flag carriers' if you are spending taxpayers' dollars.

Once back in the UK, I found there was no such restriction in the opposite direction, so I could stick with Trans World and my precious Gold Card that got me a free upgrade to Business Class, with its steak dinners and unlimited champagne. In the very early days there used to be a small pack of complimentary cigarettes on the tray, but I had given up my light smoking habit with alacrity when the serious nature of the health risks emerged, and smoking on aircraft was phased out not long after. I was sorry when TWA went bankrupt in 1992, but perhaps it wasn't surprising given their munificence. I took my business to British Airways, who were good but not nearly so generous with the upgrades, or the champagne.

Ozone depletion

The point of the Upper Atmosphere Research Satellite and its payload of ten scientific instruments, with ISAMS the only non-American one, was to study the science of ozone depletion. We knew the ozone layer was there, we knew there was less of it each year, but we did not know why, although there were plenty of theories, most of them pointing the finger at the effluent from the growing industrialized society that was already beginning to overpopulate the surface. But what exactly was humankind doing that was threatening to destroy its shield from the deadly rays from space? And what were the options for fixing it? Unusually for a topic involving some fairly esoteric science, there was a lot of media discussion of questions like these, and the phone often rang. The *Observer* newspaper explained the ISAMS experiment to its readers quite well, in an article with the rather obtuse headline, 'Oxford Team for NASA trip'.

The best theoretical work, by Paul Crutzen and others, had pointed the finger at oxides of nitrogen, 'NOx' for short, simple gases that were released into the atmosphere in small amounts by the use of soil fertilizers and other agricultural activities. Later studies suggested that traces of certain compounds of chlorine attacked ozone even more vigorously than NOx. These were building up in the atmosphere through the increasing use of chlorofluorocarbons ('CFCs') in air conditioners and aerosol spray cans. In spite of the long name, CFCs are simple molecules and were thought to be very stable, and near the ground they are, but under the conditions in the stratosphere they decompose and affect the balance of the chemistry in the ozone layer. However, how all of this actually works in detail was a mystery that UARS, and ISAMS, set out to address and hopefully to solve.

Building a satellite instrument

After a couple of years of detailed studies and painstaking negotiations, NASA finally confirmed that they had selected the Improved Stratospheric and Mesospheric Sounder for flight and the UK government, through its Science and Engineering Research Council (SERC), had agreed to pay for it. We were in business.

ISAMS was a huge project. The total estimated cost was well over 20 million pounds when all of the labour, facilities, and other 'invisible' costs were factored in. About a third of this was allocated as cash, to pay for hardware, subcontracts, travel and so on. The Oxford department was to provide scientific leadership, some critical instrumentation including the coolers and the pressure modulators, and the facility to test and calibrate the instrument before it went to the USA. The space division of British Aerospace would build the flight structure and the other sub-systems and assemble the complete instrument. Rutherford Appleton Laboratory, the large government research facility at Harwell (coincidentally and handily, not far from Oxford) would design the optics and the electronics, and manage the budgets and the schedule for the entire project. The completed instrument would be tested in a large vacuum chamber at Oxford before we delivered it to the USA where it would be mounted on the UARS spacecraft at the General Electric plant in Valley Forge, Pennsylvania.

The British Aerospace plant was at Filton, near Bristol, originally the home of the Bristol Aeroplane Company that developed such classic aircraft as the *Blenheim* and the *Bulldog*. It was also home to Bristol Motors, who produced a few hand-built cars every year, and on my frequent visits to Filton for progress reviews on ISAMS and to inspect the hardware, I would watch for *Brigands* and *Beaufighters* (names that also used to belong to aircraft built here) being test-driven around the site that the carmakers shared with the aerospace company.

One early visit to BAe led to a curious anecdote. It was September 1981 and that summer England had suffered a string of race riots in several major cities, including Bristol. The taxi driver who picked me up at Bristol Parkway railway station for the short drive to Filton was black and seemed rather moody. We drove in near-silence for a while, and then he turned on the radio to listen to the news. This was grim—there had been fresh disturbances in the Bristol suburbs not

Members of the ISAMS team at Cape Canaveral, with the Space Shuttle that was to launch our instrument on the Upper Atmosphere Research Satellite. From the top, Marie Corney, Data Manager; John Whitney, Project Engineer; Fred Taylor, Principal Investigator; Ken Davies, Project Manager.

far from where we were driving and most of the bulletin was about the seeming hopelessness of the situation, although I did note with sadness a short item about the death of Bill Shankly, who had been the very successful manager of Liverpool football club when I lived in the city and sometimes I used to watch the team play from the terraces at Anfield. The driver turned the radio off and there was another long silence while I sat gloomily in the back of the cab morbidly picturing burning buildings and riot police with helmets, shields, and body armour. Then the driver finally spoke: 'It's terrible that, isn't it?' he said. 'Well, yes', I replied, somewhat at a loss what to say. 'Bill Shankly', said the driver, 'he was so good'. We chatted cheerfully about football for the rest of the journey.

Problems and reviews

It was perhaps not surprising that it did not take very long for the complicated ISAMS project to develop problems. After a couple of years, almost every part of the hardware programme was behind schedule, and costs began to rise. The bills from British Aerospace, in particular, were relentless, and it soon became clear that the reserves in the kitty were going to run out before the instrument was complete. NASA was worried about schedule; they needed the instrument on the spacecraft when it became time to launch, but since ISAMS was paid for from British sources they could not exert the same level of control as they could over the rest of the mission, nor could they pump in extra money from their own reserves. As our team laboured to try to keep on top of the massive complexities of the project, the funding problem got worse.

Money for ISAMS was at that time being channelled from the British Government through a body called the British National Space Centre, BNSC. This was set up in 1985 to run Britain's affairs in space by controlling a partnership of no less than ten government departments and agencies. Most of the ten (which included the parts of the military with interests in space) wanted to keep their own budgets and were less than thrilled at the prospect of turning them over to this newcomer, and before long the arrangement collapsed. The name BNSC was retained for a much smaller body that specialized mainly in marketing the UK commercial space industry abroad. The UK did not get a real space agency until 2009.

In its heyday BNSC had a Science Projects Board of which I was a member, and as it happened it was chaired at the time by my predecessor John Houghton, now in his role as Deputy Director at Rutherford Appleton Laboratory. The Board was scheduled to discuss the ISAMS problem at its meeting in April 1986, not for the first time, but this time it was coming to the crunch. I arrived early for the meeting at the BNSC's headquarters in Milbank Tower, overlooking the Thames in central London, and was surprised to learn from one of the secretaries that a subset of the Board was already in session, and that the topic was ISAMS. Nobody had told me about this.

I went to the side room where the group—those with a direct interest in ISAMS, and the administrators—were meeting, with JTH in the chair. I apologized for being late, and made to sit down, but was told to leave. The matter would be discussed with me at the main meeting in 30 minutes' time.

What the hell was going on? Soon, the full Board was in session and the topic of ISAMS duly came up. The mystery of the pre-meeting was solved when the chairman proposed a solution for the financial black hole that the project had fallen into: all other grants to my Oxford department were to be cancelled and the money transferred to ISAMS. This would mean not only that the cash did not have to be found elsewhere, they said, but also that the Principal Investigator (me) would be able to devote full time to ISAMS and prevent further slides into debt and delay.

I had not expected this, and managed only a few inarticulate protests to the effect that although as PI I had overall responsibility as head of the project, I was only notionally in charge of the ISAMS budget. I had no insight into its detailed expenditure, especially at the biggest spender, which was British Aerospace, funded from the contract managed by Rutherford Appleton. I could devote 24 hours a day to budgetary matters without any hope of resolving the problems. And if I did, I would then have to neglect my main job, which was to ensure the excellence of the science and the validity of the technical and project design. Furthermore, everyone agreed that those aspects were in excellent shape, thank you very much. Finally, I was personally committed not only to ISAMS but also to several other activities in important related scientific areas that they now proposed to starve of funds.

Measuring pollution from space

These supporting ventures I spoke of, now in jeopardy as far as Oxford involvement was concerned, included the planetary missions *Pioneer Venus* and *Galileo*, but were mainly stratospheric studies closely related to ISAMS. I was a co-investigator on teams led from JPL by my former colleagues Barney Farmer and Reinhard Beer. Both were building large, sophisticated spectrometers that worked in the infrared to find and measure many of the species involved in ozone chemistry, including those ISAMS was designed to observe but also many others that might be important. They therefore complimented my own investigation perfectly, and I was not about to give them up completely, whatever the committee said, although I could not have the large involvement I wanted without their financial support.

Barney's experiment, called ATMOS for Atmospheric Molecular Spectroscopy, worked by viewing and following the Sun to measure the absorption of sunlight in the atmosphere along the line of sight. This restricted the observing geometry a lot, to sunrise and sunset when in space, but gave extremely high sensitivity because of the very bright source. ATMOS did a series of test flights in the stratosphere on a prototype of the *Concorde* supersonic airliner, and plans were being made for Space Shuttle flights. Reinhard's instrument, the Tropospheric

A technician makes final adjustments to the ISAMS instrument in our laboratory at Oxford. The large vacuum chamber in the background was used to simulate conditions in space so the instrument could be tested before delivery and launch.

The *Upper Atmosphere Research Satellite,* with ISAMS visible as the large cube on the underbelly of the spacecraft at bottom left. UARS was a huge satellite – about ten metres long, not counting the solar power panel at the top. It weighed over six tons.

Emission Spectrometer or TES, could observe anywhere day or night but some of the very rare species that ATMOS could see were not detected by TES. ATMOS was inhibited by clouds and worked best on the upper atmosphere; TES delivered its best data on the lower regions. Together they were extremely powerful. Combined with ISAMS as well, we could rule the world.

Decisions, decisions

Sitting in Millbank Tower that day in January 1987, at a large table in a room high above the River Thames, I had to decide what to do. I had to act there and then, with no warning and in a state of shock and sitting surrounded by high-powered people from business, industry, academia, and the military, none of whom showed any dissent at the proposal or any support for my position. After all, it did solve the financial problem, for the time being at least, without hurting their own pet projects. My first reaction was to resign as ISAMS PI, on the grounds that my position was untenable. However, it wasn't; I knew what we needed to do (if not what it would cost) and I had a huge personal stake in the scientific results I confidently expected to obtain when the mission flew. Not only that, but my Oxford department was so heavily committed to ISAMS that it would have collapsed as a world-class research centre without it.

In a flash of brilliance, or stupidity, I do not to this day know for sure which, I resigned not from ISAMS but from the BNSC Space Science Board itself. In a voice that was probably not as firm and decisive as I would have wished, I said that the solution just approved was unfair, highly damaging, and would be ineffective. Worst of all, as the nominal leader of the project I could not live with the fact that I had not been part of the discussion, nor even told about it until confronted with a fait accompli across the boardroom table. If this was how they did business I was off. I might lose knowledge and control without my board membership, but I would have more time for science, and it would be good for my blood pressure not to have to sit around that particular table any more.

I then left, but not without taking on board the message that I personally should do something about ISAMS finances and not just let it continue as it was until it came off the rails altogether. After a maze of negotiations, and some subterfuge, I wrote a memo proposing a new management structure. Ray Turner, a jovial and hard-drinking engineer from RAL with an excellent reputation for trouble-shooting recent projects became the new Project Manager. Ken Davies, who had been doing the manager's job up to then, would continue on ISAMS, but focus his energies on the two main trouble spots: the electronics development at RAL and the industrial subcontract at BAe. A new post in my department, with the title of Space Projects Manager, would be created to support someone to look after the schedule and the budgets for the instrument work at Oxford. Clive Rodgers would become Deputy PI and run the ISAMS science team meetings and the development programme for the data analysis software; he would also represent me at some of the many UARS team meetings held in the USA.

The 'subterfuge' involved in setting this improved and enlarged organization up was mainly to do with the space projects management post at Oxford. Although Ray and Ken were excellent managers, I could see that something on the scale of ISAMS was not going to work under remote management from RAL, 20 miles away, no matter how many meetings we held. RAL was a huge outfit, with problems and priorities of its own. This was a strength in many ways: it brought us the drive of Ray Turner, and an excellent instant replacement for the test and calibration manager in John Ballard when the incumbent walked out in search of ephemeral opportunities in the USA. But I needed a component of the management team at Oxford, always there, and working for me.

This, of course, would require even more money, and quite a lot of it to create a senior position sufficiently attractive to a person of the right calibre and experience. Both the University and the Science and Engineering Research Council (the latter having taken back responsibility from BNSC for funding ISAMS in the meantime) were loath to shell out. Instead, SERC set up an ad-hoc review group to come and sort us out. This is a time-consuming and fairly painful way of making progress, but I was glad of the attention. Several months and meetings later, the panel reported.

They praised our 'highly respected' science, saying we were widely seen as very successful. However, they said that I was trying to do too much, with my academic duties, departmental responsibilities, and widespread research interests involving a lot of collaborations and travel (I didn't tell them I was also trying to write a book). The way they put it was 'Through his undoubted skill as a scientist, Dr Taylor had been able to attract funds for a larger number of projects than his group had resources to carry our effectively'. They recommended 'that, as a matter of urgency, a Space Projects Manager is acquired . . . funded equally by the SERC and the University'. I had asked for this post as long ago as 1981; this was December 1988. I was reminded of Brer Rabbit and the briar patch. Still, I had my Oxford manager at last, and the rescue plan was complete.

As Principal Investigator I remained responsible in an overall sense for ISAMS, of course, but with this extra support I had more time to hold on to our involvements in the other key activities from which the British National Space Centre had withdrawn support before BNSC itself went belly-up. The money the Board had reassigned did not come back, of course, so the work was harder for me to handle, although I did have access to other resources. Chief among these was research students: we were training some of the nation's most successful physics graduates who had chosen to go into careers in research, starting with a doctoral degree. Not only were these young men and women extremely bright, they had seemingly inexhaustible energy and resilience. Also, of course, the permanent staff of the Department were not reliant on BNSC funding. The planetary projects survived while BNSC did not.

Launch

The Improved Stratospheric and Mesospheric Sounder was shipped to America and attached to the Upper Atmosphere Research Satellite at the plant where the General Electric Company had built the giant spacecraft under contract to NASA. Valley Forge is near the site of George Washington's headquarters during the revolutionary war of 1776, and American history is one of my hobbies, but there was no time for sightseeing. Once everything was thoroughly tested, the complete UARS assembly would ship to Cape Canaveral in Florida and be tested again. Then, it would be mounted in the payload bay of the space shuttle *Discovery* and tested yet again. Finally, on 11 September 1991, it blasted off the launch pad on its way into low Earth orbit.

As a Principal Investigator on the mission, I again had a grandstand seat for the launch, and a boring but nail-biting several-hour wait for the action to start. I thought about trying to get closer, but unlike the *Pioneer Venus* launch from nearly the same spot in 1979, this time I held back. The stubby, bulky appearance of the *Discovery s*huttle, and the knowledge that there were crew on board, contrasted dramatically with the slender profile of the unmanned *Atlas-Centaur* rocket that had launched us to Venus. Something about this made me want to keep my distance this time. Also, I was 12 years older, with my 50th birthday just three years away, and maturity makes one less interested in seeing what can be done by way of flaunting authority. Anyway, I knew now that they would catch me.

The most spectacular thing about the launch, apart from the fact that it went flawlessly, was not so much the sight of thundering rocket motors raising their massive payload skywards, but the actual *sound* of it. I had seen plenty of videos of shuttle launches and thought I knew what to expect, but nothing had prepared me for the intense low frequency rumbling and loud crackling noise that shook the ground and battered the ears. Amazing. And all our work and pain had not been in vain: after yet more tests in orbit the shuttle bay doors opened, and the manipulator arm dangled the six-ton satellite at arm's length. It stayed attached until the solar panels had been deployed and electrical power surged through the circuits that would power our instruments. Then, and only then when everything had been checked out one last time, the *Discovery* let the satellite go, closed up, and returned to base. We were on our own now, and we were in business.

Results

This short account of the build-up to the launch, the event itself, and the immediate aftermath is of course a telescoped and sanitized overview of a long and often agonizing period, around a year, of intense meetings, setbacks, reviews, crises, recriminations, and more meetings. It had its glamorous moments, and those are what one tends to remember: sitting at the long console at the Cape with the mission controllers, wearing a headset and looking at a screen to see what the readouts from ISAMS said, and giving the final 'Go' command that was an essential part of the chain leading to the final countdown. I had seen this on television and in movies so many times, and had looked forward to being part of the drama.

The reality is rather more prosaic than the movies, with the tension and delays draining the glamour away. Even the party following a successful launch is an edgy affair, because the scientific instruments stay switched off for ten days or so to allow traces of water and other volatiles trapped in various parts of the satellite to evaporate into space.

I celebrate with Ray Turner, the ISAMS Project Manager, beside a one-tenth scale model of the *Upper Atmosphere Research Satellite,* at the party held by NASA following the successful launch of the spacecraft.

Failure to do this can result in deposits of ice and other nasties appearing on the mirrors and other important surfaces in the instrument, which can impair its performance and even cause failures. The team leader cannot relax until he sees the first data coming down and the team has looked at it and confirmed that everything is working as it should.

There is not much time for celebrating even then. Once the instrument is switched on and checked out, the data flows continuously day and night, seven days a week, and has to be processed, analysed, stored, and disseminated to users. The expectation that something dramatic will be discovered and published in prominent journals, after presentation at international scientific symposia, adds more pressure. This is the fun part, however. The scientist is in control, and even funding is not much of a problem because the sponsors, having spent tens of millions of pounds on the instrument, are not going to stint on data analysis spending, not at first, anyway.

For several months ISAMS worked spectacularly well. The complex and temperamental pressure modulators, the newly invented coolers, and the sensitive fragile detectors were all virtually perfect in their behaviour in orbit, scanning the Earth and sending back data of excellent quality. The ozone layer was laid out before us; we saw the trace gases of manmade origin as they ebbed and flowed across the globe, changing with the time of day and with the seasons. We could see how they interacted with each other under all sorts of different conditions, and used this information to sharpen up the computer models of the atmosphere so they represented the processes better and made better forecasts, which we could then test. As a bonus, there was a major volcanic eruption at this time, which threw gases and dust up to a great height in the stratosphere, where we could observe the interaction of these with ozone and each other as well. We were in demand at conferences around the world to explain all this to attentive audiences.

Tea with Mrs T

One of the audiences interested in our new insights on the stratosphere was unexpected. I was at a smart cocktail party in the university for some long-forgotten reason and met George Guise who worked as a scientific advisor to the Prime Minister, Margaret Thatcher. When I told him what I did he asked me did I know anything about ozone? I said yes, I knew a great deal about ozone, and I mentioned the progress being made to address the effects of pollutants on ozone using satellite measurements. Great, he said. The Prime Minister was concerned about the ozone depletion problem and was determined to find a political solution. Would I help him to brief her?

A week or so later I was walking alone up Downing Street to confront the policeman at the door and be admitted to No. 10. Just inside the door there are several more police, with a mass of screens and electronics monitoring equipment, but once past that, the Georgian splendour one would expect to find in such a place asserts itself. I threaded my way up grand staircases lined with portraits to the science advisor's small office at the back of the building, and there worked with George on the briefings that would get the ministers up to speed on the basic nature of the ozone problem.

I went back to Downing Street several times, once for a special meeting involving most of the Cabinet with myself and the other scientific advisors forming an audience that they could consult. I have never seen anyone run a meeting as well as Mrs Thatcher did (except that she didn't notice, or affected not to, when the Home Secretary fell asleep). By mid afternoon we had reviewed the science and the politics of ozone depletion and formed the foundations of the Montreal Protocol that was signed by 46 delegates in 1987, and would be finally ratified by all 197 nations of the world when South Sudan was the last to sign up in 2012.

The best remembered feature of the protocol was the banning of halogen-based propellants from refrigerators and aerosol cans, these having been identified by the science community as the biggest culprits in ozone destruction. This and other measures taken in Montreal, with Mrs T in the driving seat, greatly ameliorated the threat to the ozone layer and it gradually faded from the headlines. The problem was not yet completely understood or resolved, of course, so it stayed on the scientific agenda, but we did feel among the UARS team that we had made a contribution to saving the world.

Back at Oxford, politicians and other celebrities showed up at the various official parties that mark major events such as the appointment of a new Vice-Chancellor, or the opening of a new building. One always met interesting people at these. Apart from the one with the Science Advisor, I particularly remember chatting to Quentin Hogg when he was Lord Chancellor, Peter Mandelson shortly after his retreat from government and subsequent ennoblement, the warm and friendly actor John Voigt, and most delightfully, Ian Hislop, the editor of *Private Eye*. I was invited to coffee one morning to meet the evangelist Billy Graham, only to learn that he had injured himself by slipping on soap in the shower and was indisposed. I was sorry about that; I was not particularly keen to be converted, but I did admire his eloquence and passion.

Bugs in space

After nearly a year in orbit, ISAMS had accumulated a huge mass of data, but was no longer operating. A mechanical fault in one small part had wiped out the whole instrument. Instead of a minimum of 18 months of data, and an expectation of a lot more, we had operated for only ten months. The gremlins, like those that caught us short at Venus, had struck again.

The failure was not in any of the sophisticated equipment, like the coolers over which we had sweated tears of dread for years. They went on to work perfectly for many more years. The culprit was the 'chopper', a small mechanical device that is a standard part of all infrared optical instruments and which should be completely reliable. It wasn't, and when it stopped all of the data from the instrument ceased to be of any use. After a few days it started up again, but after a few more days of normal operation it quit permanently. This was a major embarrassment, but not a total disaster—ten months worth of data spanned nearly the whole seasonal cycle, and covered the entire Earth hundreds of time. We could still do lots of science.

Terror on the ground

The real disaster came nine years later, and was of a totally different character. Exactly ten years after launch, NASA called a three-day meeting of all UARS investigators for a decadal review of what we had achieved, to be followed by a celebration. The party was to be at the vacation home of the chief scientist for UARS, Carl 'Skip' Reber, on Chesapeake Bay, about an hour's drive away from NASA's Goddard laboratory where we were meeting. Maps and instructions for finding the party after the meeting ended had already been issued when

Commander Creighton, who had piloted the shuttle that carried UARS into orbit, started to speak. Then all hell broke loose.

The pilot, a Texan and an ex-Navy pilot, was showing some slides and movies he and his crew had taken during the flight, when the screen went blank for an instant and then restarted showing a tall glass-faced tower building with smoke streaming from its side. What on Earth was this? The first reaction was that it was a movie, *Towering Inferno* perhaps, that had been introduced by mistake. However, the picture quality was far too poor for that. It was in fact the CNN News channel, which the shocked and confused Goddard projectionist has seen on his monitor and patched through to our screen, but in the auditorium we didn't know that.

In blissful ignorance, we were relaxing as we waited for normal service to be resumed when armed, uniformed guards from NASA site security rushed in and shouted for everybody to get out. Out of the building, we asked? No, completely off site, the entire facility was being evacuated. Where should we go? Anywhere, anywhere, just get off lab, it's an emergency. Later we learned that the Pentagon had just been struck, only a few miles away, and Goddard was a fairly obvious candidate to be another possible target. Nobody knew then how many rogue airliners were in the sky. With admirable presence of mind, Skip Reber

The launch of the *Upper Atmosphere Research Satellite,* 12 September 2001 on the space shuttle *Discovery.*

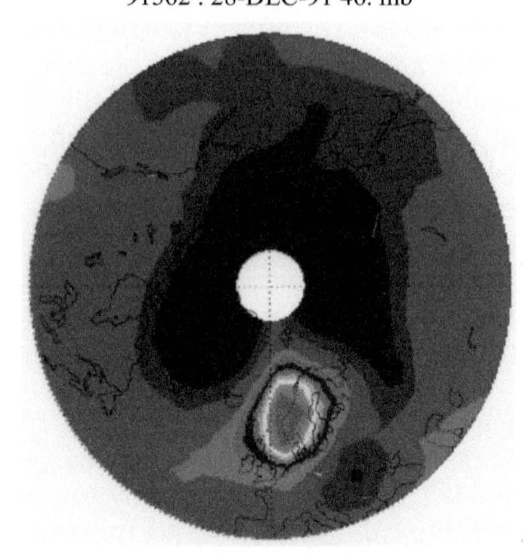

A map by the *Improved Stratospheric and Mesospheric Sounder* of the aerosol density in the stratosphere over the northern hemisphere, with the North Pole at centre. The compact feature over Scotland and Scandinavia is a rare example of a northern polar stratospheric cloud, similar to the larger versions which intensify the Antarctic ozone hole over the South Pole each winter.

Chatting about climate change to William Waldegrave, then in government as Chancellor of the Duchy of Lancaster, at the Smith Institute in 1993.

shouted 'Come to my place. The party starts now'. We fled to our cars clutching the maps and directions.

It wasn't the usual kind of party. Once assembled at Skip's house at the beach, we crowded on the floor clutching beers and hot dogs and watched the TV. The situation was still very confused, and would be for days and weeks to follow. But by the end of the day, it seemed safe enough to drive back to Washington and our hotels. The next day, the meeting resumed in a much smaller auditorium borrowed from an electronics company near Greenbelt. Goddard Space Flight Center, like all government buildings in the area, was still sealed off.

I was finally able to give my talk about ISAMS. We had achieved a lot, despite our early setback and maybe even because of it. The team had had more time and incentive to mine the large but finite data bank we had already gathered before the instrument died. In an informal league table of the number of papers published in professional, refereed journals up to that time, ISAMS came top. The experiment had disappointed in terms of lifetime but done enough to meet its principal objectives.

These objectives were, I explained to my diminished and displaced audience, firstly to obtain measurements of atmospheric temperature as a function of pressure, from the tropopause to the mesopause, with good accuracy and spatial resolution, and hence to study the structure and dynamics of the region. For illustration, I showed a spectacular picture derived from ISAMS data of the temperature structure of the whole middle atmosphere, all the way from the central southern hemisphere to the North Pole. Such coverage was unprecedented, and I had chosen a time when the data showed an impressive example of the mysterious phenomenon called a 'sudden warming'. This occurs when the circulation around the pole changes from one seasonal regime to the next. Sometimes, this change is so quick that massive amounts of kinetic energy are released as heat. In the example we had captured in detail, the temperature changed by more than a hundred degrees in a single day. Simultaneously, the instrument tracked the dynamical behaviour that was responsible for the warming by observing the behaviour of trace gases. I showed measurements of carbon monoxide, which is produced high in the atmosphere from carbon dioxide dissociated by sunlight, propagating downwards in a pattern that varied dramatically as the tracer was carried by the wind field as it changed during the warming event.

We had also investigated atmospheric chemistry, especially the behaviour of water vapour in the middle atmosphere, following up on those pioneering studies with aircraft that my mentors Brewer, Houghton, and Goody had so painstakingly undertaken many years earlier. The stratosphere really is astonishingly dry, and had it not been, the Earth would have lost some, and possibly all, of its oceans long ago, as Venus had done. But our major focus remained on the global distribution of oxides of nitrogen and their origins and their roles in the chemical cycles which control the amount of ozone in the stratospheric ozone layer. Again, the data was spectacular, with huge changes in concentrations from day to night as the solar-driven photochemistry changed, and subtler changes with latitude, season, and weather.

The makeshift UARS decennial meeting started to break up as people found ways to go home. We were depressed by having witnessed mass murder, and we knew that the world had changed forever. The whole nation was still in shock. As I drove back to my hotel, people were standing by the side of the road waving flags and holding up home-made placards reading God Bless America. Some of the team members from California took their rental cars and drove all of the way back, nearly 3000 miles. Driving was not an option for me: I had to live in a virtually deserted Holiday Inn for another week and a half before flights resumed and I could get a seat back to London.

A hole in the sky

The most newsworthy aspect of the work being done by the *Upper Atmosphere Research Satellite,* certainly as far as the general public were concerned, was the study of the ozone 'hole' over Antarctica. Dobson had recognized the loss as a basic property of the dynamics at the South Pole in winter, when a stable vortex forms and makes a 'containment vessel' hundreds of miles across in which production is curtailed by the darkness and ozone levels are low. Decades of observations by the ozone monitoring facility at Halley Bay revealed that the ozone loss in the hole was getting greater year by year, until there was virtually no ozone there at all at some levels before spring came and the vortex broke up. It was important to prove, or disprove, that this change was due to pollution of the stratosphere by anthropogenic activity. The finger had been pointed at things like high-flying aircraft, the working fluid in refrigerators and air conditioners, and the ubiquitous aerosol spray cans. These everyday items were a huge part of modern life and commerce and could not be changed or done away with unless there was indisputable evidence that they were a threat to civilization on the surface.

A key factor, it turned out, was the formation of clouds of frozen nitric acid at the very low temperatures in the polar night. ISAMS was contributing to the study of these polar stratospheric clouds, PSCs as they were called, and investigating why they only occur at the South and not the North Pole. The basic reason is the difference in the continental land masses, which affects the nature of the containing vortices, so that the north is typically ten degrees or so warmer in midwinter at these heights. This is usually enough to prevent PSC formation in the Northern Hemisphere, but remarkably we did spot one with ISAMS lying over the north coast of Scotland. It wasn't very large, and it only lasted a few days instead of months, so there was no question of an Arctic ozone hole developing that was anything like the one over Antarctica, but it was a very interesting phenomenon. I chose it as an ISAMS highlight with the others that I showed to my UARS colleagues and the NASA hierarchy who had put their faith in us that we had delivered.

There was much that remained to be done. The situation in Antarctica is interesting but unique to that essentially uninhabited region, and the politicians and the public were more interested in what might be happening to ozone over the rest of the globe. After the initial excitement, the data analysis phase of a large project diffuses into the wider community and can go on for decades. Once the main processes had been identified, the main task of quantifying global ozone depletion and predicting future trends, with and without legal steps to control the offending emissions, falls to the research groups that build and run large computer models. As someone with a lot of other interests I was not too sorry to publish the early results, finish archiving the data set, and then move on a little.

The legacy

ISAMS, the UARS satellite, and all of its other instruments fell out of orbit and burned up in the atmosphere on my birthday in 2011 after more than 14 years in orbit. As would also be the case at Mars, the pressure modulators were among the most robust individual pieces of hardware on the spacecraft and they may have survived the descent. At Oxford, we joked that they might come through our roof at the speed of sound and end up back in the lab where they were built. In fact they are now at the bottom of the Atlantic Ocean.

Long before the scientific work with ISAMS data was completed, the question of what we should do next had to be addressed. We wanted to build more instruments, and the pathway to design, selection, and funding was, as ever, long and uncertain. At the time, our workshop was busy with hardware for *Mars Observer* (Chapter 9) and *Cassini* (Chapter 10), but this was not enough to keep the whole technical team busy. We needed another big Earth observing project, if we were to stay in business as pioneers in space.

An intriguing prospect had come along, in the form of something called the *Infrared Atmospheric Sounder*. This was a project for the US Air Force, being developed by Hughes Aircraft Corporation at one of their facilities in California, although not the one at Long Beach that I knew so well from *Pioneer Venus* days. We didn't feel very comfortable working on a military project, especially a foreign one, but the Americans had come to us for information about making temperature measurements in the upper atmosphere, since that was one of our specialities. They wouldn't tell us why they wanted such data, although it was not hard to guess that it had something to do with the effects of atmospheric density variations on the paths of missiles re-entering from orbit. They said we could have full use of the data for scientific studies, and they had plenty of money, so we decided we were interested. The planning process was going well, when suddenly everything went quiet from the American end, and that was that.

We were much more comfortable with the friendly, open nature of NASA, anyway. Most of the work on new ideas for an advanced sounder had been done by John Barnett, and when NASA announced a new, big satellite series called the *Earth Observing System* to follow on from UARS, John came up with an improved version of the Air Force project that he called *Dynamics Limb Sounder*. The name was meant to emphasize that a unique approach to scanning the atmosphere would reveal a lot more detail than before in weather systems in the upper atmosphere. Advances in instrument sensitivity, and in computer models for data analysis, indicated that this experiment could make huge progress in understanding wave motions in the stratosphere and the way they affect ozone transport and mixing. This was something we had long been interested in, and was the key to new discoveries in stratospheric physics. We decided to see if we could make this our next big project.

We returned to the complexities of proposing, and winning, and then funding, a big space project. As always a massive commitment and effort is required, and often the result is disappointment if the competition is lost. Not only that, but the background conditions were not set fair; ISAMS had cost more than it was supposed to, and had caused many ructions and much pain along the way. Worst of all, for me, was that I was so badly traumatized by the experience that I was loath to start over again on a project of a similar nature and magnitude. Anyway, I wanted to give more time to our planetary projects.

Of course, I did not have to be the Principal Investigator again, although our long-standing practice was that the PI had to be the Head of Department in order to make sure all possible resources were mobilized. The Dynamics Limb Sounder was John Barnett's idea, but I thought it very unlikely that John would want the responsibility of the Principal Investigatorship, although he certainly had the ability. He was a quiet, gentle man, very intelligent but without the aggression and win-at-all-costs mentality that experience said was required for the PI to survive the slings and arrows of a large project.

I was surprised, therefore, when I asked John about it and he said he was not only willing but actually keen to be PI. He would have the support of an experienced team, including the essential John Whitney as chief engineer and now our own project manager, Ray Carvell, to provide local management. I would be a Co-Investigator and provide all the support I could muster as Head of the Department. Rutherford Appleton lab would provide the usual services, although there were fresh difficulties there. Changes wrought by JTH when he was Appleton director had led to a culture shift; so rather than the lab's traditional role in support of university research, it had started to build its own programme.

The flagship project RAL adopted was another big Earth observation experiment, the *Along-Track Scanning Radiometer*, which we

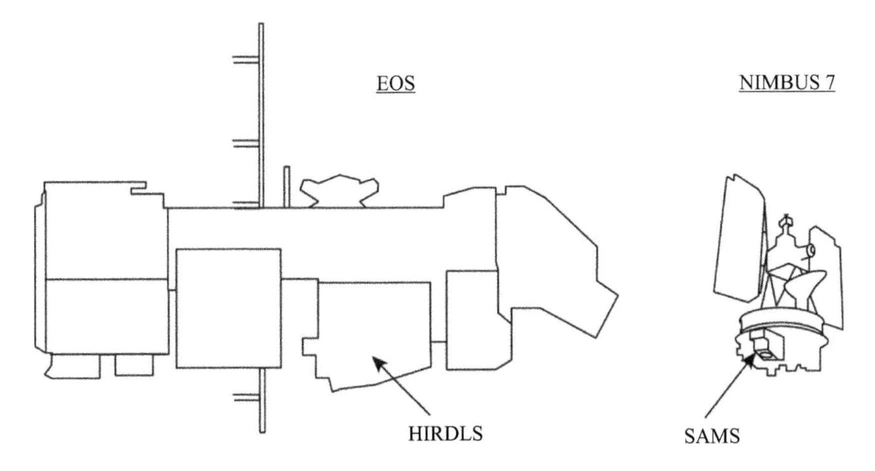

These outlines of the two of the spacecraft, *Nimbus-7* and EOS-AURA, that carried our experiments into Earth orbit side-by-side to the same scale shows how far scientific satellites have advanced in size and complexity during the last quarter of a century.

at Oxford helped to develop and support. Once the European Space Agency had selected ATSR for flight, and the workload grew, it did not make a lot of sense for us to provide key personnel from our small group to a big national facility, when we were already overstretched on our own projects. Conversely, we were already absolutely reliant on support from the Rutherford for crucial things like optical and electronic design, and it did not help when most of that became redirected into their new project.

Hurdles and HIRDLS

Nevertheless, John Barnett put together a very strong proposal, and NASA liked it. The research council in the UK indicated a predilection for coming up with the funding if the instrument was selected for flight. ISAMS might have burned them too, but so did almost any sufficiently large scientific project on their books and they were used to problems, and they knew we could deliver something special.

There was one more difficulty, however. An experienced satellite experiment group at the National Center for Atmospheric Research in Colorado had submitted an almost identical proposal to ours, which they called HIRS, for *High-resolution Infrared Sounder*. What both instruments had in common, apart from the same scientific goals, was a capacity to scan the atmosphere rapidly in the horizontal direction in order to greatly increase the resolution of the data. Older instruments did not have the size or sensitivity to do this, and it made an important difference, because we now suspected that the chemistry and the dynamics of the stratosphere were inextricably linked. John Barnett knew this, and so did John Gille, the team leader for HIRS. Gille was well known to us, and had in fact been one of my team for the VORTEX experiment at Venus 20 years earlier.

NASA could not fly both instruments so there was for a while a standoff and some friendly rivalry. It was actually a suggestion by John Harries at Rutherford Appleton that led to the solution: merge the teams, and produce one instrument built partly in each country. Oxford had worked alongside NCAR before, and the groups knew each other and we were on good terms. The merger might even save money, although I knew from experience that such schemes rarely do. NASA would have to accept two PIs on one experiment, which is normally anathema, but they did, and the *High-resolution Infra Red Dynamics Limb Sounder* was born.

HIRDLS was built, eventually, with many of the trials and tribulations that were by now a familiar part of such big projects, with the added complication of the hybrid, international approach to engineering, funding, building, and testing the hardware. John Barnett turned out to be an excellent PI, totally dedicated and hardworking, forever rushing off to the USA for meetings which, in the pattern I had grown all too used to, were usually either totally routine and boring or panic-stricken emergencies. It is hard to say which I preferred, but there is no choice as to what you get, and John took them in his stride.

It was all going so well, then tragedy struck, not once but twice, and on such a scale that ISAMS came to look like a walk in the park. EOS, with HIRDLS on board, went through the routine of delivery, testing, launch, and deployment without a hitch and everything was going wonderfully. Then, after the usual few days of outgassing, HIRDLS was switched on. Something was wrong. When it scanned from horizon to horizon to collect high-resolution data, the signal went blank over most of the scan. The reason, it turned out, was far more prosaic even than the failure of a simple mechanism that had cut short the lifetime of ISAMS. A piece of insulating fabric, bonded to the inside walls of the HIRDLS instrument to keep the temperature stable in space, had detached itself and was flapping around inside the front aperture. Nothing could be done to fix it, short of a visit by an astronaut with a tube of glue, or bringing the whole thing back to Earth, neither of which was remotely feasible.

It is hard to imagine anything more demoralizing than that, after so much effort and expense, but much worse was yet to come. Whether or not the stress of leading the project had anything to do with it, John Barnett fell victim to a brain tumour in 2009 and died less than a year later, at the early age of 62. The sense of loss for those of us who knew, loved, and worked with him was unimaginable. And there was no way back for our Earth Observation programme, at least anything on the scale of ISAMS and HIRDLS.

9
Mishaps at Mars

'Why do you want to go to Mars, Professor Taylor? It's just a bunch of old rocks there, isn't it?' John Humphrys was interviewing me in typically aggressive style on the BBC Radio 4 *Today* programme, and I was somewhat nonplussed. What do you say in answer to a question like that, on a live broadcast at 8 o'clock in the morning, with six million people listening in? I mumbled something about Mars rocks being different from those here on Earth (actually, they are not very different). Then I pulled myself together and remembered the old politicians' trick of giving the answer I had prepared, quite regardless of what the question had been. By the time the interviewer realizes you are not answering the question you are in full flow, and he has to either rudely interrupt you or let you finish. In this case he let me finish, probably because he knew it wasn't a fair question in the first place. I told him we had built an instrument that was about to arrive at Mars and unlock the secrets of its climate, which once was wet, warm, and Earthlike and for some unknown reason had dramatically changed. I didn't know it at the time, but that was only partly true as well; we would arrive, but not unlock anything. The date was 23 September 1999.

A better way

I was not the leader of the Mars project as I had been for VORTEX on *Pioneer Venus*, SAMS on *Nimbus 7*, and ISAMS on the *Upper Atmosphere Research Satellite*. The reason I was now just one of the team was partly that we needed to have an American Principal Investigator in order to get the project funded, built, and flown. Dan McCleese at JPL took on all of these things and drove the project forward. Meanwhile, I had had a defining moment following the elation and the trials of the heavy responsibility for those three earlier projects. Basically, I had realized that being Principal Investigator on a major space mission was no longer something that appealed to me. Of course, it is great when you are young, unknown, and struggling, to be catapulted into the top ranks of your field by winning such a leading role in a high-visibility project. At first, you are willing, indeed glad, to shoulder the constant strain and often terrible burden of the responsibility for succeeding in what, at times, seems a near-impossible task. And of course the chance to take the lead in defining and then pursuing exciting new scientific objectives that follow your own interests makes it all worthwhile when the results come rolling in.

My new insight was that there is another, better way that opens up to you once you have made the initial breakthrough and have a foot in the door. Each Principal Investigator, when he or she submits a proposal for a new experiment, assembles a team of co-investigators to carry out the project if it is successfully selected. Once you have been a PI yourself, you have enough contacts and a bit of a reputation (hopefully mostly good), so that invitations to become a Co-I on an experiment led by someone else also come along. Co-Is have the right, but are not usually required, to attend meetings connected with the project, and generally do not have to assume responsibility for difficult schedules, baulky subcontractors, and large but sometimes inadequate budgets. You do have to make a valuable contribution to the success of the mission, of course, but unlike the team leader or project manager you have a lot of freedom in how you deliver this.

I calculated that, with the time and effort I had expended as ISAMS PI, I could have been a Co-I on about six separate missions simultaneously, and reaped a wider range of scientific rewards with far less inconvenience, pain, or blame. Having six eggs in different baskets would also provide insurance against missions failing or being cancelled, as did happen not infrequently (as we shall see). Once enlightened, it was a no-brainer: henceforth my ambition would be to join teams on interesting missions as Co-I rather than PI. The first application of this principle was on HIRDLS, which was in any case John Barnett's concept; the second, I hoped, would be on a long-awaited mission to Mars.

Onward to Mars

Long before I moved back to the UK, well before we were finished with analysing the data from Venus, and while we were still working hard preparing for the Jupiter mission, we started talking about a sending a version of our Venusian VORTEX instrument to Mars. For a while, the concept was known as MORTEX although for semantic reasons that name did not catch on (sounds like 'dead instrument', said Fred Vescelus). The idea of using similar devices to map out the weather and climate on all three terrestrial planets was irresistible, however. The chance to inter-compare the data would surely reveal the similarities and the differences between the members of the Earth-like triad. It did, but it was a very long slog, and featured several heart-breaking failures before we got there in the end. And along the way, we succeeded, albeit unintentionally, in putting the first British hardware on the surface of Mars.

The relict of the VORTEX team at JPL, now under the leadership of Dan McCleese, and its collaborators in Oxford now led by me, applied itself to the task of defining a follow-on experiment for Mars. The first problem we had was that, at that time, there was no suitable mission to the Red Planet in the planning cycle. For me, the *Galileo* Jupiter mission came along, and was so interesting it was possible to forget about Mars, or at least leave it in the background while slowly continuing to develop plans to do it later. Dan was also distracted when he came up with a successful proposal to use pressure modulator spectroscopy to measure winds from space, although the project would later be cancelled before it flew. For him, too, Mars was an itch that would need to be scratched sooner or later. We waited for a suitable mission opportunity to come along, and also thought about how to create our own.

The way that new missions were born was, as ever, to work hard on the planning teams and management committees of the space agencies. First one had to acquire membership of one of these exalted bodies, which was by invitation only, based on reputation and always much sought after. The reason scientists are keen to join these panels is not so much that it is fun to sit for days discussing and arguing around a large table, usually in a hot windowless room, although it is never less than interesting, and it does make you feel somewhat important. More to the point, however, is that expensive scientific projects can never get off the ground without working their way through the planning cycle and the management structure of the responsible agency. If you succeed in helping to drive something through, you not only get the mission you want, you are also in pole position to be involved with it as an investigator. A lot of time can be dissipated introducing or endorsing something exciting and seemingly irresistible, writing endless reports about it, giving talks, and lobbying colleagues, often for years, only to have the project flounder at some stage and never see the light of day. But a few do go forward, all the way to fruition, and this can make it all worthwhile.

Getting to Mars isn't easy

The 1980s were frustrating years for those of us wanting new Mars missions. The legacy of the *Viking* landings in 1976, intended to find signs of life but instead showing that there did not seem to be any, had left NASA's Mars programme with nowhere to go. The climate was better in Europe, where ESA, flushed with success from its rewarding *Giotto* mission to encounter Halley's comet, had shown it had the capability to mount a modest but capable mission to Mars. A group of us, led by Ulf von Zahn, a professor at the University of Bonn in Germany, devised such a mission and called it *Kepler*. The spacecraft, designed by Messerschmitt and carrying a payload similar to that on *Pioneer Venus*, including our radiometer suitably modified for Mars, was described in a proposal which we submitted to ESA in October 1980. The story of *Kepler* and its scientific goals is told more fully in my book *The Scientific Exploration of Mars*, along with more depth on all the Mars missions discussed here.

Kepler was successful in the first round, the prize being some substantial funding for a full 'Phase-A' study. Phase A is the stage at which a concept is turned into a detailed mission design, with funds to get all the technical risks and implementation costs defined by experts. As we did the work, flying back and forth to the ESA Space Technology Centre in Holland to consult with the mission planners and engineers, we were confident that our relatively mature designs, low cost, and alluring target would win the final go-ahead at the decision-making convention that was scheduled to take place soon in Paris. At this massive event, ESA sounds out the whole European space science community before it makes the final mission selection, and a complex web of politics and self interest comes into play.

The problem for *Kepler* was that few European scientists were actively involved in studying Mars in those days. Instead, nearly everyone was a deep-space astronomer, studying stars and galaxies, while the relatively small number of groups that were doing Solar System research were focussed mainly on particles and fields, or maybe on small bodies such as comets and asteroids. The reason for that was simply that these targets were accessible with the technology of the previous generation, such as small rockets and satellites orbiting in the Earth's radiation belts, or fragments of comets available as meteorite collections in numerous museums and collections. My peers were the former students of those pioneers, now grown to maturity.

The team under von Zahn tried again and again to get approval for *Kepler*. To us, it seemed like a great idea whose time had come. Mars might not be on many European research agendas, but it had huge popular appeal. It was also now possible to get a small payload to Mars using a European launcher, in the form of the *Ariane* rocket that had been developed mainly in France. It was worth persisting, therefore, and after a rare second 'Phase A' exercise once again led to non-selection (this time we lost to *Cassini*, see Chapter 10) we gave up on the purely European approach. The last chance to find a way forward was to team up with the Americans, not just for experiments like MORTEX but for the planning and execution of the whole mission. This idea was not unique to Mars; so much of the Solar System was ripe for exploration, and for a while everyone thought pooling resources seemed to be the best way to go.

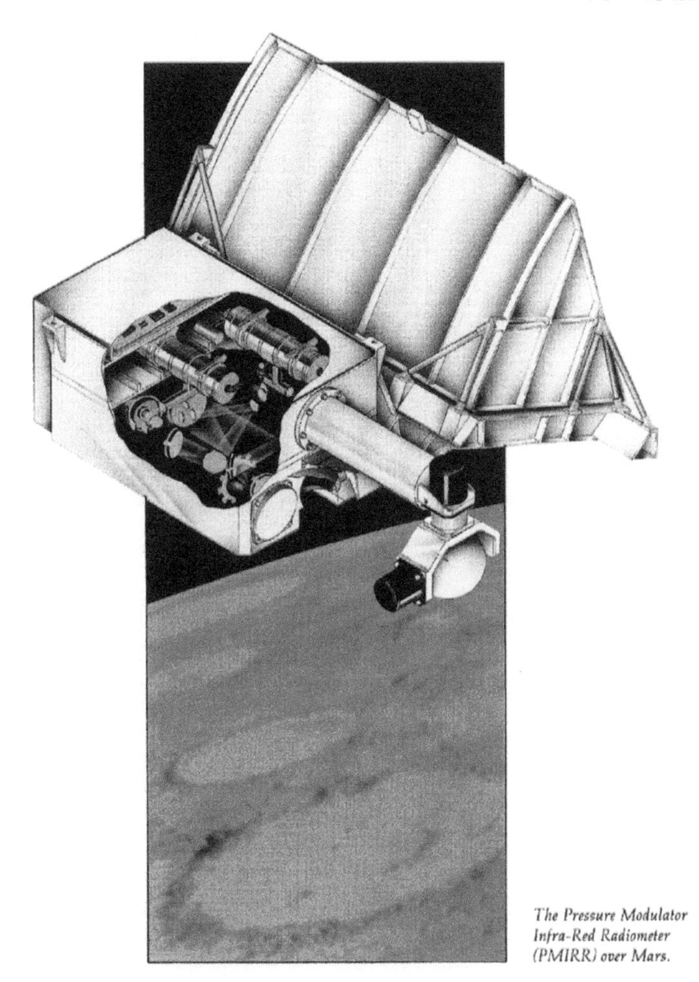

The Pressure Modulator Infra-Red Radiometer (PMIRR) over Mars.

An artist's impression of the *Pressure Modulator Infra-Red Radiometer* over Mars (of course, in reality, it was mounted on a spacecraft and not free-floating!) The large structure at top right is a radiator to keep the infrared detectors cold; at bottom right is the scan mirror that directs the view of the instrument at the atmosphere and surface following a programmed sequence. The cylindrical devices inside the cutaway view are the pressure modulators we built at Oxford.

The US–European Joint Working Group

Leaders in the planetary science communities on each side of the Atlantic decided to grasp the initiative by setting up a joint study that would make hopefully irresistible recommendations for joint missions, leading with any luck to at least one real collaborative project. Through the auspices of the US National Academy and the European Science Foundation they appointed six representatives from each side to come up with ideas and study them in detail. I was lucky to be one of the European members, being young then, but my transatlantic experience

The Mars experiment was an international partnership led by the three of us shown here, at Cape Canaveral for the launch of *Mars Climate Orbiter* on 11 December 1998. The American Principal Investigator, Dan McCleese is on the left, with the leader of the Russian team, Vasily Moroz in the centre. The spacecraft launched successfully but went on to crash into the planet instead of orbiting as intended. Thus, it inadvertently delivered the first British hardware on the surface of Mars.

was a factor. The Joint Working Group began to hold regular meetings, alternating between continents. There is an element in this sort of work of each side wanting to impress the other so the venues tended to be attractive; Woods Hole on Cape Cod, for example, when the Americans hosted, and a castle in Bavaria when we did.

The top-level membership of the working group split up into three specialized panels, dealing with terrestrial planets, outer planets, and small bodies respectively. We set up a task force for each that would coopt additional experts and work with the space agencies and the aerospace industry to flesh out the details of mission concepts, including their likely timescale and cost. I was put in charge of the terrestrial planets panel, and at our first meeting we decided to focus on Mars. It was a wrench for me to give up on Venus so early in the study, but it was clear that more of the people we had to convince would be excited by going to Mars than by any other goal we could advocate. Some favoured a mission to Mercury, but the innermost planet is difficult to get to and in any case less appealing as an objective to most of us.

The rules we set for ourselves called for each panel to define low, medium, and high cost mission profiles with a specific price range attached to each. For the low cost terrestrial planets mission, we came up with *Mars Dual Orbiter*, which was basically *Kepler* from Europe plus NASA's planned *Mars Geochemistry and Climatology Orbiter* working together at Mars with matched payloads. Each partner would deliver its own satellite to Mars but the orbits and science objectives were to be worked out together. The two satellites would be in radio communication with each other, as well as back to Earth. During the occultation events, when the beam between the orbiting spacecraft passes through the Martian atmosphere, we could perform some unique investigations of temperature and pressure profiles at very high vertical resolution, less than a kilometre. This, we asserted, would reveal many important features of the atmospheric structure and dynamics that were hard to detect by any other remote technique. Both satellites would also use cameras and spectrometers to observe interesting features on the surface of Mars and its natural moons Phobos and Deimos.

The 'intermediate' cost mission, which was our favoured option because the price and complexity were about what we reckoned the politicians and paymasters who had the final say would buy into, was a small rover that could traverse Mars and examine different terrains. This would be controlled from the Earth and so would not require any advanced robotic techniques, which were then in their infancy and probably too risky. One of the simple ways to make a rover was with a large inflatable ball, that pumped itself up after deployment on the surface. It could move by being blown by the wind across the surface, or in a slightly more controllable fashion by making it in segments and alternately inflating and deflating each one to make the whole thing crawl along. A group of our colleagues in Arizona actually made a prototype of one of these so-called 'boules' (the idea had come from Jacques Blamont in France) and deployed it on the lawn at a conference. It was ten feet high and amazing to behold as it hissed and slouched its way slowly across the grass.

An alternative mission at a similar cost would be a network of small, stationary surface stations using penetrator technology. Penetrators are torpedo-like devices that are hurled out of orbit without a parachute to strike the surface at high speed and bury themselves into it down to a depth of ten metres or more. It sounds terrifying, and indeed whenever the idea was presented to an audience someone usually asked whether we had considered the possibility that the Martians, if there are any, might see this as an act of war. We did not believe that was a serious risk, and our detailed studies and consultations with experts from the military sectors of industry showed that it is possible to build hardware, including scientific instruments, that would survive the massive shock of the impact. A single probe carrier in orbit could deploy a network of hard landers distributed all over the planet, to explore diverse terrain and make meteorological and seismological measurements.

In the final report of the Joint Working Group in 1984 we included, as the high-cost option for a Mars mission, a full-up robotic laboratory to roam on the surface and explore and sample geological foundations. With this we were looking ahead to something similar to the *Curiosity* rover that did finally make it to Mars, 30 years later. We could argue that our simple rover concept, although not a boule, also eventually saw the light of day in the form of the *Sojourner* rover that landed in 1997 on *Mars Pathfinder*. There have been no coordinated pairs of orbiters, however, and no Mars network, using penetrators or otherwise. The latter idea showed itself to have wide appeal, however, and networks keep reappearing as mission proposals and as part of long-term planning for NASA's Mars programme. The fact that it has yet to be implemented in reality is perhaps mainly a reflection of the stochastic nature of success in the rarefied field of endeavour that is planetary exploration.

During one of the wrap-up meetings of the Joint Working Group, somewhere in the USA, I shared a car back to the airport with George Haskell, who was the head of the Solar System Division at ESA and therefore obviously had a key interest in the progress of the study. We were chatting about prospects for Mars when I noticed that he had become distracted by something just off the road. 'Ooh, there's a *Dunkin' Donuts*' he said. 'Do we have time to stop?' It turned out this rather distinguished individual had a minor addiction to their sugar-loaded treats, and I must admit I enjoyed them too, with a big mug of hot coffee. One of our American colleagues referred to the doughnuts dismissively as 'fat pills', and he had a point, although George managed to stay slim all the time that I knew him.

We caught our flight on that day despite the distraction, but another ten years would pass before Europe finally did carry out a Mars mission of any kind, and one joint with NASA is still elusive. Today the European Space Agency is working with the Russians instead.

Mars Express and *Beagle 2*

When *Mars Express* was proposed to become the first European mission to Mars in 1996, it also had its origins in the Soviet Union. A few months earlier, the large and ambitious *Mars-96* mission had crashed back into the ocean shortly after take-off from Kazakhstan, taking with it a number of European instruments that were on board. I had been invited by the Russians to be a participating scientist on the mission, which would have provided privileged access to the data had there been any, so I was as shocked and saddened as any of the team at this great loss. However, ESA bounced back, and made plans to recover the European component with a much more modest spacecraft that could carry spare or rebuilt versions of the lost instruments.

The mission was assembled quickly and cheaply, hence the 'express' tag. At a meeting I explained rather pedantically to my European colleagues that 'express' does not literally mean fast or cheap, but rather it means 'specific'. An express train was one with just one destination, so it might be fast, but that wasn't what the word meant literally. I was met by blank stares and decided to let the matter drop.

Mars Express launched on 2 June 2003, around five years after it received final approval, whereas ten years would be a more typical gestation for most space missions. Probity demanded that there be a proposal and selection process for the payload, but it was railroaded through rather than built from the ground up in the usual way. I was on the selection panel and helped provide a solemn endorsement for what was certainly a good set of experiments, although largely pre-defined by their state of readiness, this being officially designated a 'flexible' mission.

Things got really interesting when it came to considering whether to restrict the project to a simple orbiter, or to include a small package to land on the planet. The additional excitement of landing appealed to everyone, and a quick study showed that it would be possible within the expected envelope of mission cost and launch vehicle capability. The cost to the agency was that involved in carrying and deploying the lander, and relaying its data; the spacecraft itself would have to be funded by the European or other nation that provided the hardware. The Russians and Americans had both developed small landers that had yet to be successfully deployed at Mars, and both offered to provide one for *Mars Express*. The third viable proposal we received was from the UK. This was the famous *Beagle-2*.

Beagle looked good: it was proposed by a strong team of excellent scientists, and backed by the superb engineering team at the Stevenage plant of British Aerospace (soon to be sold off by the short-sighted powers-that-be to EADS, the Franco-German-Spanish aerospace group, and renamed Astrium). It was also the only viable lander on offer that originated in an ESA member state, an important consideration, especially considering the fledgling state of planetary missions in Europe. There was just one problem: the British didn't want to pay for it.

Back in 1998 at the meeting in Paris to approve the payload, I was the only Brit at the table. The ESA officer in charge turned to me and asked what I thought were the chances that the UK would come up with the readies (estimated at about £50 million) if faced with the *fait accompli* of selection by ESA of *Beagle 2* to fly on the mission. I crossed my fingers, swallowed my reservations, and said something to the effect that I did not think the relevant body, the Particle Physics and Astronomy Research Council, would turn their back on such an exciting opportunity. This is what the rational part of me really thought; the realistic part of me knew it would be a struggle, but I was not going to wreck the carefully assembled case for a real European Mars mission at last, even one without MORTEX.

The rest is history. Colin Pillinger, who led the Beagle project, told the story of its genesis and fate in lurid detail in his fascinating book *My Life on Mars* before his premature death in 2013.

Mars Observer

NASA decided to return to Mars in the mid 1980s. Originally, the plan was for a low-cost mission to map the planet in more detail from an orbiter based, for economy, on an existing terrestrial weather satellite design. Dan McCleese and I developed a JPL–Oxford collaboration to propose an infrared instrument that would measure temperature, water vapour, dust, and cloud properties in the atmosphere of Mars and define how the present-day climate operates. We dropped the MORTEX design based on VORTEX in favour of something more sophisticated using pressure modulators and limb-viewing optics; basically a smaller version of the Earth-orbiting ISAMS radiometer discussed in the previous chapter. Dan would be the Principal Investigator; he named the experiment PMIRR, which stood for *Pressure Modulator Infra-Red Radiometer*.

We were delighted when PMIRR was selected for flight. It was brilliant science, and we would be able to achieve our longstanding ambition to compare weather and climate on all three Earthlike planets, using our own data from similar instruments. Also, it meant we could continue the Oxford–JPL collaboration. This was vital for my plans for the Oxford department, located in a country that did not even have a space agency or very much political enthusiasm for space, which meant limited funding. The UK government would eventually see the light on the value of having a capability in space, but not for a couple more decades. They would never have funded the total cost of PMIRR, but we did not have too much trouble getting research council support for the smaller costs of building the modulators and for setting ourselves up to work with JPL to analyse the data when it came in.

The newspapers and live media picked up the story and requested interviews and explanations. 'Scientists look at life on Mars' was a typical headline. The writer knew perfectly well that we were not looking for, let alone at, life on Mars, but apparently 'looking at weather' was not sufficiently eye-catching and the headline had to be sexed up. This sort of thing happens a lot; I have yet to understand why excellent journalists who would not dream of putting a deliberate untruth in their article have no such compunctions when it comes to the headline. Often the answer is that the headline is not written by the author, but by some harried editor who glances at the article and has to make something up quickly.

In fact, articles I have written myself have appeared with headlines that made me cringe. The worst of these was 'The Moon and Jesus', where the headline writer seemed to think that anything in space must be about the Moon, and the reference to Jesus meant my college, not our Lord. Not far behind for cringeworthiness must come 'Doctor Fred's rescue mission to Mars', which appeared in the Oxford student newspaper *Cherwell*. In the interview that led to this I had replied to yet another 'why bother?' type question with a long explanation of how understanding climate on Earth-like planets would lead to a better understanding of the climate change issues threatening life on Earth. The official *Oxford University Gazette* did rather better with 'Oxford scientists help to build Mars probe' as the front-page headline in September 1996.

OXFORD'S ROLE IN SPACE

UK scientists join hunt for ice on moon

Finding water could be the key that opens a gateway to the stars for astronauts

British equipment heads for Venus

The Pioneer spacecraft, launched to Venus from Cape Canaveral by the US National Aeronautics and Space Administration last Saturday, carries the first piece of British hardware to travel beyond the Earth's gravitational field. When the Pioneer Venus Orbiter reaches the planet in December one of the projects on board will send information about the temperature of the planet's upper atmosphere back to Earth. The infrared radiometer that will do this work was designed and built by the Jet Propulsion Laboratory with the aid of the Department of Atmospheric Physics at the University of Oxford. The Oxford team, led by Professor J. T. Houghton, produced some of the hardware that is now on its way to Venus. Fred Taylor, the leader of the project at JPL, was one of Professor Houghton's students.

The Pioneer Venus mission will be carried out by two separate spacecraft, the Orbiter and the "Multiprobe". The Orbiter will survey Venus's atmosphere and surrounding environment with 12

Oxford eyes on Venus

By LUCY TENNYSON

Galileo goes into orbit by Jupiter

'There are plenty of little chunks of ice raining on the moon all the time'

Venus probe will carry Oxford made [ha]rdware

First British mission heading for Mars

BBC NEWS

All aboard the Venus Express?

The European Space Agency (ESA) is planning its first mission to explore the planet Venus in 2005. Professor Fred Taylor of the University of Oxford, UK, tells BBC News Online why he thinks the UK should back Venus Express.

Scientists look at life on Mars

Some press clippings from the 1990s. One of the remarkable features of the last 30 years has been a gradual but substantial increase in interest in space missions by the popular media.

Mars Climate Orbiter: the first British hardware on Mars

Whether or not it betrayed an understanding of what we were actually doing, nearly all of the publicity was good. The funding flowed, so the PMIRR instrument was duly built, with the parts from Oxford nestling inside the US-built optics and electronics. After the usual rounds of testing, it was installed on the *Mars Observer* spacecraft. That summary glosses over the usual crises and delays with the instrument and the spacecraft, but these were largely Dan's worry this time. Co-investigators like I now was could just turn to one of their other projects when there were delays. *Mars Observer* ended up costing about five times the original estimate, but NASA stuck with it. To no avail, as it turned out.

In line with my reduced responsibilities and with other irons in the fire I did not travel to Florida this time but stayed at Oxford for the launch. I was also at my home base when, nine months later, the spacecraft arrived at Mars after a flawless flight. I was able to keep track of everything using the communication links we had with NASA and that we intended to use for accessing the data. But there wasn't any data. When the time came to fire the rocket motor to guide *Mars Observer* into orbit around the Red Planet, the communications downlink from Mars to Earth went dead.

Unlike the problem we had when VORTEX stopped communicating from Venus, this time there was no transmission at all from any part of the spacecraft. Even the carrier link was gone, which indicated either that the transmitter had failed, or that the antenna was no longer pointing towards the Earth. This crisis had come shortly after the command was sent to pump liquid fuel into the rocket motor, prior to firing it to send the spacecraft into Mars orbit. Apparently, something had gone wrong in this process and in the post-mortem analysis the experts concluded that the fuel may have exploded, fatally damaging the whole spacecraft. The wreckage drifted on past Mars and into a distant orbit around the Sun, taking our dreams with it.

All was not lost however. NASA announced quite soon that it would commission copies of the *Mars Observer* instruments and fly them on a new spacecraft. This was the era when the agency had come under a dynamic new chief, Daniel Goldin, who felt a fresh approach was needed. He ordered a new focus on smaller missions, to be carried out more quickly and for much less money—the so-called 'faster, better, cheaper' mantra. The *Mars Observer* payload was duly split into three, and assigned to three different missions, all using much smaller spacecraft and cheaper launch vehicles than *Mars Observer* had.

So, by 1998 we were off to Mars again with *Mars Climate Orbiter*. This time I did go to the launch, which was on a fairly modest *Delta II* rocket rather than the mighty *Titan III* that had been used for *Mars Observer*. The spacecraft was so much smaller that PMIRR was actually the only science instrument on board, except for a small camera; we virtually had the mission to ourselves, which was wonderful. The investigators on multi-instrumented spacecraft are always fighting over what the orbit should be and who gets the lion's share of scarce resources like mass, power, and data rate. The launch and flight to Mars went so well that I relaxed and, instead of crouching by the monitor as before, when the time for the encounter with the planet came I went off to hike in the familiar and beautiful country around the Cheviot Hills in Northumberland. As I wasn't Principal Investigator for this experiment, I had no formal duties during orbit insertion until the data started to flow the following week, so I took a break before the action started.

I stayed with an old friend from my schooldays in Alnwick, Tony Danby, who had looked after my old cars when I moved to America in 1970. Tony had moved on from Ilderton Station to a new home in Berwick upon Tweed, almost right on the border with Scotland. When the two of us came back from our hike, I expected to have to make phone calls to hear how it all went, but instead we found Tony's very elderly father sitting by the phone with a pile of scrawled messages. He hadn't a clue what the press men who kept calling had been saying, but he managed to write down some of their numbers so I could call them back. I had done a lot of media interviews before I left and was not expecting this kind of attention again until after we had looked at the first data. With a sinking feeling, I called Oxford first and sure enough, we had had another failure. Instead of going into orbit, the spacecraft had crashed onto the surface of Mars.

The reason, to the delight of the media who love a good disaster, especially one clearly attributable to human error, turned out to be confusion between imperial and metric units between the contractor that had designed and built the spacecraft and the NASA mission control engineers. The difference was small enough to evade obvious detection, but large enough to cause the orbiter to miss the narrow window between avoiding the planet altogether on one hand and crashing into it on the other. Thus we lost the second PMIRR instrument as well as the first. I couldn't believe it. Losing one instrument is a tragedy, but losing two begins to look like carelessness. The waggish part of the JPL notice board said 'Faster, cheaper, better—two out of three ain't bad.'

It was scant comfort when we realized that our Oxford-built modulators were sufficiently robust to have survived the crash, and were now strewn about on the surface of Mars where some future explorer might find them. This was the first British hardware on the surface of Mars, more than three years before *Beagle 2*. In years to come, when we were both at least superficially emotionally recovered from our respective losses, I would joke publicly with Colin Pillinger that I crashed on Mars before he crashed on Mars.

Scientifically, we were high and dry with still no data from the Red Planet at all. However, we had been busy developing tools for data analysis that included an advanced general circulation model for Mars that was interesting and useful, even without the new measurements that we had planned to compare with the model's predictions. A small team at Oxford under Peter Read and Stephen Lewis built up the Mars model by modifying the software developed by the Meteorological Office for forecasting the Earth's weather. The latest technique involved feeding the measurements straight into the model as they arrived, so the computer could make instant comparisons between the forecast and reality. The model would modify its predictions when it saw the data and generate diagnostics that helped the scientists to improve the model.

Groups in France and Spain joined in the modelling effort and the European Space Agency came up with funding, looking ahead to the wind and dust forecasts they would need when their plans to land on Mars came to fruition in a decade's time. And of course, we would need the model for data analysis when we finally got a working instrument into Mars orbit, which we were still determined to do someday.

Still California dreaming

Because of my ten years of employment at JPL, and with an ongoing involvement through the Mars programme that had me travelling there frequently, the lab offered me an enviable position on their Distinguished Visiting Scientist programme. This meant an office, a parking space (very important), and a budget that I could use to visit whenever I chose. After the original flurry on the PMIRR experiment, I settled down to regular visits of two weeks each, three times a year. I didn't give up this delightful arrangement until I retired from my chair at Oxford in 2011, so I remained an honorary Californian the rest of my working life.

Sometimes I shared an office with visiting scientists far more distinguished than I, such as Jacques Blamont, the chief scientist of the French space agency CNES, Richard Goody from Harvard University, and Wes Huntress, who for a time ran the planetary programme at NASA Headquarters. Goody, in particular, was known to me since student days as a legend in the field of atmospheric physics, and his 1964 textbook *Atmospheric Radiation* had guided my early efforts to get my head around the idea of the structure and behaviour of the atmosphere as something that could be understood using the laws of physics. Ilias Vardavas and I used that book as a model and a starting point when we wrote *Radiation and Climate* nearly 50 years later.

In his own biographical writings, Goody notes that he was intrigued, as early as the 1940s and 1950s, by the fact that the climate, a structure of immense and dynamic complexity, nevertheless appeared to have stable statistics. In his efforts to understand why, he laid the foundations for the remote sensing techniques based on the inversion of spectroscopic data that were taken up by others, eventually including myself. In the 1960s he even built a pressure-modulated radiometer with some features similar to the one I was struggling with at about the same time, but by then he had moved to Harvard and I did not find out about his work until much later. In any case, it was a massive device designed to work from the ground looking up and would not have been suitable for mounting on a satellite.

When I was young I little thought that I would share an office with such an iconic individual, although I had met Richard a few times, particularly at meetings of the *Pioneer Venus* project of which he was one of the architects, and I had also visited him in his beautiful retirement home on Cape Cod, when he entertained a group of us who were participating in a meeting of the NASA–ESA Joint Working Group at the nearby Woods Hole Oceanographic Research Institute. Later he visited us in Oxford while I was in charge, sometimes for extended periods, and we enjoyed many discussions. There was also one traumatic episode, which occurred during one of our departmental seminars.

In those days our meeting room was furnished with cheap chairs, made of a bent metal tubular frame to which was attached a moulded plastic seat. Goody was rocking back and forward on his as he listened, and just as the speaker was reaching the climax of his discourse there was suddenly a mighty bang as the plastic seat separated from the frame and the great man went flying onto the floor. Fortunately he was fine and took the whole thing in good spirit. I ordered some better chairs.

10
Operation Saturn

As a ten-year-old schoolboy my favourite reading was a comic for boys called *The Eagle*, which featured on its front page the exploits of Dan Dare, Pilot of the Future. This intelligent and beautifully drawn serial broke down many barriers, not least in that it had a multicultural cast that, like the storylines, was way ahead of its time (and ground-breaking, historians of *Star Trek* please note). Dan himself was an educated Englishman and Academy graduate, who in an earlier era would have been a Spitfire pilot, but his assistant Digby was a blunt-speaking, working class northerner and his Navy buddy, Lex O'Malley, was archetypically Irish. The two smartest members of the crew were a woman and an extraterrestrial, and the most skilled were a Frenchman and an American. Dan led from the front and was a pillar of virtue and courage.

In the April 1954 edition Dan and his crew are on a mission to Saturn, set in about 2005, and have just arrived into orbit around the planet, close to the rings. Here, the formidable and beautiful female scientist Professor Peabody (in whom Dan never showed any hint of romantic interest, although she was a valued colleague) asks 'Why is it so hot suddenly?' Another member of the team, Pierre, who is French in case his accent leaves any doubt, replies 'Perhaps eet ees zat ze spaceship ees on fire!' This elicits a response from Sondar, a wise and gentle member of the otherwise fierce Treen population of Venus: 'No, O Lafayette! The great hotness arises only from our nearness to Saturn. Saturn's rings are made up of a revolving mass of zesto particles—they give out a heat nearly as great as the Sun's. Any solid body which entered one of the rings would melt and vanish in a split second!'

No one knew then what Saturn's rings were made of, and Frank Hampson who wrote and drew the strip was noted for a realistic side to his imaginings. I remember wondering therefore if Sondar was citing a real possibility. (In case anyone reading this is still wondering, there are no such things as zesto particles, and the rings are made of chunks of very cold ice.) I also wondered if I would ever get to study the rings myself, and amazingly, eventually I did.

Perhaps even more than Mars, mighty Saturn, with its great size, beautiful girdle, and extensive family of moons, is everyone's idea of the archetypical planet. Lying far off in the outer Solar System, ten times further from the Sun than is the Earth, Saturn was not at first a realistic target for a large, fully equipped spacecraft like those now regularly flying to Venus and Mars. NASA had sent a small 'pathfinder' spacecraft, *Pioneer 11*, in 1973, to arrive at Saturn in 1979. This showed that it was possible to fly safely through the obstacle-strewn asteroid belt, and close to Saturn's rings without wiping out, but *Pioneer* had only basic instrumentation and took relatively little science data. It was followed by *Voyager* in 1980, a larger spacecraft with better instruments, but still on a fast fly-by trajectory that afforded little more than a glimpse of Saturn and its rings and moons. A more powerful mission combining a Saturn orbiter and an atmospheric entry probe began to seem feasible after the success of *Galileo* at Jupiter, and the prospect finally became real in October 1997 when *Cassini–Huygens* was launched.

Cassini was the name given to the Saturn orbiter, which was developed by NASA, while *Huygens* was a European-built probe which would ride with the orbiter to Saturn and then be separated to descend onto the surface of Titan, the planet's largest and, in every sense, most atmospheric moon. The international approach to the mission had its origins in 1984 when the US–European Joint Working Group, which featured in the previous chapter in connection with its proposals for Mars, also made its recommendations for missions to the outer planets, naming Saturn, and especially Titan, as the most exciting targets, and proposing an orbiter–probe combination. Saturn is twice as far from the Sun as Jupiter and it was clear that the mission would be challenging and expensive, so international collaboration and cost sharing were desirable and even essential.

Saturn backlit by the Sun: a view never before seen by human eyes, as it requires a vantage point beyond the orbit of the giant ringed planet. The Sun, and the Earth, are behind Saturn's disc in this picture taken by the camera on Cassini as the spacecraft passed through the planet's shadow on 4 September 2011. The CIRS experiment we had helped to build at Oxford is also on board, measuring the temperature and composition of Saturn's atmosphere.

The joint study actually recommended that the European Space Agency, ESA, should provide the orbiter and the more experienced NASA the Titan lander, which we considered to be a more complex challenge. However, when the agencies did a more detailed technical study, reasons were uncovered why it would be better to reverse the roles. The arguments for this included the fact that the orbiter would need nuclear isotope power sources that were not available in Europe. It was also found that the orbiter was a lot more expensive than the probe, and the Americans had more money to devote to the mission. Although not an official reason, another factor was that the Americans thought the Europeans might not stay the course, and it would be possible to fly the orbiter without the probe if they had to, but not the other way round. ESA did not worry so much about NASA defaulting, although they should have done because they nearly did, cancelling *Cassini* at one point before relenting and quickly restoring it. The European engineers relished the challenge of working out how to land on Titan more than they did the relatively mundane task of building a satellite, even one that was going to orbit Saturn. They named the probe *Huygens*, after the Dutch astronomer who discovered Titan in 1655. Giovanni Cassini (1625–1712), after whom the orbiter was named, was the first director of the Paris Observatory, where he studied Saturn and discovered several of its other moons, Iapetus, Rhea, Tethys, and Dione.

If you can't beat them, join them

Cassini–Huygens was not yet an approved mission, and it faced stiff competition for funds on both sides of the Atlantic before it would get the go-ahead to fly. It was very clear that obtaining and maintaining political support for what was obviously going to be a very long and expensive project, costing several billions of dollars, was fraught with risk. Joint planetary projects with NASA had been proposed before, and indeed since, and nearly always fallen through sooner or later. At JPL and in my interactions with colleagues among the planetary scientists in Europe I kept in touch with the campaign to develop the mission concept and advance its case for flight. I liked the science but at first I was not really enthusiastic about the mission's chances.

There were two main reasons for this, in addition to the high cost and the risky politics. The first was that studies had shown that to get such a large spacecraft all the way to Saturn, even with the most powerful launch vehicle available, would take a very long travel time, more than seven years. This not only meant the chances of a technical failure while on the way were high, possibly losing everything, but also that other new planetary mission opportunities were likely to be held back for a whole decade while this one was in progress. The second reason was that I was heavily committed to a different project that was competing within the ESA committees for the same block of funding.

My rival interest, the *Kepler* mission to Mars, could be carried out for a small fraction of the projected cost of *Cassini–Huygens,* and was a simpler machine with a much better chance of succeeding without technical problems. When the two projects came head-to-head in Paris at the gathering to choose the next ESA mission, I was scheduled to speak on behalf of the Mars mission. *Kepler* would get to Mars in just seven months, would be all-European, and plainly made much more sense as a choice for the fledgling European agency's first planetary mission than a wild and risky trip to distant Saturn. However, common sense does not always apply when ESA selects its projects; it goes mainly by popular acclaim instead, and *Cassini* had a couple of aces in the hole.

One of these aces was cost sharing. NASA was keen to go ahead with *Cassini*, and said it would cover the lion's share of the cost. *Kepler* was an all-European mission and ESA would have to pay for all of it. This greatly reduced our cost advantage. In fact, *Huygens* on its own, with a free ride to Titan from NASA, was arguably cheaper than *Kepler*.

The other ace was Titan itself. The results from *Voyager* had confirmed that this was a moon with a thick and remarkably Earth-like atmosphere. There was even talk, if not of life itself, of 'pre-biotic activity' on Titan. This rather meaningless expression, along with tantalizing glimpses of a surface covered with remarkable features including possible oceans, combined to raise interest in Titan as an irresistible target for further study. The *Cassini* proposers described how they had designed the orbit around Saturn to make multiple close encounters of Titan possible, and how instrumentation much more advanced than *Voyager's* would reveal Titan's secrets. To too many people, *Kepler* looked tame by comparison and *Cassini* duly won the competition.

To me and to many others *Cassini* still seemed like a massive gamble, and it was. International partnerships on this sort of scale are fraught with problems, and the huge distances to be travelled in space meant that almost unprecedented levels of reliability would have to be built in to the design, and of course the technology would be obsolete by the time it got there nearly a decade after launch. With the mission now approved, all these reservations were cast aside, by me as much as anyone else, in the scramble to join the teams that would participate by providing scientific instruments.

Our Oxford group had a head start, with our participation in the Joint Working Group, our experience in the outer Solar System with *Galileo*, and our expertise in infrared spectroscopy and remote sensing that would surely be a requirement on the mission. The biggest problem was the usual one that the top-level paymasters in Britain were more parsimonious than their counterparts in the other large European nations, especially France, and they did not see an imaginative space project like this as a fit way to spend large amounts of taxpayers' money. If we were going to build something at Oxford to go to Saturn, it would have to be as part of an international consortium with costs shared and only a small British financial contribution.

We looked to our friends in the French group at Paris Observatory that was pushing *Cassini* hard from the European side. In America, the leading group intending to propose an infrared experiment for the mission was the same one at Goddard that we had worked with on *Nimbus*, where their spectrometers sat alongside our radiometers. They were also the team that had beaten our JPL group for the opportunity to fly on *Voyager*, so they already had valuable experience with a Saturn mission. The original leader of the group, Rudy Hanel, had retired, and the Goddard proposal for *Cassini* was being led by his younger colleague Virgil Kunde, with whom I was on friendly terms. Pretty soon we were part of a US–French–UK consortium proposing a huge instrument called the *Composite Infrared Spectrometer*, known as CIRS.

After the usual tooth-and-nail competition with other instrument teams, CIRS was selected for flight, with Virgil as the Principal Investigator and myself as one of three dozen co-investigators. We negotiated for Oxford to provide the cryogenic system for the cooled detectors, a difficult but essential component of any infrared instrument seeking high sensitivity when observing very cold objects such as Saturn and Titan. Simon Calcutt led the design of a system based on a passive radiator, that used views to cold space to produce temperatures as low as that of liquid nitrogen inside the instrument, without carrying any cryogen, and without using mechanical refrigerators like those that were beginning to be employed by ourselves and others in Earth orbit. This not only saved weight and power, but also gave us the reliability that we would need on a 20-year-long mission.

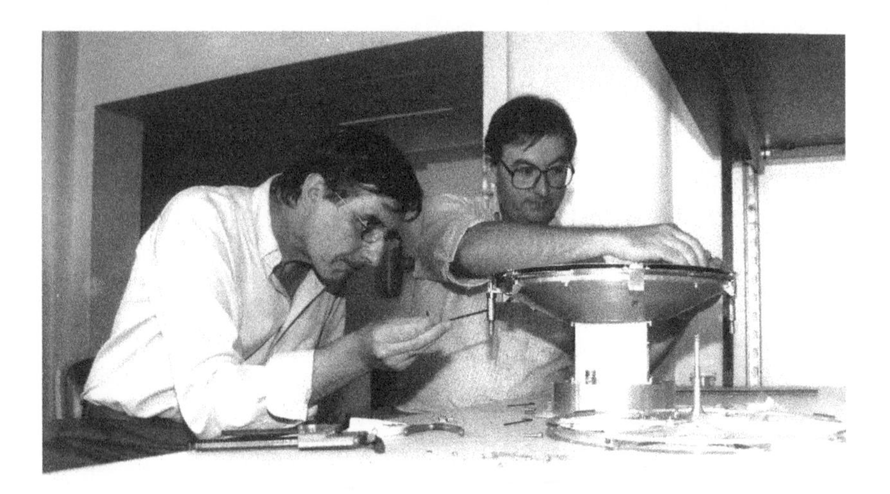

Simon Calcutt and myself with the cooler for the CIRS instrument under development in our lab at Oxford for NASA's *Cassini* mission to Saturn. More than twenty years later, it is still operating perfectly in orbit around the giant ringed planet, an achievement for which most of the credit belongs to Simon. Photograph by George Retszetter, by kind permission of Oxford Mail/Times (Newsquest Oxfordshire).

Orbiting Saturn

Cassini launched on 15 October 1997 and went into orbit around Saturn on 1 July 2004, having a few weeks earlier released the *Huygens* probe on a trajectory to land on Titan, which it did successfully on 14 January 2005. The CIRS instrument was turned on after a long hibernation and the cover was released on the Oxford-built cooler; the detectors obligingly cooled down to the desired operating temperature. We were in business.

The superb rings aside, Saturn is a smaller and colder version of Jupiter, and having close-up observations of two examples makes it a lot easier to understand the very deep, very dynamic atmospheres on the outer planets. With no entry probe for Saturn, *Cassini* was limited to studies of the highest part of the atmosphere above the clouds, and the waves and storms that show up most clearly in the clouds themselves. Noteworthy examples of meteorological activity include the weird pattern around the north pole which forms a near-perfect giant hexagon. *Cassini* observations of this, and experiments with tanks of rotating fluid in the lab at Oxford, led Peter Read and his students to conclude that it is caused by the nonlinear equilibration of barotropically unstable zonal jets (if you are not a dynamical meteorologist, don't worry about that description, few people really understand barotropic instabilities in any depth; I don't).

Equally exciting was the monster storm that erupted on Saturn in late 2010. Apart from its great size, big enough to swallow the Earth, and its long duration of several years, the interesting thing about this was the way it changed the composition of the clouds by stirring up

The *Composite Infrared Spectrometer* (CIRS) instrument in the laboratory at NASA's Goddard Space Flight Center prior to launch. The Oxford-built cooler is inside the conical structure to the left, and the ghostly figure standing behind the table is the science team leader, Dr Virgil Kunde.

Dr Athena Coustenis, my collaborator for studies of Titan, Saturn's huge moon, and co-author of a book on the subject, working at the telescope of the Paris Observatory. Photo by O. Borderie.

material from the warmer atmosphere deep below, including a lot of white material that the spectrometer identified as water ice.

Those glorious rings around Saturn are made up of millions of blocks of ice, with sizes ranging from dust grains to chunks as big as a house. The ice seems to be nearly pure water, with just traces of various unidentified contaminants. Like so many materials in the Solar System, we will not know in detail what the composition is until we can obtain samples and analyse them in the laboratory. That could be back on Earth; missions that would roam around inside the rings, grab some of the material and fly it back to Earth have been studied but so far not seriously considered for funding as a real mission. A manned *Ring Explorer* is a particularly attractive concept, one that would orbit Saturn, moving in and out of the different rings and analysing samples in a laboratory carried on the spacecraft. The views will be sensational. At the present rate of investment in planetary exploration that is an awfully long way off in the future, however.

The Sirens of Titan

The rings are glamorous but the real jewel of the Saturnian system, offering for many of us the main scientific rewards from CIRS and *Cassini*, has always been the planet's huge satellite, Titan. We knew long before arriving that Titan was more like a planet than a moon, because of its size, which is larger than Mercury, and its cloudy atmosphere. The *Huygens* probe operated smoothly, and confirmed that Titan's atmosphere is even more substantial than ours on Earth, with a surface pressure about 50% higher than here, and with the same main atmospheric constituent, gaseous nitrogen.

Observing from above with *Cassini* we could see structure in the clouds and haze, and an interesting distribution of minor constituents, all shifting as Titan orbits Saturn and Saturn orbits the Sun. The new data revealed a very cold but remarkably Earth-like world, with changeable weather and a hydrological cycle based, not on water (since all of the water on Titan, and there is a lot of it, is frozen as hard as steel), but on methane. We call it marsh gas on Earth, but on icy Titan this simple hydrocarbon condenses to form clouds and rain, which runs in rivers on the surface and gathers in lakes which come and go with the seasons. More complex molecules, possibly including simple amino acids, form in the upper atmosphere by photochemical reactions involving methane and traces of hydrogen cyanide, water, and carbon dioxide. These condense to form the orange haze that covers Titan's disc at all times.

Water and methane clouds can evaporate again, but the hydrocarbon haze probably does not. Most likely, it showers down onto the surface and seeps into it through cracks and fissures. We would dearly love to get our hands on some of it and examine its composition. We would also like to understand where the methane supply comes from. There must be one, because the haze-forming processes remove it from the atmosphere at what seems to be a prodigious rate. We picture a surface coated with oily drizzle from the haze, studded with 'cryovolcanoes' belching methane and other gases, and washed from time to time with methane and ethane rain. Deep below the surface, theoretical models of the interior tell us to expect to find a subterranean sea of liquid water where the pressure is high enough to liquefy the ice, which probably contains dissolved ammonia and other impurities to help the melting process.

Chilly conditions and high winds prevail at the surface, where the temperature is only a few degrees warmer than the freezing point of nitrogen, that is, of the air itself; if it got much colder the entire atmosphere would freeze out on the surface. Something similar to that has already happened on Triton, the large satellite of Neptune, which lies much further away from the Sun and therefore is even colder. It would be a bizarre form of life that could exist under such conditions, and in ten years of close-up investigation we have seen no trace of any.

I could go on at great length about the landscapes, the dynamic climate, and the remarkable seasonal changes on Titan, but I have already written in full elsewhere about the adventures we enjoyed with *Cassini*, particularly with regard to what was discovered about Titan. My co-author is a French–Greek colleague, Athena Coustenis, who is one of the world's experts on Titan and has published extensively on the subject in scientific journals and at meetings. Some of these we did together, including some review-type articles for the general reader, before deciding that Titan is so interesting it was worth a whole book on the subject.

In fact, we have now written two books, and have started to discuss a third, each one an update of the one before. This came about as the result of an invitation I received from World Scientific Publishing Company in 1994 to become the Editor in Chief of a book series on Atmospheric, Oceanographic, and Planetary Physics. They were, I soon discovered, the largest publisher of scientific books in the world, based in Singapore but with offices in London and elsewhere. However, I also discovered that I was supposed to find suitable topics and persuade the appropriate authors to write about them. The easiest place to start was at home with the volume that Athena and I were already considering.

The theme of our first book was what we knew about Titan before *Cassini* arrived. We would not wait for the new data, still many years away, but decided to write about the mysteries we hoped to address when the big new spacecraft and its load of powerful instruments finally arrived in 2004. Then when we finally had our hands on the new data, and insight into the many discoveries and clarifications we had anticipated for so long, we would write it all again. This we duly did. Between those two books, my Oxford colleague Clive Rodgers wrote a book about so-called retrieval theory—the art of converting what you can measure, such as infrared spectra, into what you want to know, such as atmospheric composition and temperature profiles. Next, my Spanish collaborator Manuel Lopez-Puertas and I worked on a book about atmospheric radiation under non-equilibrium conditions, such as those that exist in the upper atmosphere, a highly rarefied topic in more senses than one but rich in interesting advanced physics. The new Titan book, with the first results from Cassini, then became the fourth volume in the series.

At about the time that Athena and I finished the second edition of our book, World Scientific made a deal with Imperial College that turned their London headquarters into Imperial College Press. My contacts and everything else changed, and it seemed like a good time to wrap up the book series with just those four volumes. However, the question of another, third, edition of the Titan book is still in the air. We vowed we would write it when the mission was over and the last data had been studied. There would not be any more until the next Titan mission, which is still not defined but must be at least two, and probably more like three, decades away, or even longer. At the moment, *Cassini* is expected to continue until 2017, so we might write 'Titan III' by 2020 or thereabouts.

The Sword of Truth

Writing a book involves spending extended periods at each other's institutions. Athena was based, not in Cassini's old observatory in the centre of Paris but at its 'new' home (since 1877) on the site of the old Chateau de Meudon built by Louis XIV on a hilltop overlooking the great city. For light relief between writing sessions we would sometimes discuss our common interest in speculative fiction, and at one of our get-togethers I brought along a copy of a book by a new writer, Terry Goodkind, that I had picked up on a trip to the USA. The title, *Wizard's First Rule*, explains enough about the sort of story it was, actually fantasy not fiction, but we found it to be cleverly and imaginatively written, so we both became fans. Goodkind went on to produce more than a dozen sequels, all of which we bought and shared as soon as they came out in hardback.

On one memorable occasion in the summer of 2003, I was working at JPL and spotted in the *Pasadena Star News* a notice that Terry Goodkind would be visiting Vroman's bookshop on Colorado Boulevard in person the next day, for a signing session on the latest instalment in his sword and sorcery series. I resolved to go and meet him, buy his new book, *Naked Empire*, and get it signed by the author. I wrote to Athena, who was in Japan at the time, 'It's silly, but having spent so much time reading his nonsense I want to see how crazy he really is!' I duly queued for over two hours in a long line that extended outside into the car park behind Vroman's, enjoying the chat with the other sci-fi nerds. There wasn't time to talk to him, but Terry duly signed my book, writing also 'Talga Vassternich'. This turns out, when you read the book, to be the Wizard's Eighth Rule, in the High D'Haran language, which translates as *deserve victory*. The book explains this as: 'Be justified in your convictions. Be completely committed. Earn what you want and need rather than waiting for others to give you what you desire.' Great stuff. We lapped it up.

Artist Kees Veenenbos produced this gorgeous impression of the surface of Titan specially for the cover of our book, 'Titan: Exploring an Earth-like World', published in 2008. In it, we described The exploration of Titan from its discovery to the latest results from the Cassini mission, which observed Titan in detail during several close encounters while in its orbit around Saturn.

World Scientific Publishing made a thousand copies of this beautiful 3-D card to publicize our Titan book. Unfortunately, it had a similar colour and design to a rival book on the same subject and so we were reluctant to give them out. The publisher was disappointed that we were not more enthusiastic after all the trouble they had taken. The globe in the background shows the topography on Venus as measured by the *Pioneer* orbiter discussed in Chapter 4.

Travelling the World

Involvement in space research inevitably includes a commitment to a lot of long-distance travel. Each project, especially the protracted international ventures like *Cassini* will, from day one, involve attendance at team gatherings, conferences, meetings to take care of essential space agency business, or just to work on a problem or task with distant colleagues. As Principal Investigator, one has almost back-to-back meetings here, there, and everywhere as a matter of course, and often almost no choice whether to go or not. Getting up early and setting off for the airport every fortnight or so soon becomes part of your lifestyle. As a Co-Investigator, however, one has much more freedom to choose, having only the broadest of obligations to attend as many meetings as possible.

Travel is often triggered by invitations to speak at conferences, sometimes to give a 'keynote', which is a long introductory talk that sets the scene for a few invited talks and then a larger number of shorter, contributed talks that goes on typically for the next couple of days. An invitation to give a keynote talk confers prestige and is not to be missed. The invited talk is less exalted, but still an indication that the meeting organizers think you have hot new data, or something interesting to say, or a stylish way of saying it. Most of any session will be made up of the more mundane 'contributed' talks, where the speaker applies for permission to speak and may be selected or rejected. Bottom of the barrel, and worse than rejection, is being told that your intended talk is accepted only for poster presentation. This condemns you to print out your slides and stick them up on a board somewhere in a remote basement of the conference centre, there to be largely ignored while you hover around hoping someone will ask you to explain them.

The author David Lodge, in one of his satirical campus novels (*Small World*, 1984) wrote that 'the modern conference resembles the pilgrimage of medieval Christendom in that it allows the participants to indulge themselves in all the pleasures and diversions of travel while appearing to be austerely bent on self-improvement.' He was being funny, of course, but a lot of people think there is a germ of truth in this view. Some of these cynics are clearly at NASA Headquarters, where they have been known to ban travel using the funds they administer if they deem the destination to be too attractive. They also compose rules that say things like the duration of the trip must not exceed that of the meeting itself plus one day. This might suggest that business travel is fun, when your business is science, but once you get over the initial buzz that comes from being involved and in demand it rarely is.

The reality much of the time is an uninspiring story of public transport, budget hotels, and taxi rides with barely a glimpse of the attractive features of whichever city it is this time. Space Agency meetings generally take place in a characterless, windowless room that is either too hot or too cold, and always conclude with a mad rush back to the airport. There are exceptions, certainly; and meetings would sometimes be hosted in beautiful locations. But despite what NASA HQ might suspect, I know as an occasional organizer myself that the choice of an attractive venue may not be for the delight of the participating scientists, although for certain kinds of conferences that have to compete for participants it does help to ensure a good attendance. Instead, it is because a meeting in a tourist destination is more likely to enjoy good communications, cheap flights, and plenty of hotels with rooms that are available at a discount out of season. Thus, in the winter we might end up in beach communities like the Florida Keys, for example, when it is too cold to swim, and in the summer in a ski resort in the Alps, when there is no snow.

As part of an international project with participants from a wide range of countries, the *Cassini* and CIRS meetings had a particularly good chance of ending up in varied and often interesting locations. Often they were simply in one of our home bases, most often the institution of the Principal Investigator. By the time *Cassini* arrived at Saturn and started to return scientific data, Virgil Kunde had retired and the leadership of the experiment passed on to Mike Flasar, an atmospheric scientist like myself. Mike was a colleague of Virgil at the Goddard Space Flight Center, the big NASA facility in Greenbelt, Maryland, just outside Washington DC. This was the same place I had visited so often in connection with the *Nimbus* and *Upper Atmosphere Research Satellite* projects described in other chapters, and where I had been unceremoniously evacuated on 9/11 along with everyone else on the site.

Mike ran the CIRS experiment in exemplary fashion, including efficient team meetings, and he liked to bind and inspire the team by holding some of them in his co-investigator's countries. This was made easier by the fact that the *Cassini* Project Office, containing the team responsible for operating the spacecraft and making key decisions about orbits and observing sequences, had many international partners and would itself need to hold foreign meetings. The NASA Project Scientist for *Cassini* was my JPL colleague Dennis Matson, who excelled at the job. Dennis had an additional talent which was that he could sniff out the nearest cocktail bar and the best restaurant in any city in the world within hours of arriving.

Mike would hitch the smaller CIRS meetings onto those larger gatherings so that we could all attend both while limiting the amount of tedious long-distance air travel required. Once the scientific results began to flow, the emphasis shifted away from team meetings in Greenbelt, Paris, and Oxford and more towards international conferences where we could discharge the team's duty to share the excitement with the wider scientific community. Then, we would end up in niche locations like Oahu or Venice, large cities like San Francisco and Vienna, or resorts like Nice or Estoril with their large conference centres and lots of convenient hotels.

In March 2005 the *Cassini* team met in Florence. I had never been to the birthplace of the Renaissance before, and decided that I would find a quiet morning during the meeting week to slip out and fulfil a long-held wish to see Michelangelo's statue of David. It wasn't far to the Galleria dell' Accademia from the meeting location but I had been warned to expect massive queues to get in, even at that time of year, so I went early. The entrance is on an unremarkable street and when I found myself alone there it was so quiet that I wondered if I could be at the

The author sitting in the front row at a typical CIRS Team Meeting at Goddard Space Flight Center near Washington, DC, in October, 1998.

right place. But I walked in, paid, and entered the gallery to find David looming over me; I hadn't realized from pictures that the statue was so large. There was no one else there at all, I was totally alone. For about 40 minutes I examined the enormous masterpiece and associated displays in a solitary trance, until suddenly a party of noisy school children arrived and the magic was dispersed.

If this seems like an example of meetings as free vacations it should also be said that most of each day was taken up by hard work. At the time of the Florence meeting, *Cassini* had been in orbit just a few months, and *Huygens* had been on the surface of Titan for just a few weeks, so excitement was high as we pored over the early data. The schedule was packed, and I was able to have my adventure with Michelangelo only by skipping the discussions of the data on magnetic fields in the Saturn system. That data consists of long plots of wiggly lines with the occasional glitch due to some phenomenon or other, which the presenter would earnestly describe. Even the magnetometer PI prefaced her talk at a more recent meeting by saying that no one outside her immediate team was interested in magnetic field data, to general murmurs of agreement. That topic is well outside the area of Saturn's and Titan's atmospheres that I worked on, so taking a break left me fresher for the real work in the afternoon and the informal planning sessions in the evening. Honest, guv.

The Millennium Mission

Cassini, and its payload of scientific instruments including the Composite Infrared Spectrometer, is still operating as I write. We recently celebrated more than ten years in orbit and are prepared to keep going for several more years, until finally the fuel runs out in 2017. Missions that last this long cost a lot of extra money on top of that originally approved, mainly because of the cost of the manpower required to track and control the spacecraft, and of course to collect and distribute the data. The scientists working on analysing and publishing the data, including attendance at the aforementioned meetings, need resources too.

The way it usually works is for the agency, NASA in this case, to put out a fresh call for proposals from the team to continue for an addition period of, say, three more years. This has to show that there is new science that hasn't been done yet, and that the spacecraft and instruments are still healthy enough to address these additional objectives, possibly with a new orbit and additional scientific participants. This is then weighed up against the cost and usually approved, sometimes badged as a separate mission. Cassini is now on its second extension, called the Solstice Mission, and for this the main objectives centre around the seasonal changes that have been observed on Saturn and the desire to follow these until the planet reaches solstice. Then it will be midwinter in the Northern Hemisphere on Saturn, something that comes around only every 15 years, and is next due in late 2017.

Another driver is to better understand the weather and its manifestations in surface features on Titan. Monsoon-like rainstorms fill the rivers and lakes at certain times of the year, and at other times they dry out. Erosion has produced all sorts of interesting structures, including continental plateaus, sinuous valleys, and wind-blown dunes. The processes and timescales involved in their formation are hard to follow, because the *Cassini* observations of Titan are a series of short snapshots. The spacecraft approaches close to just one of Saturn's moons each orbit, for a relatively brief fly-by, and each orbit takes about two months. So we have the same problem of interpreting fleeting glimses when probing the other interesting moons, like Enceladus with icy jets escaping into space from its subterranean ocean; Iapetus with its two-tone colouration and a massive ridge nearly ten miles high running around the equator; and Mimas with one side dominated by the huge crater *Herschel*, formed by an impact that nearly tore the whole moon apart. With these and countless other features and phenomena to explore, it becomes vital to accumulate as many orbits as possible before *Cassini* runs out of fuel to navigate and is abandoned to plunge into the atmosphere of Saturn. As of now, we're still going.

11
The Days of the Comet

He gazed in a sort of rapture upon that quivering little smudge of light among the shining pin-points .... "Wonderful," he sighed, and then as though his first emphasis did not satisfy him, "Wonderful!". ... this scarce-visible intruder was to be one of the largest comets this world had ever seen .... The spectroscope was already sounding its chemical secrets ...

H.G. Wells (1906)

A Meeting in the Mountains

Long ago, in March of 1984, I was invited to a meeting in the beautiful Swiss ski resort of St Moritz to help plan an ambitious new space mission. As the little train wound its way up the mountain, my thoughts of the opportunities and challenges ahead in space were mixed with wonder for the engineers that had built a railway in such a spectacularly difficult environment. When, five hours after leaving Zurich and following a change of train to the mountain railway, we eventually reached the plateau that held the small jewel of a town and its station, the snow was lying thickly and crisply on the ground. I, and colleagues I had met on the journey, destined for the same meeting, wended the short distance to the palatial Kulm Hotel which was to be our home for the next few days.

I felt privileged to be there, not just because it was unmistakably a playground of the rich (someone said the Aga Khan was staying in our hotel, but if so I didn't see him), but also because the invitees were mostly scientific elder statesmen considerably more distinguished than I. The purpose of the meeting was to recommend a strategy for new missions to the European Space Agency, ESA, which had marked its coming of age as a big-budget multinational organization by devising a long-term plan which it called *Horizon 2000.* The plan outlined when the budget would allow small, medium, and large (called 'cornerstone') space projects to be carried out during the next several decades, but did not specify what missions they should be. That was to be our job.

The attractive location was selected, not just to make sure the bigwigs actually came, but also to provide a secluded venue, far from everyday activities and concerns, where a retreat-type atmosphere would apply. In such a setting, a focussed debate could be carried out and difficult decisions taken. The crucial part was choosing a candidate for the next cornerstone opportunity: Europe could afford these big, expensive, challenging missions at a rate of only about one per decade. I had done my homework, and was ready to advocate three stunners: a mission to the outer planets; a Mars rover; and a probe to land on Venus. Surely one of those would get the nod.

I soon discovered that working with the great and the good in my field was a mixed blessing. Unlike relative youngsters like myself, the elderly German and French professors who dominated the meeting had careers that predated the space age, and had learned their trade by observing through telescopes, theorizing or, to a remarkable degree, by analysing meteorites. The museums of Europe are full of these messengers from space that fell to Earth. Most of them are fragments from the comets that periodically sweep through the inner Solar System, crossing the orbit of the Earth. There is no denying that analysing them in the laboratory has taught us volumes about the origin and history of the Solar System, and continues to do so. That, however, was not a good reason, so far as I was concerned, for giving top priority to collecting meteoritic material in space when we already had many tons found in Antarctica and elsewhere on the Earth.

At the meeting what I thought clearly didn't matter very much. Brushing aside all suggestions of Mars or Venus, the decision was soon made that the first 'planetary' cornerstone mission would target a comet in flight. Not only that, but the spacecraft was to land, drill, and extract a core of material from the comet and store it on board. The core was to be at least one metre long, and it must be brought back to Earth intact. 'Intact' meant that the relative positions of the material in the core had to be preserved, so any layering that was present would be revealed in the laboratory when the sample was retrieved.

Comet Rendezvous and Sample Return

All concerned were well aware that comets in space consist largely of ice; the rocky meteorites that we find and collect on the Earth were once embedded in the icy comet like plums in a pudding, and they are of course the only part that can make it all the way to the ground without evaporating. Furthermore, spectrometers on telescopes had been used to make observations of the coma, the 'atmosphere' of the comet, a shell of gases blown off from the solid nucleus as the ices sublimate when the Sun heats them. The coma, and the trail of dust and ionized gas it leaves behind, are what we see in the sky; the solid part of a comet is thousands of times smaller, usually only a few miles across, and quite invisible even through a large telescope. The latest observations of recent cometary apparitions showed that the ices in the nucleus were not all water, but included carbon dioxide, ammonia, and other frozen gases as well. It would certainly be exciting to analyse a pristine sample of the ices, dust, and rock in the core extracted from the solid nucleus. But first, we would have to get it home.

Rendezvousing with a comet is not easy. They travel rapidly in highly elliptical orbits around the Sun, so the spacecraft has to match that orbit and this takes a lot of energy, time, and fuel. Once there, drilling a long core of completely unknown material (anything from hard rock to fluffy snow) and handling it so that it stays intact while it is stored on the spacecraft would require robots of exquisite skill. Worst of all, some of the ices known to be present, like frozen methane, convert to the gas if warmed ever so slightly. To preserve them, everything would have to be kept at a temperature no higher than –200 °C all the way back to the laboratory on Earth. Failing that, we would just end up with a tube full of a mixture of gases with a few pebbles in the bottom.

I was less experienced in space technology than some of the others present, but while at JPL I had contributed to a study of a comet mission as part of long-term planning for NASA. I was sure that demanding a robotic mission capable of returning a pristine core sample would be asking too much of the team that had to design and build such a miraculous piece of kit. Rather than display negativity in the meeting, during the coffee break I asked one of the most forceful senior protagonists, Professor Hugo Fechtig of the University of Heidelberg, whether he did not think the concept was too much of a challenge. He shrugged. 'The ESA engineers will find a way, it is up to them.' The ESA chiefs present did not voice any concerns, so neither did I. The *Comet Nucleus Sample Return* mission was born.

ESA's new and ambitious *Horizon 2000* programme spanning the next 20 years was announced later that year and described in a glossy booklet. As well as the comet sample return, it included a solar–terrestrial cornerstone mission (later to fly under the name *Soho-Cluster*) and two big astronomy missions, a stellar spectroscopy project (FIRST) and an X-ray observatory called XMM. The newly approved comet mission was to be the third cornerstone, with a launch in 2002; it needed a name too.

Rosetta

The community had taken to referring poetically to comets as 'the Rosetta stone of the Solar System', a reference to the inscribed slab found near Alexandria in 1799 and now in the British Museum, which led to the deciphering of ancient Egyptian hieroglyphics. The Rosetta link to comets was first made by my former JPL colleague Ray L. Newburn in the 1970s, when he, and I with him for a while, worked on planning possible missions to comets for NASA. Those missions remained on what was euphemistically called the 'back burner', but the classical allusion was widely taken up, first by NASA then by ESA, who named their comet mission *Rosetta*.

The managers and engineers at the European Space Technology Centre, ESTEC, in the Netherlands got to work on planning the new projects in detail. It was not long before alarm bells began to ring over the estimated cost of carrying out the sample return part of *Rosetta*. Behind closed doors, the planning teams were saying that it could not be done at any price. The following year I received a two-page letter from the space agency headquarters in Paris saying in effect that, in order to proceed further, the mission would have to be scaled back to just a comet rendezvous, without any sample return to Earth. As a member of the committee that had planned it, they required my agreement to this scaling back (called 'descoping' in the business) by signing at the bottom of the letter and returning it. I was itching to write 'I told you so' as well, but of course I didn't.

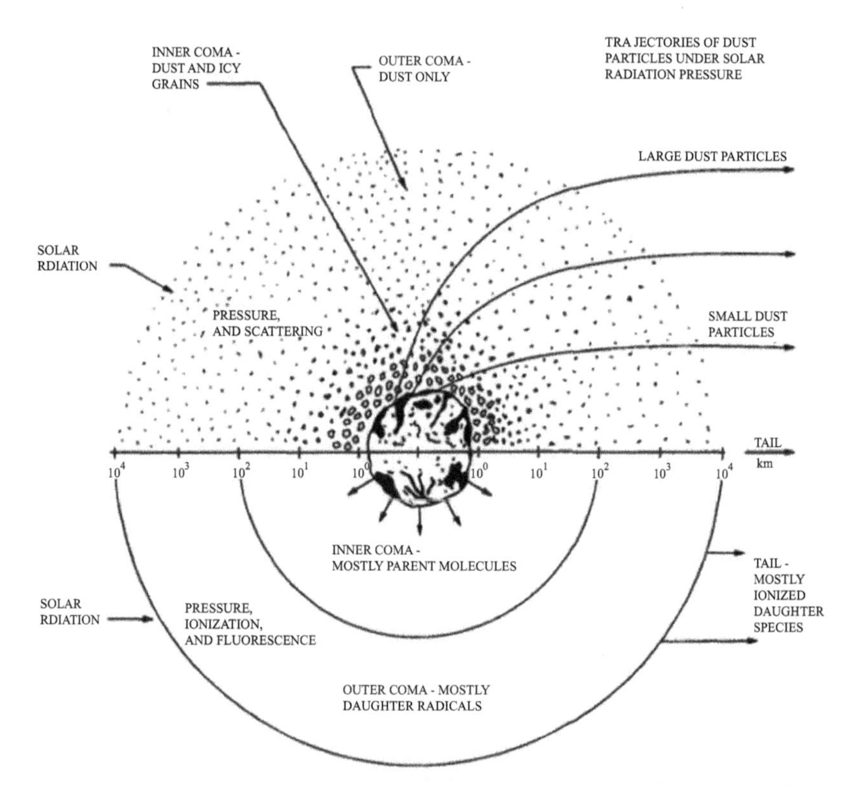

I made this sketch of the predicted structure of a 'typical' comet in 1973 for a JPL technical report entitled 'A model of the physical properties of Comet Encke'. At the time, Enke was to be the target for NASA's first mission to a comet, but the project got cancelled and never flew. Still, the work came in useful forty years later when *Rosetta* reached comet Churyumov–Gerasimenko, which does resemble the model in many ways.

The VIRTIS saga

In November 1994, ESA released a call for letters of intent to propose experiments to be carried on the *Rosetta* mission, saying it would launch in 2003 and encounter a comet called *Wirtanen* in 2011. Comets are usually named after the astronomers who first observe them; Carl Wirtanen was an American astronomer working at the Lick Observatory in California when he made the discovery in 1948. In the long wait for the chance to participate in the mission, we at Oxford had built collaborations with groups in Paris, Rome, and Berlin to develop a large and sophisticated infrared spectrometer we called VIRTIS, for V*isible and Infrared Thermal Imaging Spectrometer*. The Italian component of our four-part team was led by Angioletta Coradini in Rome, who was also in overall charge as the Principal Investigator for VIRTIS until her premature death in 2011. The other three team leaders were Therese Encrenaz in Paris, Gabrielle Arnold in Berlin, and myself in Oxford. The preponderance of women was unusual, especially then, and the leadership team became known among the troops as the 'Three Angels' as we held successive meetings in each of the centres to define what exactly we would build and who would do what.

The German Space Agency where Gabrielle worked was based in the old East German Meteorological Office, close to where the Berlin Wall had recently come down. We met in smart modern meeting rooms, but once when I took a wrong turn on my way to the bathroom I ended up in a part of the building that had not yet been modernized, this being quite soon after the reunification of Germany in 1990. I found the toilet I was looking for, but the grimy walls were green with slime and the floor was damp concrete. I soon fled back to the newly westernized section and stayed there.

On the walls of the corridors were old photographs of leather-helmeted pilots, standing by their primitive aircraft in the time around the First World War. In one of them I recognized von Richthofen, the Red Baron, the greatest air ace of that war, with his Albatros biplane. It turned out I was standing on the historic site of Germany's first airfield where pioneers like Fokker had developed and tested their flying machines before and during the war. The meteorological office probably dated back to the days when von Ricthofen was learning to fly there, supporting him and the other aviators with weather forecasts and storm warnings.

From the coffee lounge near the top of the tall building, we could look across to former West Berlin and see the scar stretching right across the city where the wall between East and West had been before it was torn down. Although he wasn't involved in our comet project, in a coffee break during one of these visits I ran into Gerhard Neukum, a scientist local to Berlin with whom I had worked on various Mars studies. As we stared at the view together, he pointed out a space between the streets, what in England we would call a square, with some modern tower blocks along one side. There used to be a beautiful palace there, he said, the finest building in all of Berlin. When the Russians took over they demolished it, saying it was bourgeois and such symbols of elitism should be completely eradicated. A great shame, said Gerhard. Then we talked for a while about the Berlin Wall; I suggested that they might put up markers to show where it used to be, for the information of future generations. Oh no, said Gerhard, it must be completely eradicated! So it was: when I saw that view again after more than a decade had passed, no trace of the Wall could be seen.

On a later visit, I had a free Sunday and spent it touring the centre of Berlin, locating various historic buildings and sites like the Brandenburg Gate. When it came to the Reich Chancellery from which Hitler directed the final phase of the Second World War, there was surprisingly little information. I knew the building had been demolished by the Russians some years earlier, but the site was not marked and seemed if anything to be a secret. It turned out that the policy of the German government at the time was to subdue the history of anything connected with the Nazis, particularly if it might encourage renewed interest. It was not difficult, however, to overlay an old map on the current one and soon I was standing in the former garden above the bunker complex where Hitler spent his final days. This site, too, had been levelled and the tunnels filled in, so all I could see was a car park and blocks of shoddy Soviet-era flats. The flats screened the main road, and the side roads were empty, so I was completely alone, standing on the spot where my research said they burned Hitler's body, with only a few parked cars for company.

It was eerie, and reminded me of the time a few years earlier when I was at a meeting in Rome and an American colleague remarked that he had spent the weekend researching the exact location where Julius Caesar had been stabbed by Brutus and the other conspirators, and claimed to have found it. He marked it on my map, and when I had a free lunch hour I walked over there and found myself standing on an unremarkable back street. There was nothing like the same sensation of being surrounded by ghosts that I experienced in Berlin, and I wondered if he had got it right. Since then, professional archaeologists have looked at the question and pinpointed the likely location, well away from where my amateur historian colleague had said it was.

Not so cool

At Oxford at the time we had a world leading position on designing and building miniature refrigerators for use in space. We had developed these for our ISAMS Earth-orbiting instrument (Chapter 8), and shown that they could outperform any other way of keeping payloads and detectors cool in order to increase their sensitivity. They could keep working almost indefinitely with high reliability and low power consumption. Everyone was interested in them, and the secrets of the design soon leaked out to industry all over the world. To this day, I don't know whether to feel pride that we 'spun out' such an important device from our research, or remorse for the fact that we lost out on the fortune we could have had if we had but marketed it properly. Instead, we turned to a branch of the government that had been set up for this sort of purpose, and watched the opportunity sink in a slough of neglect and desultory pen-pushing.

It was agreed that we would build a cryogenic system for VIRTIS as our largest contribution to the instrument, and we incorporated that in the flight proposal. After the usual tense competition, we emerged successful with a letter from the Space Agency saying VIRTIS was selected for the *Rosetta* mission. To get that far, you had to have submitted a pledge from your funding authority saying they were willing to support you financially if you were chosen. However this was not a guarantee; if the total involvement of a nation that ESA selects costs more than the money it has available, there can and will be problems and backtracking. This is rare, however, since winning an opportunity to participate in a major mission is such an achievement that some degree of support is nearly always provided by the happy nation that is home to the successful proposer.

The relevant budgets in the UK at this time were very short of money because of a large and sustained investment in ground-based telescopes demanded by the numerous and politically powerful astronomy community. The Science and Engineering Research Council could only afford five million pounds for the entire UK involvement in *Rosetta*, and they decided to give all of it to one experiment. The rest of us were politely told to get lost. The winner was an instrument called *Modulus*, a mass spectrometer for analysing the composition of the comet from samples obtained when the lander module reached its surface. This experiment had been proposed by Colin Pillinger and his team at the Open University but ironically, despite this victory, Colin soon lost interest in *Modulus* and passed his role on to someone else, so he could concentrate on his passion for sending *Beagle 2* to Mars. Sadly, it did not end well, when the UK experiment crashed on the comet in November 2014 along with the rest of the *Philae* lander and never delivered the essential data on the composition of the nucleus that we had all been hoping for.

I protested that, having got this far, funds must be found to build our part of the VIRTIS instrument, now that the mission was approved with us on board, but this time to no avail. 'We held a competition. You lost', said the director at Swindon headquarters of the research council brusquely when I visited to plead with him. He had problems of his own: an anonymous quote that now appears on Wikipedia sums it up well: 'The Science and Engineering Research Council struggled to combine three incompatible business models—administratively efficient short-term grant distribution, medium-term commitments to international agreements, and long-term commitments to staff and facilities. Given a lack of control over exchange rate fluctuations and the need to meet long-term commitments, cuts regularly fell on the short-term grants, thereby alienating the research community.' I knew about all of this from my work on the other side of the fence as well, being a member of some of their committees charged with evaluating applications and dispensing funds from a pot that would expand and shrink (mostly shrink) rapidly and often without warning.

Resigned to our fate, I had to call the Angels and tell them we could not provide our quarter of the instrument after all. They were kind but I don't think they understood; their countries were delighted by their success and prospects and funded them gladly. What was wrong with the UK? I was wondering that, too. But without much trouble the Italians came up with the extra money to plug the gap. For a while I thought they might use this to fund us in Oxford to build the coolers after all, and there was a flurry of travel and correspondence, but instead they went for one of the companies that had commercialized a version of our design, in Israel of all places. They did a good job.

Encounter with Lutetia

Years later, in 2010, I was working late in my hotel in Paris for a presentation I was due to give at a meeting the next day. My laptop pinged and an email came in from the new VIRTIS Principal Investigator, Fabrizio Capaccioni. *Rosetta* was now in space, but delays in the launch had made the original target comet Wirtanen unreachable. Instead, the spacecraft was heading for *Churyumov–Gerasimenko*, a periodic comet discovered in 1969 by two astronomers in the Soviet Union. *Rosetta* had just made a close fly-by of the large asteroid *Lutetia* on its way to the comet and made some interesting observations.

Lutetia had puzzled astronomers for years with its unusual spectral properties, suggesting its makeup was not like most other asteroids. *Rosetta* found that it is a very dense object, suggesting a high proportion of metals, although it also has a low radar cross-section which would argue otherwise. The answer appears to lie in the thick layer of dust, possibly several kilometres deep, that coats the 100-kilometre-wide planetoid. The dust layer prevented VIRTIS from probing the bulk composition of Lutetia, and itself shows hardly any spectral features at all, except a slight reddening.

Fabrizio had drafted a paper about these findings for submission to the journal *Nature* and was inviting the team to read it and make contributions. As I looked at the draft, late at night, I found myself idly wondering where the name *Lutetia* came from. The internet soon reminded me that it was the Roman name for the town on the south bank of the Seine, the precursor to Paris, located approximately where the 6th arrondissement is now. I was sitting right in the middle of it. So, according to the old records, was the astronomer Hermann Goldschmidt when he discovered the minor planet in 1852, and this presumably explains his choice of name for it.

So, still on the VIRTIS team, despite not building hardware, I attended as many team meetings as I could over the years and intended to enjoy the science and contribute to the analysis of the results. Lutetia was interesting, but it was the comet that promised the most excitement. For obvious reasons, *Churyumov–Gerasimenko* is more often referred to as just Chury, or by its catalogue number 67/P (Wirtanen is 46/P). The P indicates a periodic comet that returns to the inner Solar System regularly, about every six and a half years in the case of 67/P. Previous sightings meant that the orbit of the comet is well documented, which is obviously important for purposes of navigation and rendezvous. On the other hand, many previous encounters near the Sun would mean the nucleus might be depleted in ices, especially the interesting ones with a low sublimation temperature, making it less active now. We were not sure what we would find; even the size and shape of the nucleus remained to be discovered when *Rosetta* got close enough to resolve such details.

At the Comet at last

Answers started to come in when the data flowed as *Rosetta* nosed in close to 'Chury' in August 2014. The relative speed was now less than a metre per second, and a series of burns with the spacecraft's jets were used to start to fly around the comet at a distance of only about 25 kilometres. We called this an orbit, but it wasn't really; the nucleus is so small, just a couple of miles across, that its mass provides only a tenuous hold on the spacecraft by gravity, far less of course than the grip the Earth holds on weather satellites, the space station, or the Moon. Instead, on arrival *Rosetta* had to use its thrusters regularly to steer around the comet in a roughly triangular path. Later, a gravitational orbit was in fact achieved for a while, by moving in close and allowing the spacecraft to travel very slowly in an eccentric path around the nucleus, but when the comet started to blow gases from its interior this was enough to perturb the orbit and send the controllers reaching for the thrusters controls again.

So, fully 30 years after the planning meeting in San Moritz, *Rosetta* had matched trajectories with its cometary target and was manoeuvring around it. The cameras had revealed 67/P to have a remarkable shape, likened in media reports to a rubber duck or an alien spacecraft. The 'neck' is produced by a deep fissure near the middle of the object, on one side of which is a spectacular cliff a kilometre high.

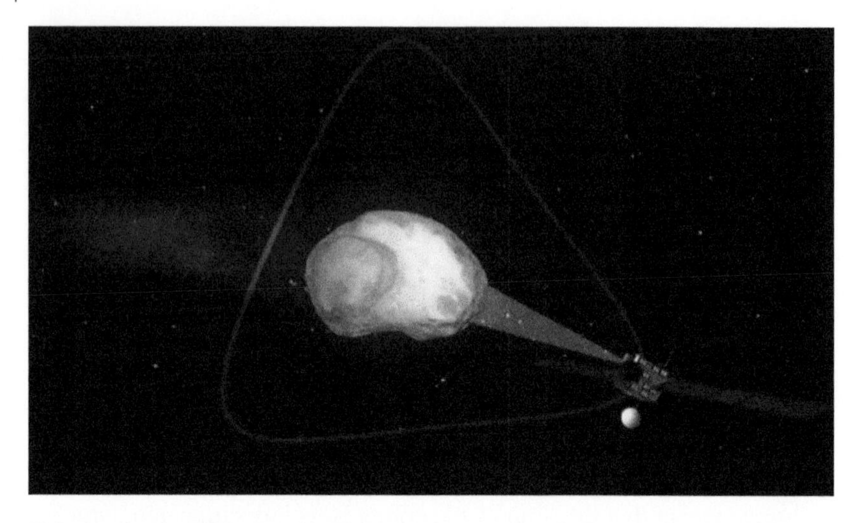

This drawing by ESA shows how the *Rosetta* spacecraft flies around the nucleus of 67/P in a triangular pattern, using gas jets for propulsion. The low gravity generated by the small mass of the comet (relatively speaking; it weighs in at about ten trillion tons) means a normal orbit is difficult. The beam down to the nucleus represents the view of VIRTIS and other instruments as they scan the terrain, before and after *Rosetta* dropped the Philae lander on to the comet's surface.

Amid the general shock and speculation about how and where on this curious terrain the lander should be deployed, we on the science team were pondering how it got that way. The most likely guess is that 67/P was a double comet, with two or maybe three large objects apparently fused together to make the nucleus. This is how we think the planets formed, with kilometre-sized hunks of rock and ice bumping into each other and sticking until we got Jupiter-sized and Earth-sized objects nearly five billion years ago. It is encouraging to see the 'planetesimal' theory apparently supported by this example, exactly the sort of thing expensive space missions are for.

The comet probably started its existence out in the cloud of similar objects that we see orbiting the Sun at a great distance, beyond the orbit of Neptune. It could have been more or less spherical initially, before something, a collision, or a gravitational tug from a chance alignment of the outer planets, or something else entirely—some comets may come from outside the Solar System altogether—pushed it into its present orbit. This is a huge loop that takes the comet from beyond Jupiter, but not as far as Saturn, to a point just outside Earth's orbit, every six and a half years. During its repeated passages close to the Sun, the icy parts could have evaporated away leaving more of the rocky or claylike material which happened to have the strange shape we were now observing.

The key to resolving these and other possibilities could belong to our VIRTIS spectrometer, because spectral imaging of colours at hundreds of wavelengths, most of them beyond the range of the human eye, should

The nucleus of comet 67/P *Churyumov–Gerasimenko* in a picture taken by the camera on *Rosetta* in August 2014, when the spacecraft was just 100 kilometres away from the object. It is less than five kilometres across in the longest direction.

be able to tell us whether the two main pieces of the object are different in composition, as they might be if they formed separately and joined together later. As the inevitable pressure grew to publish the early results, the team looked into this key aspect. In my new role as elder statesman, I was reading emails from younger colleagues in Italy, France, and Germany as they beavered away on the new results, looking for answers. Sometimes I lamented the decision that had failed to fund the UK part of the instrument, so long ago; but sometimes it is enough to bask in the privilege of being in a position to be able to see and discuss findings that were not yet out on the street, and adding opinions and interpretations.

The early indications were not promising for resolving the 'binary' question. There were differences in composition, but rather than the 'head' and 'body' looking different, the contrasts were scattered around both parts. So we couldn't say that they were different, but we couldn't say they were the same, either. More work was required. What was more remarkable was what didn't show up in the spectra, and that was very much water ice. The surface of the nucleus seems to be very dry, and also, from the way its temperature changes as the body rotated, quite loose and porous. That would fit with the low density of the nucleus, which can be measured by its gravitational tug on the orbiting spacecraft. It turns out to be only a little more than half the density of water. Probably, the solid-looking object that we see has big voids in it, like a sponge. At one time these cavities might have contained a volatile ice like frozen methane which long since evaporated away.

Photographs taken by the camera on *Rosetta* revealed a cliff nearly one kilometre high on the surface of the comet. The large boulders at its foot are about 20 metres across.

The surface of the comet is blacker than coal, apparently from a thin coating of some sticky dark substance. The instruments on the small probe that *Rosetta* deployed onto the surface might have analysed this and told us what its composition is, had it landed properly. Since it crashed, all we know is that the distinctive signs of tarry organic materials are there in the spectra. Everything that drifts around in space for a while seems to get coated with dark material, which is apparently a component of the interstellar dust that fills the universe, so it was no great surprise to find it on *Chury*. Some of it may be due to the decomposition of methane by the Sun's ultraviolet radiation as it diffuses out of the comet, followed by recombination of the fragments into larger organic molecules.

The missing water and porous surface were not a total surprise, either, since the near-surface volatiles would have been baked out during previous trips around the Sun. We observe the comet as it becomes active again on its current orbit, with the ices deeper in the crust evaporating as the interior also gets warmer. As I write, the comet and its new companion are still out beyond Mars, and the surface temperatures we measure are as yet well below freezing point. The first weak jets of gas from the interior have begun to be detected, with water vapour and carbon dioxide prominent but in different places. Pieces of the comet have started to be blown off into space as the gas flows outwards through the fragile structure. Small avalanches have occurred, revealing pristine material that, sure enough, contains a lot of water ice. A lot of the action is near the relatively thin neck; as the erosion continues we may even see the nucleus split in two. Whatever happens, we are sure to end up with much more information about these primitive members of the Solar System family to which we Earth-dwellers also belong.

12
Return to the Silent Planet

Knocking on NASA's door

On 27 June 1979, following the success of the *Pioneer Venus* mission that delivered the first British hardware to another planet, the Oxford University Public Orator was delivering the annual Crewian Oration, in Latin as usual, in the historic Sheldonian Theatre. He said:

> Quae Terram ubicumque, Venus bona, iungis amantes
> Amplexu stellae cingeris ipsa novae.

which translates as

> Kindly Venus, thou that linkest
> Lovers all in time and space
> Art thyself henceforth encircled
> In a satellite's embrace.

This turned out to be the high water mark for Venus exploration for many years to come. Those of us that had been involved in *Pioneer* in 1979 and *Magellan* in 1991 grew more and more frustrated as time wore on and there was no new mission to Venus to follow up and dig deeper into the unresolved mysteries we had uncovered. I hadn't been involved much in the *Magellan* mission, because it had a single objective, which was to map the surface using radar to reveal the features hidden below the permanent cloud layers. The results were surprising, with a multitude of volcanic structures of different kinds and no evidence for continental plate tectonics like that which shaped the surface of the Earth. Deep, dry river beds were found snaking hundreds of miles across the baking hot landscape. The geologists were as keen as the rest of us to follow up.

By the turn of the millennium, I was a seasoned member of the space science community and had regularly joined my colleagues in the United States to carry out studies, write proposals, and form pressure groups, all with the aim of persuading NASA to resume the exploration of Venus. The Russians, who had led the charge in the early days, had political problems and were no longer very active, and it was more than 20 years since they or anyone else had flown to Venus to study the atmosphere and climate. I was sure the hiatus must be temporary and continued to work steadily on a variety of initiatives for new missions. These produced attractive proposals for projects on a whole range of cost and time scales, and now and again one of them would seem to have gained support among scientists, planners, and administrators right across the board. Hopes were raised but somehow it never went any further.

We were banging our heads against a brick wall as far as NASA and Venus were concerned. This was in stark contrast to the new programmatic strategy that the agency now had in place for the exploration of Mars. Instead of the usual piecemeal approach, NASA had persuaded its sponsors in the US Government to approve a whole series of missions to launch one after the other over a period of a decade or more, until a complete scientific strategy had been satisfied. Encouraged by this, we set to work on an exploration strategy for Venus for the years 2000–2010 to see if we could achieve the same goal for the cloudy planet. Our US colleagues would present this to the Solar System Exploration Subcommittee and seek to get it into NASA's plan, or 'roadmap' as they liked to call it.

Why Venus?

What was driving us was the urge to understand why the planet most like Earth in terms of its size and composition seems to have diverged so much from our own during the course of their common history. The two planets formed together from the cloud of gas and dust that surrounded the young Sun about four and a half billion years ago. Why had Earth evolved conditions suitable for the development of life while the Venusian greenhouse makes the presence of life as we know it impossible? Sure, Venus is closer to the Sun, but not, on the face of it, by enough to make such a difference. We talk about the 'habitable zone' around a star, but we clearly do not really know what that means, even for our own familiar star. If we could understand why the two neighbouring planets developed such differing environments, it would have implications for the likelihood of finding Earth-like planets elsewhere in the Galaxy.

We were also still a long way from understanding how basic processes work on Venus, and again this is especially interesting because Venus is so Earth-like. I was interested in things that held the key to understanding the hot Venusian climate, such as the atmospheric circulation and cloud formation, and the input from volcanoes that vent gases into the atmosphere from sources in the interior of the planet. In both cases, what we see happening on Venus is obviously not the same as on Earth, and yet they have to be related through the same laws of physics. We didn't yet have enough information about atmospheric and surface composition, or curious surface features like the long, deep river valleys discovered in pictures from the imaging radar on *Magellan*. We don't even know what it was that flowed so long and fast to create these and other erosional features, but they are there. Add to this questions on geological processes in the interior, magnetic field generation, volcanic activity, bulk composition of the solid planet, molecular composition of the atmosphere, generation and maintenance of the global cloud layer, and chemical weathering of the surface, and it all starts to get quite complicated.

A Venus shopping list

Scientists are inveterate list makers. No meeting on mission planning, progress in developing hardware, scientific goals, results, in fact anything to do with planetary exploration is complete without at least one, or probably dozens, of bullet lists. These are brief statements of critical items or activities, each preceded by a large black dot—the bullet. It is a convenient way to break down a complex task or problem into bite-sized pieces, and if done properly is a useful tool and a persuasive way of presenting a plan. I've avoided them so far in this book because they do not always make easy reading for the non-specialists I hope are in my audience; but this is a good place to give a single example.

Before I do, here's an anecdote that shows bullet lists at work. Many years ago, I was sitting in a small group in Washington listening to a NASA official describe the requirements for participation in a new mission they were planning. Scientists who wanted to become involved had to submit a proposal by a certain date, promise to attend progress meetings, commit to deliver hardware on time, and so on. Each of these appeared on a bulleted list which he addressed for some time, before concluding that we would be successful only if we took all of these 'NASA bullets' to heart. What about achieving the scientific goals, shouldn't that be on the list? enquired John Gille, who was sitting near me. That's up to you as investigators, that is not a NASA bullet! responded the official. Do NASA bullets have to have teeth marks in them? asked Gille sweetly.

Here's the list we collectively compiled to summarize the things that were keeping us awake at night thinking about Venus, just before the end of the twentieth century. I've simplified it slightly to make it clearer to the general reader what is meant in each bullet point.

- What is the level of volcanic activity on Venus, past and present?
- How has volcanism shaped the surface of the planet?
- What gases and magmas are produced by volcanoes on Venus?
- How does volcanism affect atmospheric composition and the global cloud layer?

- Is Venus geologically 'alive' like the Earth or 'dead' like the Moon?
- What difference from the Earth leads to Venus having no magnetic field?
- Was there a magnetic field on Venus in the past?
- What chemical processes ('weathering') occur between the atmosphere and the hot surface?
- What dynamics and chemistry maintain the Venusian cloud layers?
- What drives the atmosphere to circulate at cloud-top level 50 times faster than the solid planet?
- What fluctuations occur in the atmospheric circulation on different scales in space and time, and what causes them?
- Can we explain the meteorology revealed in the cloud patterns at different altitudes?
- What determines the characteristics and variability of the polar vortices?
- How deep does the equator-to-pole circulation extend?
- How do the waves in Venus' atmosphere interact with the global circulation?
- Did Venus ever have oceans of liquid water? Could life have developed there?
- Are ancient rocks preserved in the highlands of Venus, above the lava plains?
- Why are some common surface minerals on Venus different in detailed composition from their counterparts on Earth?
- What sort of layering of rock and soil exists in the lava-filled lowland plains? Could we extract the geological history from them?
- Are there any geological features on Venus that might contain a record of ancient life?
- Do 'venusquakes' occur, and what are their characteristics?
- Why does Venus not have plate tectonics ('continental drift') like Earth?
- Is there a crust–mantle boundary and other deep structure like that on Earth?
- Why does Venus have slow retrograde rotation and no satellite?
- Why are the noble gases, such as neon and argon, present in such different proportions on Venus and Earth?

We could go on and make a much longer list, but there comes a point where it is futile to keep listing objectives without thinking about the limits to what we can actually do in the difficult Venusian environment. Some sort of aircraft or floating station could be designed move around in the deep atmosphere and land periodically on the surface to take and analyse samples. Such vehicles go by the generic name of 'aerobots', although we sometimes refer to them as 'submarines' because the high pressures where they have to operate are like those deep in the Earth's oceans.

When we start to think about this approach, it immediately becomes clear that we will need a technology validation mission to test the aerobot flight technologies, high temperature electronics, and autonomous instruments required to address the science questions. Ideally the probe would be capable of deciding for itself where to land locally and how best to carry out measurements, and this will need advances in robotics somewhat beyond any capability we have now.

Another useful approach to new missions to Venus, one that would be easier and cheaper than submarines, could use a modern version of the *Pioneer Venus* multiprobes. A lot of probes, at least 16, and preferably even more, would be great for measuring atmospheric temperatures, cloud layering, and wind fields right across the globe. If they could not only measure atmospheric profiles, but also go on to become short-lived surface stations targeted to specific areas, they would provide information on the range of crustal compositions on Venus and the nature of surface–atmosphere interactions.

The probes might use some version of the penetrator technology that had been proposed for Mars, although in the case of Venus there is a problem in hitting the ground hard enough to dig in after passing through the dense atmosphere, which slows them down. They might have to carry drills to get sufficiently far below the surface, where they can hope to find unweathered material to compare with that on the surface that has spent long periods reacting chemically with the atmosphere. Getting still more ambitious, if a way could also be found to give the probes a long life, they could perform the first seismological measurements. This is just about the only technique we have for studying the deep interior of a planet. Does Venus even have venusquakes? We don't know, but it probably does, and we'd like to listen in to them, preferably from multiple locations so they can be timed and triangulated.

The most exciting and promising mission concepts for Venus are undeniably technically difficult and expensive to carry out. Robot geologists, seismic networks, and the rest are much easier to deploy on Mars, and that planet has the added allure of the search for evidence of past or present life. In spite of regular surges of optimism generated by some realistic proposals for new Venus missions, nothing has reached the top of NASA's 'must fly' list since the *Magellan* radar mapping mission a quarter of a century ago, and this remains the situation in the USA right up to the present day.

Europe offers hope

NASA might not be interested, but in the global scientific community we felt that there was still plenty that could be done at Venus with the least expensive type of mission, such as a new orbiter with the latest instruments on board. This, we thought, might appeal to the European Space Agency as it gradually grew in stature and confidence and began to embrace deep space projects. Remote sensing instruments were getting better and better, so new satellites orbiting Venus and making long-term studies of the atmosphere could offer better insights into the dynamics of the puzzling Venusian climate. We knew now how to observe the surface right through the clouds without using large, heavy radar equipment and could do things like map active volcanoes with relatively simple optical instruments. So, in 2001, I was one of a small group of European planetary scientists who wrote to ESA urging the agency to fit a mission to Venus into their programme. We wrote as follows:

> 'Venus is the brightest object in the sky after the Sun and Moon, the closest and the most easily recognizable planet. Venus is the twin sister of Earth with similar size and mass and yet a drastically different atmosphere, featuring a high concentration of carbon dioxide, sulphuric acid clouds, extremely high surface temperatures, hurricane winds circling the planet, giant vortices, and a direct interaction with the solar wind in a way typical of comets. It is therefore a natural laboratory for the study of atmospheric dynamics, chemistry, radiative balance, and plasma effects. Perhaps the most puzzling aspect of Venus' history is the evolution of its atmosphere, which one might have expected to be similar to the Earth's. Yet at some point the evolution on each planet took drastically different paths. Was it just natural evolution or the result of a catastrophic event in the planet's history?'

> 'Venus has a powerful greenhouse effect that sets the absolute record in the Solar System. The increase in the surface temperature, over what it would be without any atmospheric greenhouse, is about five hundred degrees centigrade. This is due to the presence of the same gases and clouds that are also present on the Earth, although in different proportions and in smaller amounts. For Earth, the corresponding number is about thirty-five degrees. We know this will increase and change the future of our planet if atmospheric pollution continues to grow. What we do not know is how stable is our own climate system, and hence what the detailed response will be. How likely, or unlikely, was it for the Earth in its evolution to manage to squeeze between Scylla and Charybdis, as represented by cold Mars and hot Venus, to gain the delicate balance that supports human life? '

We went on to end with a flourish:

> 'Many of the questions that we would like to solve for Venus apply to Earth as well, and popular interest in these issues is high. So, as well as pursuing the exciting exploration of Earth's nearest and most similar planetary neighbour, the further study of Venus represents a key step in our understanding of requirements for survival on Earth.'

Going for broke

I had a very strong personal interest in seeing a new mission to Venus take off. Ever since my VORTEX instrument discovered the great double-eyed storm at the north pole on Venus more than 20 years before (Chapter 4), I had been pondering how I could get back there and take a closer look. *Pioneer Venus* had opened up so many questions about the atmosphere and the climate on our neighbour that were then left unanswered. In particular, I felt almost desperate to understand the huge, circulating 'dipole' feature at the north pole that we had discovered with *Pioneer* all those years ago. We had observed many interesting phenomena for the first time then, but still hadn't unlocked the secrets of the climate on Venus. A planet much like the Earth and so close by should not be in such a defiantly different state without us having some clear idea on why it is like that. I knew my scientific life would not feel complete without some answers.

By 2003, time was running out; I would retire in less than a decade and lose most of any influence I might have to help start a new mission to our sister planet. I had already tried various proposals through all the various mechanisms that existed: the US Committee on Planetary Exploration, COMPLEX; informal pressure groups; official study groups; calls for ideas from the space agencies; papers at planning sessions at scientific meetings; everything, but so far to no avail. Then the Japanese Space Agency, which already had a Mars mission in the pipeline, announced that they were also thinking about mounting their own mission to Venus. I sent the project leader a letter of encouragement accompanied by a dozen reprints of some of my papers on Venus mysteries.

Then there was a further stroke of luck. ESA had sent a small orbiter to Mars in 2003 and was very pleased by the results and good publicity it had produced for a relatively small cost, just 300 million euros. It can cost as much as that to make a movie these days. The agency launched a competition, inviting proposals for a new mission that would use a copy of this *Mars Express* spacecraft to do something different. The re-use of the existing designs would mean that the cost was even lower.

Most of the ten or so proposals they got back involved going back to Mars with a new set of instruments, but we were able to show that it would be just as easy to send the spacecraft to Venus, with very few changes needed. Some of the Mars instruments could bc adapted easily for Venus as well. Others were available off the shelf as spares from current missions like *Rosetta*. *Venus Express* was born.

In order to win the competition, it was necessary to put together a team of supporters across Europe who would campaign for the mission and express an interest in participating if ESA chose it for flight. Fortuitously, Dmitry Titov, who had participated in the Russian *Venera* missions to Venus as a young prodigy, and whom I had met many times at conferences, had moved to Germany after the break-up of the Soviet Union. From his new home in the Max Planck Institute for Solar System Research, Dima led an inspired campaign for *Venus Express*. Emmanuel Lellouch rallied essential support in France, and together the three of us organized the writing of a detailed proposal with contributions from several other countries, especially Italy. On 4 November 2002 we learned that we had been successful, when ESA's Science Policy Committee gave its approval.

Love thy neighbouring planet as thyself

Here's what we had promised to do. The top-level summary of the proposal had rather vaguely offered a 'global investigation' of the planet's atmosphere and plasma environment from orbit, and to address 'several important aspects' of the geology and surface physics. You have to be more specific than that, so we went on to list some detailed scientific goals.

Top of the list was the vertical temperature profile, sounded repeatedly to build up a picture of the whole atmosphere in space and time. The temperature contrasts that emerge are those that drive the atmospheric engine, with its strong winds and wave phenomena. Global measurements of the composition include tracers that also delineate the motions, and provide the ingredients for cloud formation. The Venus clouds are not simple condensates, but rather are part of a chemical cycle that includes the sulphur-containing gases from the volcanoes on the surface. Compositional variations in the clouds are obviously present and need to be investigated.

With all that, we said we would study the energy balance and show how the extreme Venusian greenhouse effect produces the high temperatures at the surface. The high pressure is a major factor; this is some kind of a balance between the gases pumped into the atmosphere from the surface and interior and the escape into space from the top of the atmosphere. We would study all of the processes that are involved: try to quantify volcanic activity, the surface properties, and the plasma environment where atoms, ions, and electrons are swept away by the solar wind.

Most of the atmosphere of Venus is under cloud cover and hard to observe. *Venus Express* had an ace in the hole: it would be the first mission to attempt global monitoring of the lower atmosphere since near IR transparency 'windows' were discovered. These are wavelength regions where a suitable instrument can 'see' right through the clouds and down to the surface; their existence was unsuspected so previous orbiter missions like *Pioneer Venus* had been restricted to mapping the atmosphere above the clouds. We could do a lot better now.

Finally, we noted that together with the *Mars Express* and *BepiColombo* missions to Mars and Mercury, the proposed mission to Venus, through the expected quality of its science results, would ensure a coherent programme of terrestrial planets exploration and provide Europe with a leading position in this field of planetary research. The *Venus Express* orbiter would also play the role of pathfinder for future, more complex missions to Venus, and the data obtained would help in planning and optimizing future investigations. In support of these promises and assertions, we listed 50 potential collaborators from ten different countries and 18 potential payload instruments. These would be designed and manufactured in ten European countries plus the USA, Japan, and Russia.

Venus Express takes shape

The new spacecraft used the same design and some of the spare parts from *Mars Express*, and was put together by the same team of engineers. This meant the assembly was done at Stevenage in England, by the company that was formerly British Aerospace, by then foreign-owned and renamed Astrium. (Today the company has had another name change, to the even stranger *Airbus Defence and Space*.) The payload of scientific instruments was mostly scraped together from *Mars Express* and *Rosetta* spares, or copies where there was no spare, which kept the cost down without compromising the science by very much. Venus had been neglected for so long that even second-hand instruments offered the chance to make massive progress on many of the key questions.

The visible and infrared thermal imaging spectrometer VIRTIS was one of the instruments we had borrowed from *Rosetta*. It was so versatile that a straightforward copy with no design changes could also do a great job at Venus. Most of the existing science team, including myself naturally, were enthusiastic about extending their involvement to Venus. The only downside to re-using everything was that the funding crisis that had so damaged us on the comet mission meant there would be no Oxford hardware on *Venus Express*, either.

From the scientific viewpoint I would have been satisfied with a VIRTIS co-investigatorship, as on *Rosetta*, but there was a greater prize to be won. Following recent practice in NASA, ESA was offering a small number of appointments to the mission as an Interdisciplinary Scientist, which carried the same status and privileges as a Principal Investigator for an instrument, but with access to the whole payload and not just a single experiment. This was a very attractive proposition and competition for the three places on offer was massive, especially since ESA allowed Americans, with their long experience of Venus missions, to apply. I wrote a proposal reminding ESA that I was one of the architects of the mission, and promising to address the climate question, bringing in my long experience of the Earth where similar processes are at work.

Theoretical expectations are that Venus should be warmer than the Earth, but not by anything like as much as it is, I said. Just as Mars is the planet with the best evidence for climate change, Venus is the global warming champion of the known universe. Earth sits in the middle with relatively modest versions of both, but they are all out of the same box. Understanding them all at once should be more effective than trying to pick them off individually. This was the theme of my Interdisciplinary Scientist project, which ESA did select as part of the mission in 2006. The necessary funding for travel to meetings and computing, something of the order of £50,000, still had to be raised, but since this is a relatively small amount (not like the millions needed to build an instrument), on this occasion it was relatively straightforward to get the backing of the latest incarnation of the UK funding outlet, the Science and Technology Facilities Council (sic). Apart from the fact that Science, Technology

The *Venus Express* spacecraft under assembly at Stevenage, England in 2004. The author is in the centre of the group of engineers and managers from Astrium Ltd (formerly British Aerospace) who built the spacecraft.

After the arrival of *Venus Express* at the planet, my colleague Dmitry Titov and I were interviewed at the European Space Operations Centre in Darmstadt for German television about the objectives of the mission, accompanied by a 1/10 scale model of the spacecraft and a lot of dry-ice generated smoke.

I am speaking at a press conference at the European Space Agency's headquarters in Paris following the successful arrival of *Venus Express* into orbit around Venus. Next to me is Dr Hakan Svedhem, the Project Scientist for the mission, then Drs Jean-Loup Bertaux, Giuseppe Piccioni, Mats Holmstrom, David Grinspoon.

and Facilities Research Council would make more sense, no-one seems to have cared that the organisation, based in Swindon, ends up having the same initials as Swindon Town Football Club.

Astrium did a first-class job of building the spacecraft, basically copying the Mars Express design they had built earlier with just a few changes to suit the new destination. These included a smaller communications dish, since Venus is closer to Earth, and better thermal protection, since it would be operating nearer to the Sun. There were some of the usual panics—ESA actually cancelled *Venus Express* at one stage, not because there was anything wrong with it but because they had problems elsewhere and needed to save money. Fortunately, just then there was a change in leadership of the Science Directorate at ESA and the resourceful David Southwood, a professor from Imperial College in London, took over. One of his first acts was to restore *Venus Express*.

With no further crises, everything was ready on time and on budget for launch in November 2005. Lift-off was also perfect, on a Russian rocket fired from the Roscosmos launch site in Kazakhstan. This was a purely commercial arrangement, since it was cheaper for ESA to pay the Russians to launch their spacecraft than to use our own European launcher, *Ariane*. I decided not to travel to see the launch, it was a difficult journey and I suspected that Baikanour was not as much fun as Coco Beach, the village near Cape Canaveral where I had several times enjoyed staying for NASA launches. All went well at Kazakhstan without me, and five months later in April 2006 the short journey ended with a flawless insertion into orbit.

I travelled to Germany where ESA put on a big press conference to coincide with the arrival at Venus and the first signals received by ground control in Darmstadt. It was bated breath time again, but once again there were no mishaps. When we were sure the spacecraft was safe, the champagne was opened and the party began.

More on science goals

While the spacecraft was travelling on its way to Venus, media interest was high and there were many requests for quotes, articles or appearances to spell out its objectives for the public. I wrote a standard screed:

> 'First and foremost is the scientific challenge to understand the climate on Venus. Much effort and ingenuity had been expended, but we still didn't know why Venus is so different from the Earth. It really shouldn't be if we just consider distance from the Sun. Secondly, it is a long time since the last mission specifically to explore the Venusian environment—Pioneer Venus in the late 1970's—and ideas about what we should be looking for, and what techniques to use, have moved on. In particular, thanks to a lucky discovery in the 1980s, we now know how to probe the atmosphere and surface below the clouds, using 'windows' of transparency in the near infrared part of the spectrum. Finally, the growing awareness that the Earth is evolving to a warmer more polluted state—one more like Venus, in fact—has added renewed interest in understanding the greenhouse effect and other aspects of climate physics common to both planets.
>
> So, off we go. Can we expect any surprises—scientific or otherwise—when the spacecraft arrives at Venus in March 2006 and commences full-time observing? What about the possibility of life, that great driver of the imagination whenever planetary exploration is mentioned? From what we already know, the chances of life of any description on or below the surface of Venus are remote indeed. Some have speculated that microbial life might hang out in the clouds, at levels some 50 kilometres above the ground, where temperatures and pressures are indeed quite Earth-like, basking in sulphuric acid and soaking up the energy of the Sun, twice as intense as it is on Earth. Strange, variable, dark patterns in the clouds could be testimony to a living population there, but realistically these are much more likely to be due to cloud chemistry and meteorology (it would be nice to understand those, too). There is a more profound link with life on Venus if we think of the distant past and future: the isotopic ratios in atmospheric hydrogen suggest that the planet once had an extensive ocean of water, while model studies hint that Venus could once again revert to an Earth-like climate in a few million years if and when volcanic activity subsides. These possibilities are well within the capability of Venus Express and its instruments to research and elucidate.'

A popular astronomy magazine, *The Sky at Night*, challenged me to summarize what was exciting about the scientific investigations it would carry out in less than 250 words. I came up with a bullet list:

- How does the greenhouse effect work? Should Venus be warmer than the Earth, because it is closer to the Sun, or cooler, because its brilliant clouds reflect most of the sunlight and Venus actually absorbs less heat than Earth does (and not much more than chilly Mars)? In fact, Venus is baking hot—why?
- Venus rotates slowly, every 243 days, but the bulk of its atmosphere whizzes around the planet every 4 days. This is called 'zonal super-rotation', a common phenomenon on the outer planets, but a mystery on Venus, especially since it is so fast.
- Venus has weather, like the Earth. Even deep in its atmosphere where the pressures and temperatures are very high we see major weather systems that are hard to understand. At the poles, there are giant vortices, like hurricanes, but stationary and permanent—and with *two* 'eyes'!
- There is evidence from the deuterium abundance that Venus used to have a global ocean, at least several tens of metres deep. This must have been lost, leading us to ask how Venus interacts with the solar wind, and at what rate are atmospheric molecules lost by this process?
- Venus is almost certainly lifeless now, but was it always so, especially if it used to have oceans in the past? Will Venus become more Earth-like in the future, if the volcanoes that stud its surface, thought by many to still be vigorously active, finally subside?

Viewing Venus

By October 2007 Venus had made two trips around the Sun and the nominal mission was over. ESA granted an extended mission of another two Venus years, and the science teams busied themselves writing up the first results for a special issue of *Nature*, which appeared on 29 November 2007, supported by a press conference.

The press is impatient and keen to remind us that we are spending public money, so they want to know right away what we are doing, what discoveries have been made, and what problems encountered. The investigators invariably protest that we have not yet had enough time to analyse even the earliest data, that Venus is turning out to be much more complicated than anyone had expected, and that we do not want to jump the gun with premature conclusions. Before long, however, we were reporting new findings on the dynamics of the atmosphere, and the weather patterns deep below the cloud tops featuring lightning discharges; the rapid (although not as rapid as we expected) escape of gases from the top of Venus' magnetically unprotected upper atmosphere in the solar wind; compositional variations in the atmosphere that traced some planetary-scale processes including the fate of Venus' oceans; and, of course, the 'runaway' greenhouse effect that has taken such toll of the tropical dream that mankind once had of the environment there.

At the same time, Venus's atmosphere revealed new complexity. The features that are seen in the clouds at ultraviolet wavelengths are due to an unknown absorber that still defies identification. In the near infrared, the clouds show even more dramatic and complex structure, some of which correlates with the ultraviolet markings and much that doesn't. The reason is that the ultraviolet markings are the result of absorption near the cloud tops; in the infrared, the radiation is emitted in the hot deep atmosphere and transmitted through the clouds, its intensity modulated by the structure in the most dense part of the cloud layer, which is also the deepest, lying about 25 miles above the surface. Below that, the atmosphere appears to be relatively clear of particles, probably because all the candidate cloud materials are vaporized by the high temperatures.

The structure in the cloud patterns shows remarkably high levels of meteorological activity, turbulent and wavy at the low latitudes near the equator where the solar heating is strongest. It is probably here that the lightning *Venus Express* detects is generated, in the regions of very dense cloud that also signal strong upwelling, both requirements for lightning production on Earth. It's also possible that the volcanic eruptions generate lightning in the dust plumes they emit. Further towards the pole there is a sharp transition to a smooth 'laminar' flow, characterized by long bands of cloud that extend large distances around the planet. These are produced by the zonal super-rotation of the atmosphere—very high winds, parallel to the equator, that mean the atmosphere rotates more than 50 times as fast as the surface below. Theoretical models show that this behaviour is generated in the turbulent region by the deposition of solar energy in the main cloud layer, well above the moderating influence of the surface, where most of the solar heating occurs on the Earth.

The Tempest

The much-anticipated high-resolution views of the polar vortex were high on my and everyone else's agenda as the first data started to come in. The VIRTIS team had prepared software to produce a picture of the 'dipole' on the first day, while the press was still in attendance and looking for early results. They got them—a spectacular picture of the structure that was definitely the feature we had found 30 years earlier with VORTEX, but now seen in beautiful detail. Actually, it wasn't quite the same feature: just as *Pioneer* could only observe the north pole, the orbit of *Venus Express* was such that it could only see the south. But we expected Venus to be symmetrical, and it is, more or less. So both poles have giant, double vortices.

In spite of the much clearer view, it was not easy to explain what we were looking at. The vortices are definitely vast, circulating air masses that the team initially described as a third dynamical regime, along with the turbulent tropical and the laminar mid latitudes. More recently, the laminar region has been found to by dynamically linked to the vortex and better considered to be part of it. A fourth regime, sited on top of the other three at altitudes above about 100 kilometres above the surface, has also been confirmed. This is a classical subsolar to antisolar flow pattern, characteristic of slowly rotating planets, that we could see taking over where the super-rotation, which falls off with height, reaches a low enough value for the global circulation to be dominated by warm, rising air near noon and cold, subsiding air near midnight.

The most remarkable feature of the polar vortices is the detailed structure of the 'eye' at the centre of the hurricane-like rotating mass of air and cloud. Working at much lower resolution than *Venus Express* achieves, *Pioneer Venus* saw the eye as a dipole—not one, but two relatively cloud-free regions where the spinning air descends and recycles back towards the equator. Early results from the new mission tended to confirm this, capturing images of the dipole that confirmed it was actually S-shaped, two vortices either side of the pole linked to each other by a graceful sweeping band of bright clear air, forming a feature as large overall as the continent of Europe.

Some elementary theory, imported from terrestrial atmospheric physics, can explain this behaviour as a consequence of the equator-to-pole overturning Hadley cell, combined with the super-rotating zonal winds, and a wavenumber-two instability near the poles. What *Venus Express* went on to see, however, was something much more complicated. The 'dipole' changes its shape on an almost daily basis, sometimes becoming a monopole and sometimes a tripole, moving on and off the rotational pole, speeding up and slowing down its rotation rate, and sometimes assuming shapes of quite amazing complexity; on one notable occasion resembling what one team member called a 'ladder down to hell'. Venus meteorology really is quite amazing, and it will be some time before we understand the details.

So what is it? We have been pondering it ever since, and slowly working it out. As I mentioned in Chapter 4, it was not surprising that a vortex exists at the pole, nor really that it is so big and dynamic, given the high cloud-top winds that are carried toward the poles by the solar heating gradient. Even the 'dipole' aspect could be explained, as an instability in the rapidly rotating flow; theory predicts that a double maximum is the most likely characteristic to appear, and it does. It isn't always double; when the new data showed one, three, or even four maxima on occasions, it was not unexpected and it does show that the feature is unstable even though it is always present in one form or another.

The big puzzle now is the detailed structure. The two 'eyes' are not circular but a sort of chevron shape, linked by a narrow linear feature that connects opposite ends of the two chevrons. Pondering this, I was reminded of the 'strange attractor' described in some very famous early work on the Earth's climate by Professor Ed Lorenz of the Massachusetts Institute of Technology. Lorenz pioneered the construction of computerized climate models, on the primitive machines available in the 1950s and 1960s, and found a family of mathematical solutions in which two sorts of climate were possible. These might be thought of as

The cover of my 2014 book about Venus featured an artist's impression of the surface of the planet, with its baked lava plains, mountainous continents, and sulphur-loaded cloudy atmosphere.

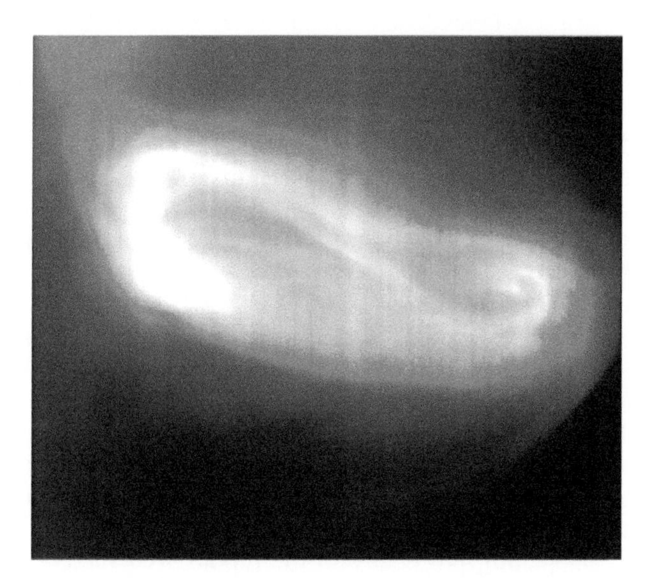

An infra-red image of the eye of the Venusian south polar vortex, as observed by the VIRTIS spectrometer on *Venus Express* in 2006. The area covered by the bright feature is roughly the same as the continent of Antarctica on Earth.

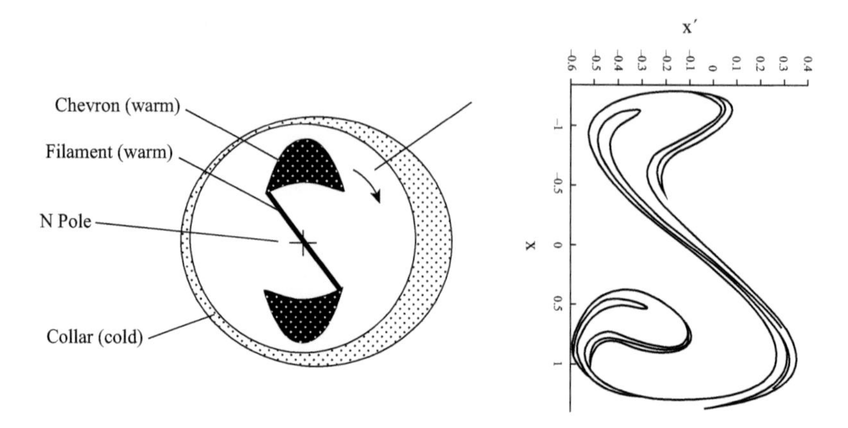

I made the sketch on the left of the Venus 'polar dipole' in 1987, at a meeting in Potsdam, when it was still part of the East in divided Germany. My aim was to provoke a discussion, which was largely inconclusive at the time, about how the strange shape might be produced. Later, I came to believe the dipole is a manifestation of the dynamical phenomenon called Lorenz's strange attractor, shown on the right.

analogous to our present benign climate on one hand, and an ice age on the other, for instance, although represented in Lorenz's model by just a few simple equations.

The model hypnotically traces out a pattern with two lobes in three-dimensional space as time progresses, occasionally making the transition between one regime and the other, and later back again. This became known as a 'strange attractor', because every solution was attracted to one or other of the two regimes, and was strange because the values never repeated. The really fascinating thing was that the 'climate' switched between the two regimes with little or no external forcing. This has become known as the butterfly effect, since, especially as the models got more realistic, the idea that huge transitions in the world's climate could be triggered by something as trivial as a flap of a butterfly's wings caught the popular imagination. The question of whether today's very sophisticated and complex models, and indeed the real climate system, might exhibit this sort of behaviour remains a hot topic of discussion.

What does this have to do with Venus? Well, it started when I noticed that a cross-section through the strange attractor bears a remarkable resemblance to the shape of the dipole structure on Venus. I read Lorenz's original papers from more than 50 years earlier and conjectured that there was a connection. I sent Professor Lorenz some details of the Venus phenomenon and its possible relationship to his 'attractor', and he replied saying it was interesting and to keep in touch about it, but he passed away a few months after that at the age of 91.

What's happening physically in the dipole, as I picture it, is that descending air is clearing the clouds away so that radiation from the hot depths of the Venus atmosphere is beaming upwards and being seen by our infrared instruments. The downward motion of the air is obeying the dynamical equations that represent a strange attractor, and reproducing its patterns. As a parcel of air migrating up to the pole from the equator enters the vortex, it is unpredictable which of the two vortex eyes it is most likely to travel towards; some small perturbation (the hypothetical butterfly's wings) ends up making the choice.

I'd like to see this theory on the cover of *Nature*, where the dipole itself first appeared, but so far I haven't tried to write it up, not even for one of the more specialized journals, and probably I never will. Speculations like this are interesting but it is very difficult to prove anything definite or to make useful predictions without a high-resolution computer model that mimics the behaviour of the dipole, and which can then be broken down into its components and analysed. This would be a challenging task, comparable to modelling near-spontaneous, very rapid changes in Earth's climate, something that is being attempted but so far not with complete conviction. However, we are pretty sure that these kinds of changes have taken place in the past and could happen again. If the polar dynamics on Venus are any kind of an analogue, they do.

Global Warming

The biggest question of all on my science agenda was to gather in all of the latest research from everyone involved and match it up with the data from *Venus Express* to get a better explanation for the hot climate on Venus. I had promised to make progress on this crucial question if I was accepted for the privileged Interdisciplinary Scientist job on the mission, and eventually I did, to my own satisfaction at least. But an important instrument, the largest on the payload, was not working. This was a versatile infrared spectrometer called the *Planetary Fourier Spectrometer* that I had expected to provide vital data on the clouds and on volcanic emissions, It had a mechanical failure in the scan mirror, another example of a simple part that should have been completely reliable, but which nevertheless let down the rest of a complicated device, which otherwise worked perfectly. The instrument could not view the planet and never produced any useful data, and this reduced the climate-related information coming back from Venus by quite a lot. Still, what we did have, plus a lot of thought, shed light on many of the big issues, and I felt we had made spectacular progress. After a lot of papers in the scientific literature and the popular press, I wrote a book, *The Scientific Exploration of Venus* (Cambridge University Press, 2014), to put the *Venus Express* findings and my final conclusions and speculations about Venus and its climate on record. For good measure, it includes the complete story of Venus exploration from the earliest times.

In brief, Venus is hot not simply as a consequence of being nearer to the Sun. In fact, because it is so cloudy and clouds reflect sunlight, when you do the sums it is easy to show that the Sun heats Venus actually less than it does the Earth. No, the reason is the enormously high surface pressure, nearly a hundred times that of the Earth and therefore equivalent to that found deep in Earth's oceans, about a kilometre down in fact. A straightforward calculation, using physics so basic that there is no controversy possible, shows straight away that melting-lead type temperatures are not only possible but inevitable, even with only small amounts of sunlight penetrating the clouds.

So we pass the buck for the hot climate to the high pressure. But why is the pressure not more like the Earth? Well, we have answers there, too. Venus' atmosphere is nearly all carbon dioxide, with nitrogen and other gases making up just only a few per cent. If the carbon dioxide was somehow removed, we'd be left with an atmosphere on Venus made mainly of nitrogen (like the Earth) at a pressure roughly a hundred times less (like the Earth).

The Earth has oceans, and water dissolves carbon dioxide quite easily (think of soda water). Once dissolved, it tends not to come out again, but to end up in the solid surface as coral reefs and thick layers of chalk. The geologists have worked out that there is so much of this material on Earth that the CO_2 that was removed from the atmosphere to make it was sufficient to produce an atmospheric pressure that in round numbers is nearly a hundred times that of the present day. So, Venus's problem is the lack of water.

But then why is Venus so dry? It must have been wet once; water is common in the universe and we know of no way the Solar System could have come together with a wet Earth and a dry Venus. Here's where distance from the Sun comes in: our calculations tell us that the upper atmosphere of Venus was warmer in the early days, which made it easier for water to escape from the planet. The water on Earth, in contrast, is trapped by the cold stratosphere and mostly stays in the lower atmosphere. If water vapour reaches the high levels of the atmosphere, the radiation from the Sun breaks it up into hydrogen and oxygen and sweeps it away into space. The sweeping is easier on Venus, too, because it has no protective magnetic field.

It all begins to fit. The picture is in some ways quite simple—Venus and Earth started out looking about the same, but Venus lost its oceans to space and retained its atmospheric CO_2 as a result. On the Earth, the opposite happened: the planet kept its water but lost its CO_2 to the oceans. Our planet is awash with water, but the equivalent to the mass of carbon dioxide that gives Venus its crushing surface pressure and searing temperatures is mostly locked up in Earth's oceans and solid crust. The White Cliffs of Dover, the Great Barrier Reef, and many other well-known features of the Earth's surface and crust are made of chalk that consists in large part of carbon dioxide that was once in the atmosphere.

Venus is still losing water to this day. *Venus Express* observed both hydrogen and oxygen, roughly in the proportion of two to one, streaming away from the planet, swept by energetic particles and photons from the Sun. The original oceans were probably removed in this way a long time ago. Their legacy is to be found in the remaining atmosphere, where the ratio of heavy hydrogen (deuterium) to the common isotope is more than a hundred times that found on the Earth. That ratio was probably the same on both planets when they formed, and by far the most likely way for Venus to become enriched in deuterium is by losing a lot of water. During the escape process, the small proportion of water molecules that contain a deuterium atom (so are HDO rather than the more familiar H_2O) escape a little less easily because of their slightly greater mass, and so undergo the process called fractionation. What *Venus Express* observed was the accumulation of deuterium-rich hydrogen during billions of years of water loss.

With the oceans long gone, what we are seeing now is probably the water from recent volcanoes escaping at a much lower rate than we would have seen billions of years ago. If we assume that the present-day water budget is in balance, we can estimate from this how the level of volcanic activity on Venus compares to the Earth. The trouble is, we do not know what proportion of the gas in a Venusian volcanic plume is water vapour. On Earth, the proportion is quite high, and the plumes we see are full of steam as well as dust and ash, and this is what gives them their explosive force. However, water from the oceans is cycled through the crust into volcanic reservoirs and obviously Venus does not enjoy the same input. The whole character of volcanism on Venus must be different, for this reason alone.

It seems likely that further research with new missions in the future will end up filling in the details of this general picture, rather than coming up with a whole new paradigm. The next step should be to get down there and investigate large numbers of individual volcanoes. However, there is one big detail that we really need to understand, and don't yet, and that is the role that volcanoes may have in controlling the climate on Venus.

Are volcanoes the culprits?

When I say that Venus has the same problem of global warming as the Earth, and for the same reason, that being what we loosely call the greenhouse effect, people sometimes say: but surely the warming on Earth that has some of us so worried is caused by cars, power stations, and factories, and there are none of those on Venus? It's a good point. The primordial carbon dioxide on Venus might have been expected to decline in abundance since the planet formed, even without oceans. The reactions to produce chalk and other minerals go much faster when the CO_2 is dissolved in water, but they still happen, slowly, directly from the atmosphere, and four billion years is plenty of time. The Earth would have essentially zero CO_2 in the atmosphere if something wasn't replenishing it.

There are natural sources of carbon dioxide, of course, and the commonest is volcanoes. These are belching somewhere on Earth all the time, and together with human activities like burning fossil fuels and the rain forests, contribute to making the carbon dioxide content of the air we breathe about one part in a thousand. Venus has a lot of volcanoes; the images of the surface show thousands of them, in all sorts of shapes and sizes. Some of them seem to be active, and although we don't know yet how many or how active the evidence is piling up that carbon dioxide and other gases are being shoved into Venus's atmosphere at least as fast as on Earth, and probably much faster.

Hot rock, unlike cool water, can emit carbon dioxide as well as absorb it. An interesting experiment made several decades ago involved putting silicate rock in an oven filled with carbon dioxide gas and varying the temperature and pressure until they reached the equilibrium point where carbon dioxide was being emitted at the same rate as it was being absorbed. This was before the experimenters, or anyone, knew what the conditions are on Venus—nobody thought it could be as hot, or as dense as it is—but the gas and the rock in the oven duly settled down at a pressure of nearly 100 Earth atmospheres and about 450 °C. This is an unstable equilibrium, but pretty much exactly what we find on Venus.

New Worlds for Old

The *Venus Express* team leaders wrote an 'early results' paper for *Nature*, in which we emphasized the similarities between Venus and Earth, with the title 'New results from Venus show a more Earth-like planet'. The point was that our neighbour was not the weird and alien place of the 'evil twin' image most people, especially in the media liked to portray, but much more obviously another Earth on which the same parameters that give us our comfortably habitable world here had worked out rather differently. The evolutionary differences involving volcanism, early dense atmospheres dominated by carbon dioxide, and warm oceans on both planets neatly explained, always with caveats about the still unknown details, why Venus does not look like a promising destination for human colonists from Earth. They do not yet explain the relatively subtle influences that make an Earth-like planet take one path or another.

Most of us vaguely realize that the rate at which we are filling up the Earth and using its resources is such that in a century or so it may be imperative to cross space to new worlds in search of pastures new. If this is so then many people assume Mars will be the destination, and in the shorter term that may be true. But in the long run Venus might be the better prospect. This rather unexpected idea was an outcome of all the thinking about Venus, analysing data, and making computer models of the climate that came about as a result of *Venus Express*.

The key role of volcanism in maintaining conditions there could mean big changes in the climate in the very distant future, when the interior of the planet cools some more and the emission of volcanic gas into the atmosphere subsides. Then the models predict that the pressure will fall as the carbon dioxide combines slowly with the crust, and the surface will get cooler. In the very long term, Venus ends up with a nitrogen atmosphere with a surface pressure just twice or three times Earth's. The temperature is what we would think of as tropical, but not too hostile or unbearable. If only we can wait that long! It might take one or two billion years, we don't really know, but Mars used to be very volcanic, and now its volcanoes are almost or totally defunct. Venus, and Earth, will surely follow, taking longer to cool inside because they are larger, but getting to a Mars-like state eventually. Loss of volcanism might not make a huge difference on Earth, but on Venus it could lead to the benign conditions and attractive destination that for centuries Earth dwellers expected.

Because we lost a crucial instrument on *Venus Express*, the actual level of volcanic activity on Venus has yet to be measured, or even estimated reliably. There are good reasons why it is probably much higher than on Earth, but a lot of my colleagues argue that it may just as likely be zero. I don't agree with them, but we won't stop arguing until there are better data. We still do not have very much information about the composition of the surface, either, so we have to be careful about blindly accepting the results, however pleasing, from simple experiments and models until we have more actual data. Since it will be quite some time before geologists can be rooting around on Venus, one of the next steps after measuring volcanic activity has to be a mission to gather samples and bring them back to Earth for analysis. Such a mission has actually been studied in considerable detail by the European Space Agency, and I was a member of the team that carried out the study. We found it to be quite feasible, although so expensive and risky that it is not likely to get approval for flight any time soon.

Bringing Venus to Earth

It is always interesting to help to plan future space missions, even very far off ones, so I was intrigued when, in early 1998, ESA asked me to be part of a team being formed to carry out a comprehensive study of a mission to land on Venus, drill to acquire core samples of the surface, and return them to Earth. This was before *Venus Express* was conceived, let alone launched, and the reason for committing time and money to advanced studies like this was not so much that they had definite plans to launch such a difficult and expensive mission, but rather that they wanted to take an imaginative look at what could be done and the technology that would be required. Some of the capabilities that would be needed for *Venus Sample Return* would become part of their long-term development programme and be available if and when such a mission actually flew.

The study was mostly carried out at ESA's technology centre near Amsterdam, the same place I had visited in a successful quest for a job offer back in my graduate study days. Now it was September 1998, and during the intervening 30 years ESTEC had grown much larger than it was when I first saw it. The tulip fields were all but gone, built over to provide buildings and car parks for the thousands of people that now worked there. For a time I travelled there every couple of months for two and three-day meetings at which we tried to figure out how to set the scientific objectives and actually carry out the mission in the searing furnace-like conditions on the surface of Venus.

We had a support team of engineers and a decent budget, so we could call on contractors from different parts of the European aerospace industry to provide specialized knowledge about the mechanisms and materials that might be needed to do the job. I was surprised to learn that it is not possible to take off from the surface of Venus using rockets alone, because the air is too dense. Instead, a balloon, made of materials that could survive the high temperatures and the sulphuric acid clouds, would float the sample return module up to the cloud tops and it would blast off from there. Once in space, it had the tricky job of navigating back to the mother ship, as it waits in orbit to fly the sample back to Earth. The whole mission will take six years to complete, including just one hour on the surface of Venus. Sample return was, we reported, something that certainly could be done, but at a price.

We are still waiting for it to happen in real life and no one is holding their breath. Being involved in futuristic exercises of this complexity always raises philosophical questions in the mind of whether a mission of this kind will *ever* happen. I have seen too many much less ambitious but equally worthwhile projects founder on a lack of aspiration, which always manifests itself as a shortage of money. However, the costs, although they sound large, are actually perfectly affordable when compared to many other kinds of human activity, some of them frivolous and others downright lethal. As John Kennedy demonstrated when he went out on a limb for the *Apollo* programme to put men on the Moon, not only can formidable sums of money be found if the will is there to do it, but that which seems extravagant at first actually pays dividends in many ways initially undreamed of. Will we, that is humankind, ever find the motivation to even attempt to found a colony on Mars, or to set up a manned research station floating above the clouds on Venus? I'm not sure about NASA or ESA, but perhaps the emerging space nations will. Or will our civilization collapse first, perhaps in a global war, or perhaps under the weight of unsupportable population growth burdened by debilitating climate change? I wonder.

Postscript: *The Sky at Night, Discworld*, and the Tardis

Occasionally at conferences, lectures, or dinners I had met Patrick Moore, the world's most famous amateur astronomer, author, and broadcaster. During his long career Patrick had spent many hours at the telescope observing Venus and had written several books on the subject, and so he was naturally interested in what was coming out of the *Venus Express* investigation. I received a letter from him inviting me to appear on *The Sky at Night*, his popular television programme, which at about this time had earned the distinction of being the longest-running TV show ever, with the same presenter. I had watched and been awestruck by him as a schoolboy and the chance to be a part of this ongoing legend was too good to resist.

Patrick was getting on a bit by the time I became part of his circle, and no longer did outside broadcasts, nor did he travel to the London studio to record his interviews. Instead, a caravan of BBC people came to his house in Selsey, on the south coast of England just across the Solent from the Isle of Wight, and set up lights and cameras in his study. Shooting started promptly the next morning, so we all had to travel down the night before. Patrick invited me to stay with him in 'Farthings', the large and comfortable thatched house he had lived in from birth, and to have dinner with him the night before.

Over dinner he remarked that, since we would be discussing Venus on the show, the recent transit of the planet across the face of the Sun in June 2004 was bound to come up as it had attracted such popular interest. He asked me if I would kindly explain to the viewers why such transits occur as they do in pairs, eight years apart, but only one pair every 126 years. I could hardly say no, or admit that I hadn't thought about it and so did not really know the details. My mind was racing as I tried to work out the orbital dynamics of Venus and Earth and why they led to this rather odd pattern. Fortunately, Patrick's house was stuffed with books and after dinner I soon found a detailed account in one of them and swotted it up. But how to put it across to a TV audience? I decided I needed to make a model that would show the orbits graphically.

I got up early and found one of the young helpers that came with the BBC engineers to set up the equipment for the recording. He went into town for me and found a shop in which he was able to find two hula-hoops of different size, a ball of string, and two table tennis balls to represent the planets. However he had forgotten the large ball that I needed to play the part of the Sun, and there wasn't time to go back. A quick visit to Patrick's gin and tonic larder produced a large lemon, and that had to do. I held this ungainly contraption up to the camera and showed how the slight tilt in Venus's orbit relative to Earth's led to the rare transits of the tiny black disc of the planet across the Sun.

This apparently went well enough that I was not only invited back on the programme four years later, but also to a series of parties in big marquees erected among the observatories on the lawn at Farthings

After a lifetime of admiring Patrick Moore's ground-breaking BBC TV series, I finally appeared as his guest on *The Sky at Night* on November 2006. Patrick was a keen observer of Venus through the telescopes in his garden; I talked about the new findings from the Venus Express spacecraft which had recently arrived at the planet.

I was photographed with Doctor Who's Tardis at the BBC Television Centre in London while participating in the filming of a 'farewell' event for Patrick Moore after his death in 2012.

to mark Patrick's birthdays, and milestones like the 500th *The Sky at Night*. One met all sorts of interesting people at these, most memorably the author Terry Pratchett, whose *Discworld* novels were much loved by many, including my wife. He had just announced that he was suffering from the onset of early dementia, but seemed quite normal as we had a pleasant chat.

The sequence of events finally ended when Patrick passed away in 2012, with a grand finale in the form of a Festspiel at the BBC Television Centre in London. We gathered for drinks behind a soundproof window on the balcony above the newsroom, looking down on the presenters with the weatherman alongside us on the higher level. Remotely controlled cameras ran on rails all around the balcony and whizzed past us like model trains. Moving into the BBC auditorium, we heard from producers and others who had worked with Patrick over the years and watched as they showed amusing clips of some of the more memorable moments in 50 years of *The Sky at Night*. Finally, some celebrity Patrick fans like Brian May and Brian Cox gave short eulogies before we all said our silent goodbyes and departed.

13
Marooned on Mercury

Mercury is the densest planet

I think Scott Adams, who created the popular comic strip *Dilbert*, must have had Mercury, the densest planet, in mind when he wrote:

DILBERT: And we know mass creates gravity because more dense planets have more gravity.
DOGBERT: How do we know which planets are more dense?
DILBERT: They have more gravity.
DOGBERT: That's circular reasoning.
DILBERT: I prefer to think of it as having no loose ends.

Most people, even most astronomers, don't spend a lot of time thinking about how things might be on Mercury, despite the fact that it is an important member of the family of terrestrial planets to which our Earth belongs. Mercury was not much on my personal radar either, until an invitation came in from the European Space Agency to join a study they were doing of a possible future mission to land on this small planet. With typical ESA bravado, they were considering not only landing, they were going to look into ways of drilling into the surface, collecting samples, and bringing them back to Earth.

Mercury is the planet that orbits closest to the Sun. It is the smallest, as well as the densest, of all the planets, apparently made almost entirely of solid metal, with just a fairly thin rocky crust. We see it with the naked eye as a morning star like Venus, although much dimmer in our skies, setting soon after the Sun. It is not an easy object to observe through a telescope, and until recently, it was very little explored. Landing there would be a challenge.

At the end of a technical study of a similar and equally ambitious mission to return samples from the surface of Venus, in which I had participated with great interest as already described in the previous chapter, the European Space Agency decided they wanted to go on and do a similar study for Mercury. These advanced technology studies did not carry any obligation to go on and fly the mission, indeed it was clear even before the study started that it was not on the cards for the foreseeable future. When I was asked to join the study team I protested weakly that I was not an expert on, nor especially interested in, Mercury science and would rather they got someone else. However, they wanted to steam ahead using the momentum built up by the same team that had just published its conclusions about returning samples from Venus, so I knuckled down with the others and got on with it. In the end I was glad, because I not only found myself getting interested in the study and able to contribute, but it was the beginning of a path that led to involvement in a real mission to Mercury.

Like the Moon but different

It was because I was interested mainly in atmospheres and climate that Mercury was not a planet that I had thought about very much over the years. Small and close to the Sun, very hot by day and very cold by night, it is essentially a barren and airless world. Gerard Kuiper, the pioneering astronomer and scientist who wrote a review of planetary atmospheres in 1950, devoted considerable space to 'repeated' observations of haze or clouds on Mercury, which he reported as being thick and persistent enough to cover surface features for several days at a time. As observations got better the haze went away, and probably it never existed, although we cannot be completely sure until we understand where Mercury's polar icecaps came from. Certainly, there is not

Mercury is a small planet, as this same-scale comparison with the Earth (bottom) and Moon (top left) shows.

much atmosphere on Mercury; current estimates put an upper limit of the surface pressure that is less than one trillionth of that on Earth.

With no atmosphere or water to erode and sculpt the landscape, Mercury's surface is still scarred by the bombardment it received in the past, mostly the very distant past, by drifting debris in the early history of the Solar System. To a casual observer, Mercury looks a lot like the Moon, which received a similar battering and also has no atmosphere to protect or scour its surface. However, on closer inspection the two are quite different. We can see using spectroscopy, the technique where we measure the amount of light that is reflected at different wavelengths, that the minerals present in the rocks are not the same, and the effects of collisions and of volcanic eruptions, both long ago, have left different records on the surface as seen in high-resolution photographs. For example, superficially similar parts of the two bodies' surfaces are found to have different colours if the contrast is stretched, even in photographs taken at visible wavelengths. Infrared maps reveal further differences in composition and physical properties that speak of a quite different evolutionary histories of Mercury and the Moon.

The fact that Mercury is much denser than the Moon, or the Earth for that matter, also reveals differences in composition. To be as heavy as it is for its size, Mercury must have a high proportion of metals, probably iron for the most part, with some nickel, sulphur, and other elements mixed in. Thus the surface we see is part of a rocky shell over a large metallic core which occupies most of the radius of Mercury. The Moon, on the other hand, is entirely rocky and has little or no metallic core. Earth, Venus, and Mars, are somewhere in between. The basic reason for this seems to be that, while the Moon was formed when a large, unidentified object collided with the young Earth and detached material from the rocky outer layers, Mercury formed like the Earth itself from primitive material orbiting the Sun. Being closer and therefore hotter, in Mercury's case the lighter rock-forming elements were less likely to condense than the heavier and less volatile elements like iron.

Mercury is not very big as planets go—smaller than Saturn's moon Titan—and the metallic core should have had no problem cooling and solidifying over the billions of years since it formed. However, the visit by *Mariner 10* in 1974 showed it has a planetary magnetic field strong enough to imply that the core is liquid. Possibly Mercury has a large proportion of very heavy metals, including radioactive elements like uranium and thorium that keep it warm by emitting heat as they decay. If so it may also be rich in the rare earth metals that are so important, and scarce, on our planet: mining on Mercury might be something that lies in our not-too-distant future.

Thus primed with some basic facts, and several puzzles to solve, our study group started to make plans to gather samples and bring them back to Earth for laboratory analysis, feeling such a mission was clearly worthwhile. The group had not been at its task for long when

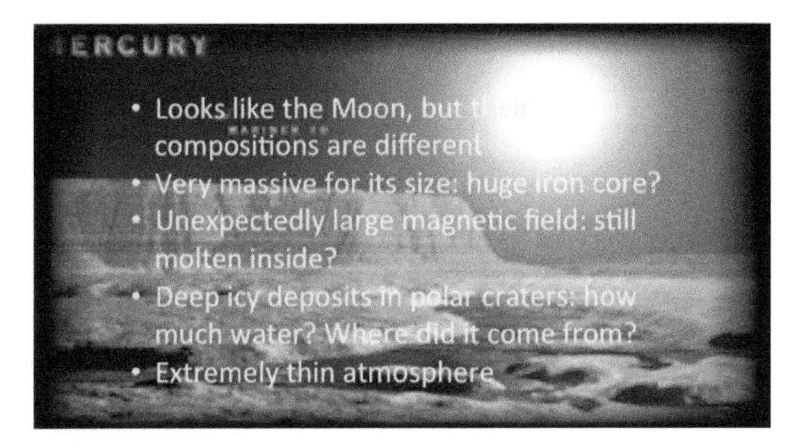

In the 1990s I gave many talks in support of a mission to explore Mercury. This is one of the slides I used, which lists my favourite scientific objectives for such a mission against a background of a NASA artist's impression of a Mercurian landscape baked by the Sun, which appears much larger in the sky than it does from the Earth.

major difficulties became apparent. The first is just getting there: flying a large spacecraft into orbit and then landing is hard when the planet is small, especially when it is close to something very large. Mercury's gravity is only about a third of Earth's, and the Sun, with its huge gravitational field, makes orbital capture quite tricky. Shielding the delicate instruments, fuel, and electronics from the searing heat of the Sun is also challenging. The absence of any significant atmosphere means cheap, lightweight parachutes cannot be used for landing; the required deceleration and steering must all be done with heavy, expensive rocket motors.

A further and very major problem for the scientists on the team was deciding where to land, since of course that would determine where the samples came from. For safety reasons it had to be somewhere relatively smooth and bland; landings attempted in mountain ranges or other rugged terrain were likely to crash and burn. Mercury has deep basins and extensive plains that look smooth and safe, but they are mostly filled with thick layers of dust or soil. This loose material is probably at least partly made up of micrometeorites that rain down from space onto the airless surface and accumulate there. We did not want to go all the way to Mercury, at great trouble and expense, just to come back with a sample of ground-up meteorite that we could have got from a science museum on Earth. How could we find and avoid the places with thick layers of this regolith material, and find somewhere smooth enough to land but with a surface of representative planetary crust?

The answer we came up with was to map various sites with an infrared radiometer that could determine the thermal inertia of the

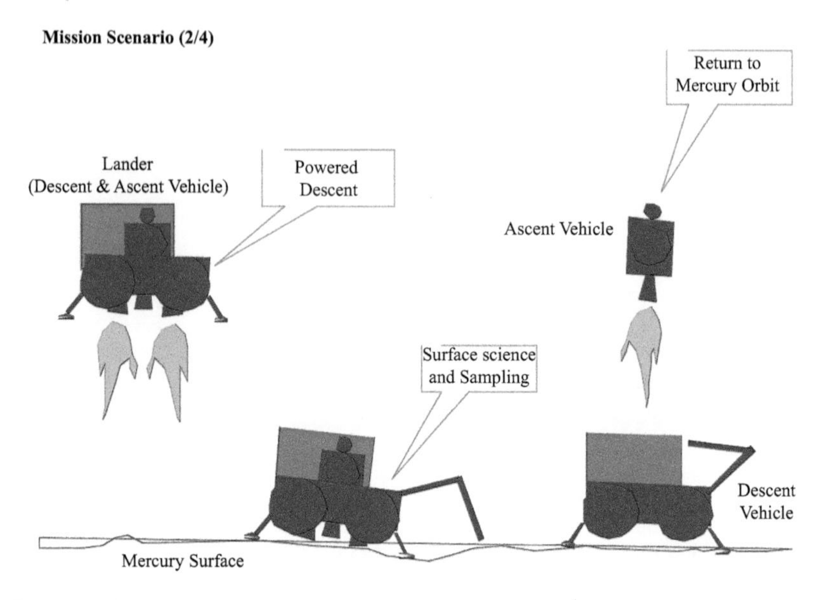

Sample gathering and return is a complicated business, even (or especially) on an airless planet like Mercury. This sketch showing how it might be done is from the final report which I helped to write as a member of the ESA study team. There was no intention, even at the time, to fly such a mission any time soon; the purpose of the study was to highlight the technical advances, and the large budget, that will be needed to do it several decades from now.

terrain. This is a parameter that describes how quickly material cools, for instance at night after having been heated by the Sun during the day. Relatively loose materials like dust and soil have a low thermal inertia, and cool quickly as the Sun sets. More compact material, like rock, takes much longer. That sort of measurement had already been made on Mars to obtain the heat-retaining properties of different types of surface such as various kinds of rock, dust, and ice. The radiometer on the *Viking* spacecraft that orbited Mars in the 1970s detected layered terrain over more than a third of the surface of the planet, where single-layer models could not explain the observed temperatures. The thermal inertia values varied by a factor of almost a hundred, strongly correlated with geological structures seen in the superimposed high-resolution images. The larger thermal cycles, the absence of a substantial atmosphere, and the almost complete lack of any prior information about the surface, means that thermal inertia measurements on Mercury will be even more useful than they are on Mars.

We worked up the design for a suitable instrument and added it to the model payload for the spacecraft design we were studying. Of course, it also went into my notebook as a candidate to be proposed for the payload of a real mission if and when one came along.

Ice in a furnace

Around the time of our work on the study, Mercury suddenly got even more interesting. Astronomers using one of the largest radio telescopes on Earth, at Aricebo in Puerto Rico, had succeeded in bouncing radar pulses off the Moon, then off Venus. Now, they had refined the technique to the point where radar imaging of Mercury was possible, showing details on the surface of the planet. In a quite remarkable further advance, the resolution of these images was improved until features as fine as a few kilometres across could be seen in the images. These features included something very bright partially filling the craters near the north pole. The strong reflections meant that the only realistic candidate for the bright stuff was water ice.

So, there appear to be large deposits of frozen water near the poles on the sun-baked planet. Interesting, indeed! How much is there? How does it survive there? Where did it come from? The answers to these three questions seemed to be (a) a lot; (b) in permanently shadowed regions; and (c) we don't know. The strong radar echoes required extensive deposits of ice many metres deep, not just a thin coating of frost. The NASA *Messenger* mission recently confirmed the presence of a lot of water ice, up to a trillion tons of it, near the pole. Permanent shadowing is possible at high latitudes because Mercury has almost no axial inclination, so no seasons like the Earth where the poles tilt towards the Sun. Given a rough enough surface, which certainly exists, there are crater bottoms and cliff faces where the Sun never shines, and has not shone for billions of years.

So Mercury has a cryosphere, to use the term that has become fashionable when discussing the icy regions on the Earth that are a crucial part of our climate system. Mercury's ice layers obviously hold clues about the amount and the migration of water in the Solar System. Had it come from the interior of Mercury, from now-dead volcanoes, or from fissures in the rocks that might still be seeping vapour, like Enceladus out in the domain of Saturn? Maybe it fell on to the planet as icy chunks of cometary material, or arrived all at once in a collision with a big comet. Are the Mercurian glaciers ancient, or still actively forming and dissipating? We must find out. Infrared radiometry could address this, as well.

Two new moons for Mercury

As we were drawing up designs for sample return, ESA already had plans for a less ambitious but still challenging Mercury orbiter mission on its books, and in 2007 this matured into a real mission, the curiously named *BepiColombo*. The Italians wanted to name the mission after Professor Giuseppe Colombo, whom I had met at meetings when he was aging and I was young. He became famous for correctly working out Mercury's peculiar orbit around the Sun. It is such, he said, that three years are equal to two days on that planet. This odd situation arose because the orbit is quite non-circular, and because of the huge tug from the nearby Sun. Giuseppe, or Bepi as he was known, died in 1984 aged only 63, and so the mission became his memorial as *Colombo*. The trouble was, too many people in the media thought it was named after a TV detective, so the professor's nickname was soon incorporated.

ESA planned a polar orbiter that could map the entire surface of Mercury in great detail, originally accompanied by a small lander. The 'surface element' could have made use of the first part of our sample return study, but it was soon deleted from the plan on grounds of cost. The Japanese space agency expressed interest in contributing a second, smaller orbiter to investigate Mercury's unexpectedly large magnetic field, so *Bepi* became a dual orbiter mission.

We already knew that orbiting Mercury would be no easy task; the small mass of the planet plus the proximity of the hot and massive Sun meant horrendous thermal design problems for the spacecraft and a long and convoluted journey. In the final design, *Bepi* is destined to

While I was working with ESA on designing a possible mission to Mercury, astronomers at the Arecibo observatory in Puerto Rico obtained these spectacular images of the planet using a radar technique. The bright, reflective deposits in the craters near the north pole of Mercury are thick layers of water ice that survive only in places that are permanently shadowed from the Sun. Investigating them, and finding out where the water came from, became a key objective for the embryonic mission.

circle the Sun many times, flying close to the Earth, Venus, and then Mercury itself five times, in order to tweak its trajectory until Mercury orbit can finally be achieved. In addition, it uses an innovative solar-electric propulsion system that has required special development and additional risk taking. It remains to be seen whether it will all work, as long delays in development have meant that the spacecraft will not see the launch pad until 2017 at the earliest, nine years behind the original schedule.

For some odd reason, the European Space Agency and its advisors relish missions like *BepiColombo* where the game is so massively difficult as to appear not really worth the candle (*Rosetta* was another example) and the mission was selected for flight. As should have been expected, delays mounted and costs escalated at the expense of the rest of ESA's programme, and all for what is exciting but not actually really top priority science. Part of the attraction was that comet rendezvous and Mercury satellite missions were both something the more value-oriented Americans had not yet done. I thought the selection of *BepiColombo* was rash and would, like *Rosetta*, absorb the funds that might be spent on several more practical and scientifically productive missions, but being right is not a deciding factor and, having been unable to prevent it happening, I wanted to be involved.

Science goals

The sample return study had fired me up about two fascinating objectives: probing the polar regions and their incongruous ice deposits, and understanding the dense, possibly molten, interior of Mercury. The thermal properties of the regolith would be interesting, too, even if we didn't have to worry about finding a landing site. We set to work on a relatively cheap and simple instrument that could address these three issues in detail, using the infrared techniques that we specialized in at Oxford. A similar concept had already looked good to ESA on the model payload of the hypothetical sample return mission.

We knew very well that a good proposal is only half the battle, with publicity and political manoeuvring to prepare the ground for the selection process also crucial. As part of that, I presented a paper at the 2001 meeting of the European Geophysical Union, an annual gathering of thousands of scientists that this year took place in the Acropolis conference centre in Nice on the south coast of France. As I presented my pitch about the investigation of the surface, atmosphere, and interior of Mercury by multispectral mapping of the planet in thermal emission, I noted with satisfaction that the ESA lead scientist for the Mercury Planetary Orbiter, Rejean Grard, was sitting in the front row and taking notes.

I already knew Dr Grard from working together on the unsuccessful *Kepler* mission to Mars, and a short-lived, earlier study of a Mercury orbiter mission that would have also delivered a probe into Venus as it flew past. As I had hoped, my talk in Nice convinced him that the measurements I described were essential for a definitive investigation of the surface and atmospheric conditions on Mercury by *BepiColombo*. After a few meetings, they became part of the reference payload for use by the engineers designing the spacecraft. Our instrument was the subject of an accommodation study by ESA and its industrial contractors, one of only two UK-led 'model payload' instruments currently being considered in this way.

It was fun to be back at ESTEC working with teams of scientists and engineers to plan a big project, and with the expectation that we now had an inside track to be selected as investigators with our own instrument on the spacecraft. Our job was to define the science goals and the relevant measurements and work with the engineers to turn these into a mission concept—spacecraft size, orbit, and so on—and to refine the model payload. The real payload would be determined competitively later using the model version as a starting and reference point.

We thought we were designing the first orbiter of Mercury, but we took so long over it that the Americans nipped in with a small, opportunistic mission to Mercury called *Messenger*, which achieved orbit in 2011, while the European spacecraft was still being designed and built. *Bepi* could still claim to be bigger and more sophisticated, and therefore still likely to deliver a worthwhile science return despite being pipped at the post in the race to orbit Mercury. On the other hand, the development was running behind schedule and over cost, and still had a host of technical problems. ESA initiated a review to consider cancelling the mission, but in the end they decided to soldier on.

The MISTERIE proposal

At Oxford we produced a proposal entitled: the *Mercury Infrared Surface Temperature, Emissivity, Reflectivity and Inertia Experiment*, which delivered the obligatory catchy acronym, MISTERIE. The subtitle of the proposal was: An Infrared Thermal Mapper to study the Surface and Atmosphere of Mercury from the *BepiColombo* Planetary Orbiter. We declared that our measurements would address some of the most important goals of the mission, including characterizing the different regions on the surface with thermal inertia measurements just like those we developed in the study for sample return.

We would also investigate the origin of the unexpectedly strong magnetic field of Mercury. According to what we knew about the interior processes needed to produce such a field, it had to imply the existence of a completely or partially molten core. This in turn means a source of heat in the interior and geothermal activity at the surface as heat from the core slowly escapes. The heat flow at the surface is most unlikely to be uniform, but rather to occur in localized 'hot spots', possibly near old and now quiescent volcanoes. If one of these emitted heat at a rate higher than the average, its equivalent blackbody temperature would be hotter than the typical night-side temperature of around –200 °C and could be detected by our radiometer. By mapping them all over the planet, we would be able to characterize the internal heat source and say something new about the structure of the crust, for instance whether there was any sign of a continental plate structure.

The third major objective was to investigate the polar ice deposits and try to work out how much ice there is and where it came from. One of the problems is that even the shaded parts of craters more than about ten degrees from the pole itself are not cold enough (according to theoretical models) to accumulate water ice faster than it sublimes away. It has been suggested that the deposits in these craters are some less volatile material like sulphur but this is unlikely on other grounds. The mystery could be resolved with measurements of the temperatures associated with the ice deposits, providing the data needed to improve the models and infer the identity of the condensate. The planned maps of thermal infrared emission and derived products would cover the planet, including the night-side and the shadowed polar regions, where neither the television camera nor the near infrared and ultraviolet reflectance spectrometers also expected to be on board could operate.

As I grafted away at the proposal, I was reminded again of how the notice boards at JPL, when I worked there in the 1970s, would sometimes carry light relief in the form of witticisms that struck a chord of truth in the minds of the hard-working scientists and engineers who worked there. One of them took the form of a strip cartoon from the popular 'Wizard of Id' series, set in a jokey version of medieval England, that appeared every day in the *Los Angeles Times*. The first frame showed a scruffy peasant reading a poster offering a job cleaning out stables, posted outside a yard with a couple of horses sticking their heads out over the gate. In the next scene, he is being interviewed by Rodney the knight who asks him if he has any previous experience of shovelling excrement. No, says the peasant, I've spent the last 20 years writing proposals. That's close enough, says Rodney, handing him a pitchfork.

The instrument we proposed for *BepiColombo* claimed heritage from three state-of-the-art flight instruments: focal-plane multilayer optical interference filters, similar to those developed in the UK for HIRDLS on the NASA *Earth Observing System*, micro-machined thermopile detectors, developed and built at the Jet Propulsion Laboratory for the PMIRR on *Mars Reconnaissance Orbiter*, and lightweight structure from the UK-built part of the CIRS spectrometer on *Cassini*. The instrument optical design and layout, however, would have to be new to allow it to be accommodated on a spacecraft of totally new design, and to assure the integrity of the measurements under the difficult range of observing conditions expected at Mercury.

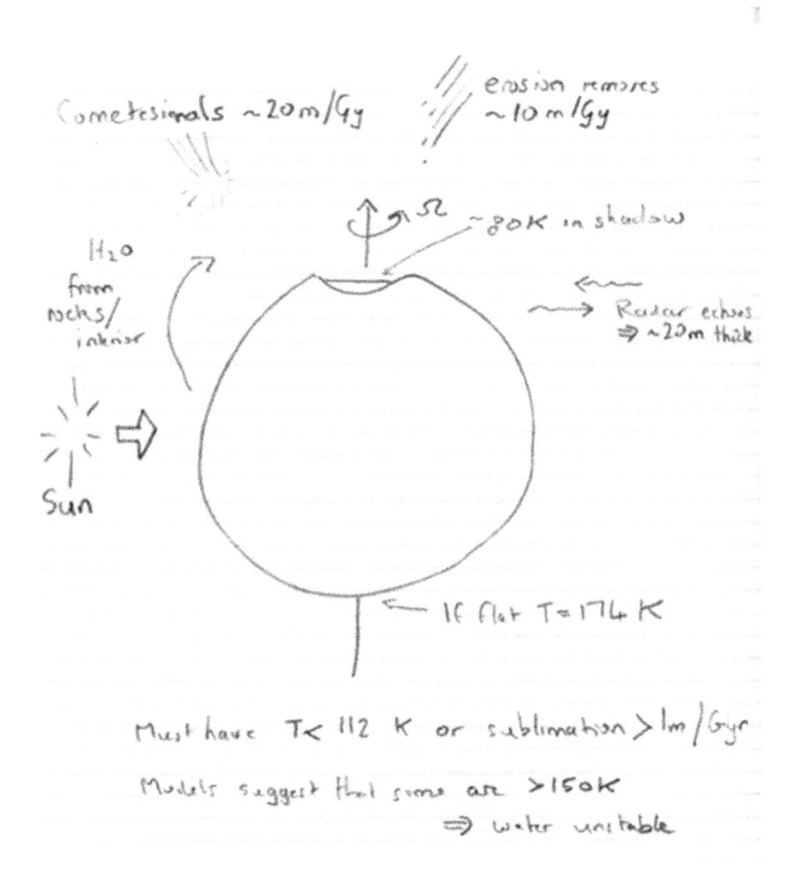

This sketch from my notes for a talk at the Royal Astronomical Society attempts to explain the ice at Mercury's polar regions. It could have been there for billions of years, surviving from a time when conditions on Mercury were very different. Alternatively, it might be seeping out of the interior from vents in the crust, even now. Yet again, it might have accumulated from small (or large) icy comets that crashed into Mercury over time. We have no idea, in other words, and might have to land at the pole and drill into the ice before we find out. If astronauts do this, they might be able to ski on Mercury during their recreation breaks.

The proposed science team included experts on planetary surfaces from the University of Lancaster and University College, London to augment the infrared and atmospheric expertise in the Oxford and JPL groups. Ilias Vardavas in Crete had spent many years studying the evolution of the Solar System and was a valuable addition. We all spent months developing plans for data processing, calibration, and software tools to allow plotting and sampling of the data by interested scientists working remotely. It is essential to include a plan for generating attractive looking maps and other products of wide interest to the media and the general public, including schools and colleges.

We worked out the instrument mass at a very reasonable five kilograms, and the development costs, we reckoned, were a snip at £2,063,000. The instrument was to be ready for launch in 2008, as then planned; we were blissfully unaware of the development problems that would push it back by an unprecedented nine years and counting.

Rejection and redemption

All of this careful planning and hard work came to naught when the ESA selection board summarily rejected the MISTERIE proposal. I was shocked, because I thought it was really good experiment, and just what the mission needed. I still think that, years later; it ranks with the unconsummated *Kepler* mission to Mars and our frustrated opportunity to build part of the VIRTIS instrument for the *Rosetta* comet mission as one of my greatest disappointments and gives new meaning to the legend of Oxford as the home of lost causes. There is always an element of randomness about these choices, as well as subterranean intrigue: the loudest voices on the committee were French and I didn't have any team members from France, in retrospect a fatal mistake.

We did have German colleagues, though, and that allowed us to salvage something and stay involved in the mission. During the lengthy study and proposal writing we worked with scientists from the German Space Agency in Berlin who specialized in infrared spectroscopy and who also had plans to propose an instrument for the mission when the time came. Their focus was high resolution mapping of features on the surface using mainly reflectance spectroscopy, to identify which minerals are present in the rocks. It also had a capability for infrared mapping, and so could address some of our Oxford objectives, although not as well as our proposed radiometer.

Our two teams were friendly enough, and our plans sufficiently mingled, by the end of the study to agree that we would add each other as co-investigators to our respective proposals. The German proposal for infrared spectroscopy was selected, and we ended up as minor partners on an instrument called MERTIS, for *Mercury Thermal Infrared Spectrometer*. The selectors around the table at ESA headquarters in Paris argued that the spectrometer would be able to address the radiometer objectives as well, but probably the sensitivity is too low to detect subtle temperature anomalies on the night side or in the icy polar craters.

MERTIS is still an interesting experiment and a chance to study Mercury from orbit is a decent consolation prize for all our hard work. For me personally it is fortunate that I didn't have a larger involvement, because the long delays to the mission would have tied me up intolerably if I were leading the team, with endless meetings to attend. Alternative projects like *Venus Express* would have been impossible, so retrospectively I can see that I nearly made a mistake by pursuing a principal investigatorship again in defiance of my earlier resolution to stick to less onerous commitments. Fate steered me into a co-investigatorship and hopefully will deliver the opportunity to make something of it.

As I write, we are still waiting for *BepiColombo* to launch, with endless technical problems pushing the launch back again and again, most recently to January 2017. Even if this is achieved, the spacecraft will not arrive in orbit around Mercury until New Year's Day in 2024, after no less than five close encounters with Mercury itself, two with Venus, and one in which it actually returns to the Earth to skim close by. These gymnastics are needed, in addition to the new ion drive motor, to supply the energy the spacecraft needs for the long and difficult flight. If everything works as planned, which is by no means certain, eventually the data will flow and the scientific work will begin. I shall be nearly 80 years old and lucky to make any kind of contribution, even as an elder statesman.

Every once in a while, during one of the occasional apparitions of Mercury from the Earth, I search it out as a dim dot in the sky just above the horizon. For a quiet moment I am like one of those early astronomer-philosophers, gazing at another world and thinking about what I wish I knew about it. The difference is that now we can look forward to a close-up view, and possibly some answers to what is going on in that strange place.

14
Mars at last

Calvin Coolidge, who was President of the United States in the 1920s, famously said that nothing in the world can take the place of persistence and determination. Getting a third crack at sending our 'weather satellite' experiment to Mars was an exemplar of this aphorism. After supporting us for *Mars Observer*, and seeing the money wasted when the spacecraft exploded, the funding people in the UK didn't want to know about starting over again to build a new instrument with fresh funds. NASA had helped us to rebuild the hardware for the recovery mission with *Mars Climate Orbiter*, but that had failed as well, and our carefully crafted instrument was now just wreckage on the Martian surface. After the crash, NASA did not have a suitable mission for us to even contemplate a third attempt, following the hard-learned lesson that the faster, cheaper approach was not necessarily better.

Third time lucky

However, NASA does not give up easily, and neither did we. It was only a year after the crash that they announced that they would be mounting a new and much larger satellite mission to Mars, to be called *Mars Reconnaissance Orbiter*. They needed this to carry out a survey of potential landing sites on the surface in great detail before sending the robotic explorers that were already at an advanced stage in the agency's planning cycle. It is always hard to find the right compromise between the most interesting places to visit and the ones offering the safest places to land. This trade-off is particularly important when looking ahead to eventual human landings, and existing surveys are inadequate.

The detailed pictures needed for a proper reconnaissance would require a camera viewing the planet from a low orbit through a large telescope. This, together with the need for a big antenna dish to provide a high data rate back to Earth, drove up the size of the spacecraft. The telescope aperture in the final design was half a metre, considerably larger than on any previous planetary mission, and powerful enough to grace all but the largest observatories on the Earth. With a mass of over two tons, the new spacecraft could also easily carry a third version of our instrument to investigate the meteorology on Mars.

To make sure that it did, we argued that investigating atmospheric conditions and making infrared maps of the surface would contribute to the objectives of the reconnaissance mission. One of the key factors in planning any descent to the surface, for instance, was strength and variability of the wind at and above the landing site. It helped our case for inclusion on the mission that we had worked on reducing the demand we would make on valuable spacecraft resources by updating the design of our instrument to something much more compact and lightweight. This was possible using new infrared detector technology that had come along since the 1980s, with much of the development work actually done at JPL. These new detectors did not require cooling, so a large and expensive part of the previous design went away.

Rather sadly, we also eliminated the pressure modulators, which were descendants of the crude devices I had built as a graduate student and flown on balloons at dawn on Salisbury Plain. Apart from some instruments on operational weather satellites developed by the Met Office, this meant the device I had laboured over for years, and that had found so many uses, had finally run out of applications. It was too heavy and temperamental, and its job of allowing the instrument to select different atmospheric gases could now be done with miniature solid-state optical filters. These filters were another recent development, also in the UK, driven along by the requirements of the HIRDLS instrument, our latest large Earth observing experiment.

The Meteorology of Mars

Dan McCleese of JPL continued to lead the team and he named the new experiment *Mars Climate Sounder*. We still built part of the hardware at Oxford, and by 2005, the instrument was ready to fly following several years of the usual drama but thankfully without any exceptional crises worthy of relating. The *Reconnaissance Orbiter* launched on schedule, arriving at Mars in March 2006, and successfully achieved orbit. Our instrument worked as expected when it was switched on and delivered great data. In fact, it is still working as I write this in 2015.

The science goals of *Mars Climate Sounder* were the same as they were for its ill-fated predecessor, the *Pressure Modulator Infrared Radiometer*. We wanted to measure temperature and water vapour profiles in the atmosphere, and to identify and map the various different kinds of clouds and haze that occur on Mars. We knew that there are two kinds of condensate cloud, one type made of water ice like cirrus on Earth but another, because Mars is colder, formed of crystals of frozen carbon dioxide. In addition to those, there is a permanent haze of windblown dust, sometimes quite tenuous and at other times chokingly thick. We wanted to study and understand them all; how and where they formed, spread and dissipated, and what effect they have on the climate. That effect is vey large, especially the effect of the dust on temperatures and winds.

Reconnaissance Orbiter is the largest spacecraft so far to be deployed successfully at Mars. The way our experiment worked was to look sideways from the spacecraft and scan the atmosphere at the limb. We could measure a complete profile of temperature, water vapour, cloud, and dust every second, and keep repeating the measurements until we covered the planet. Then we would do that repeatedly and look at what had changed, producing a dynamic picture. At programmed intervals, the instrument would look downwards and map the surface temperature as well. Because it was measuring thermal emission, that is the heat radiated from the atmosphere and surface, it laboured day and night, winter and summer, and because the spacecraft was orbiting over the poles we could cover the whole planet. The resulting global four-dimensional data set gave us the key to understanding the meteorology of Mars.

The science underlying our efforts at Mars is covered in detail in my book, *The Scientific Exploration of Mars* (2009), where I also recount the entire history of Mars exploration since the earliest times of Victorians peering through telescopes and seeing canals and vegetation.

This NASA artist's rendition of *Mars Reconnaissance Orbiter* over the polar regions of the red planet adorns the cover of my book about Mars exploration. The cylindrical object near the centre of the base of the spacecraft is part of the *Mars Climate Sounder* instrument developed at JPL and Oxford to study the meteorology on Mars.

Simon Calcutt and Sarah Harrington on the bleachers at Cape Canaveral for the launch of *Mars Climate Orbiter*. Simon became my first research student in 1979 and went on to lead the instrument development work for our planetary experiments with great success. His crowning achievement was to marry Sarah, who was my wonderful secretary for most of my years in charge at Oxford.

Slowly, the misconceptions have receded and a picture has been emerging of the present climate and the whole range of Martian meteorological phenomena. Many things are still puzzling, such as the great dust storms that blow up typically every two Martian years, and, with no rain or oceans to mop them up, go on to envelop the entire planet.

Key players: Seated at centre is Rich Zurek, the Project Scientist for *Mars Reconnaissance Orbiter*, flanked on the right by Dan McCleese, the Principal Investigator for *Mars Climate Sounder*, and on the left Dave Paige, the PI for the *Lunar Diviner* experiment described in the next chapter. I took this picture on a plane when the four of us were on the way to a meeting in Washington.

The thing that drove my interest, apart of course from the buzz that always comes from taking a fresh look at another world, was the chance to make detailed comparison to the Earth, and thereby understand the atmospheric behaviour on both planets better. This is a much taller order than it first appears, and it was never easy. The most systematic way of approaching it is by comparing the new data to the theoretical predictions of computer models of the current climate on Mars. The model represents what we think the instrument should see, based on what we already know and extrapolations from the Earth; the data, obviously, is what we actually see. Our team had already done a lot of work on global circulation models for Mars, in preparation for the two previous experiments, starting with models from the Met Office that were developed originally for terrestrial weather forecasting.

In principle we could now produce weather forecasts for Mars and see how well they performed when compared to our new observations, gradually refining the models until the differences were minimized and we understood how the processes involved compare to Earth. Getting even more ambitious, we could also extend the models backwards to conditions on early Mars, and test the various theories of what caused the dramatic changes in climate that we see etched into the geology on the surface. More prosaically, and this is what justified quite a lot of the funding for our efforts, we could make predictions of the surface wind field as required to soft-land spacecraft more safely on Mars.

Winter Wonderland

In addition to global climate studies, we could investigate specific phenomena. Infrared sounding is ideal for probing the remarkable changes that occur during the seasonal cycle at the poles of Mars. This is a fascinating business, somewhat analogous to the seasonal build-up and decline of snow and ice in the high-latitude regions of the Earth, but far more dramatic. On Mars, it is not just the water vapour, not much of it by Earthly standards, that condenses, but the main bulk of the atmosphere itself. Around ten trillion tonnes of carbon dioxide condense on first one, then the other, polar cap during the winter months of total darkness. If the season lasted long enough, almost the entire Martian atmosphere would end up as ice on the winter pole. In reality we find that the surface pressure planet-wide drops by about a third. This is a huge change, and of course nothing like it happens on the Earth.

The Martian winters, during which the arctic region is in permanent darkness, are twice as long as ours. At first glance, it seems like that should be plenty of time for the big polar cold trap to draw in all of the condensable part of the entire atmosphere and deposit it as ice, rather than less than a third which is what we see. The key to understanding this came from our early temperature measurements, which showed that the atmosphere in the dark polar night is actually quite a lot warmer than we expected based on calculating the flow of energy as radiation into and out of the region. At heights well above the icy surface, the air is actually warmer in the polar night than it is at lower latitudes, despite the fact that the latter are still receiving some sunshine.

Studying our new data, we could not only see the relatively warm temperatures but also follow changes that made it possible to demonstrate how they are produced. The agent at work is the atmospheric circulation, which has air rushing in from warmer low latitudes to descend over the pole, where some of it condenses and the rest recirculates back to the equator. The descent is so vigorous that the compression heats the gas, rather like the air inside a bicycle pump, but on a planetary scale.

Near the surface, the warming is insufficient to stop the blizzard of dry ice that covers the winter pole and continues until the carbon dioxide runs out. Locally, it can do this to some extent, because the second most abundant gas in the atmosphere is argon, which never gets cold enough to condense. We can't measure the argon concentration but it must be high, and this, plus the latent heat release from the freezing carbon dioxide, works to slow down the rate at which the ice is formed and so has an effect in determining the mass deposited over a winter. In the spring, the sunlight comes back and the snowfall stops, then is reversed as the volatiles return to the atmosphere. The atmospheric pressure all over Mars goes up for a while, and then after several months starts to fall again as the opposite pole enters the winter night.

Building intelligent models

The effort spent on computer modelling in order to make progress on all this interesting behaviour is a daunting task for a small team, and there are many formidable problems to be addressed with the data, too. Using only a single spacecraft, coverage of the planet is limited in both time and space and the choices made when the payload is selected end up restricting the range of atmospheric variables that can be measured. The coverage that is obtained is subject to the well-known limitations in the precision and effective resolution of remotely sensed observations. While it has its inadequacies, data comes in thick and fast and just handling and making use of it in any way at all requires careful planning to keep from being washed away by the flood.

The team in the USA and the UK, with collaborators in France and Spain, are still working towards a comprehensive numerical model of the dynamic Martian atmosphere, a step at a time. The ultimate goal must be approached gradually. When the model gets as close as it practically can to being able to predict observations correctly before they are made, we can then make intelligent extrapolations and predictions for regions of the atmosphere or times of the day or year which have not yet been observed. We are not there yet. Entire conferences lasting several days have been devoted to specific aspects, such as the polar energy balance question and its seasonal cycle. Modelling of ancient climates is still a speculative business.

One of the virtues of working with models is that they generate complete, physically consistent sets of atmospheric fields which can be subjected to detailed diagnoses and to comparison with new data in order to test the hypotheses on which the model is based. Of course, although they are guaranteed (unless someone makes a mistake) to be consistent with the laws of physics, they are not necessarily accurate descriptions of the atmosphere, since they rely on simplifications and approximations to exist. But as the discrepancies become better understood, the model can be improved and the cycle repeated.

This can be a tedious business if done piecemeal with the huge volumes of data generated by modern instruments like the *Climate Sounder*. It is worth finding a way to do the comparisons automatically instead, or as well. Suppose the observations are directly assimilated into the model, replacing its values for temperature, composition, or dust by the measured value whenever a measurement becomes available, thus continuously updating the model. If the difference is large, the model receives a shock and can become unstable unless the discrepancy is corrected and smoothed over by clever software. At the same time, a diagnostic report has to be issued to tell the scientists what happened, since the difference between model and data is essentially the new information being sought.

Thus we can end up with the best possible outcome in the form of a full, physically consistent analysis of all atmospheric fields, including, crucially, those not directly observed. At the same time, we identify and can correct deficiencies in the model, as well as extracting the most benefit from a relatively sparse observational record. We can gradually improve the physics in the model, which includes the hydrodynamical code that computes the large-scale atmospheric motions, the parameterization schemes for the radiative heating and cooling of the surface and atmosphere including dust effects, turbulent atmospheric motions, convection, topographical drag, gravity waves, and the condensation of carbon dioxide in the polar regions.

Modelling has long been a standard procedure for the Earth, and has now become almost an industry as weather forecasting becomes increasingly automated and climate change forecasts hit the headlines. Models for other planets lagged behind, and were little more than toys until the new data sets with vast numbers of individual retrieved temperature and dust profiles started arriving from new instruments like ours. Then it becomes irresistible to bring the latest computing and programming techniques to bear. Efforts to put a data assimilation scheme in place for Mars as part of the development of a new and more advanced circulation model started well before the loss of the *Observer* spacecraft. Getting the assimilation method to work properly was a massive task for our small team, with funding an erratic business as the fortunes of the experimental programme ebbed and flowed. But by the time the *Reconnaissance Orbiter* was on station and the data ready for processing, three Mars years of test data had been merged into the model in a way that was shown to produce promising results.

What killed *Beagle*?

It was 2003 when the first European mission to Mars delivered a small lander, *Beagle 2* onto the surface. *Beagle* had been developed in the UK, by the formidable Colin Pillinger, supported by the British aerospace industry. As related here in Chapter 9 and in more detail in my book *The Scientific Exploration of Mars*, I had been on the payload selection committee for *Mars Express*, and was involved in approving the flight of *Beagle 2*. We all waited anxiously for news after it was released from the mother ship and communication ceased until it could deploy its own antenna on the surface. *Beagle* dropped straight into the atmosphere and descended initially at great speed, before it was slowed down by a combination of friction on its heat shield and drag on a succession of increasingly large parachutes. The plan was to make a landing that was, if not exactly soft, at least survivable.

Working out the design of the shield and the main parachute required assumptions about the density of the Martian air as a function of height. Because the atmosphere is overall so thin, the calculations showed that this was critical, with barely enough air for the parachute to bite on before the probe hit the surface. The designers came to the Oxford–Paris Mars Modelling Consortium for the best possible predictions, made specifically to match, as closely as possible, the date of the landing of *Beagle 2* on Christmas Day 2003.

In the event, the probe maintained radio silence during its descent, as planned, but failed to report afterwards. Photographs obtained ten years later from the super-high-resolution camera on the *Reconnaissance Orbiter* showed that the housing that should have exposed the solar cells and the radio antenna had failed to deploy properly, probably because it was damaged on impact. It was widely reported in the press that the prime suspect for the failure was unusual atmospheric conditions that the models failed to forecast. This, however, was the result of a misapprehension based on some errors present in the processing of the first stellar occultation obtained by a French spectrometer aboard *Mars Express*. These preliminary data, showing an unusually thin upper atmosphere, were presented at a conference but rapidly revised as the correct interpretation emerged from the detailed data processing. The final version fell well within the range of model predictions available before the launch of *Mars Express*.

The *Observer* newspaper wrote that *Beagle* 'has plenty of company among the other bits of wreckage that pepper the red planet's landscape, as Oxford's Professor Fred Taylor knows only too well. He led the team that built a device for analysing variations in water vapour, dust and temperature in the Martian atmosphere: the *Mars Climate Sounder*. It was launched on the US *Mars Observer* in 1992. "Everything was going fine until the engine was fired to put the craft in Mars orbit. It blew up. We never did find out why that happened." So Taylor and his team built another sounder—for the *Mars Climate Orbiter* launched in 1998. This time there was a navigation error, caused by controllers who confused imperial and metric measurements, and the spacecraft flew too close to Mars and crashed. Taylor's *Mars Climate Sounder* was resurrected and put on the *Mars Reconnaissance Orbiter* which was successfully put in orbit round the planet in March 2006. "We are now getting lots of lovely data from it," says Taylor. "We have found that the Martian climate has undergone massive changes over time, from conditions that used to be warm and wet, like those on Earth, to the ice age that now permanently grips the planet".'

This friendly and intelligent article would make a good exam question for a journalism degree, where the task would be to identify the half-dozen or so slight inaccuracies (and one major one—embarrassingly for me it slights Dan McCleese, the actual leader of the project, I was only in charge of the British bit of what was mainly a JPL project, and even then Simon Calcutt did most of the hard work on the instrument). It gets the general points across well, however. I would have liked them to point out that our British hardware crashed on Mars three years before *Beagle* did, a point I used to make tongue in cheek in public lectures, especially if Colin Pillinger was present, to general hilarity from the audience. I was not really amused, and I'm sure Colin wasn't, either, but you have to put on a brave face and carry on in this business like any other.

Was Mars warm and wet?

The latest data and models together are revealing the secrets of the Martian weather and current climate in a very satisfactory way. They are also getting a grip on the explanation for the obvious and major changes in climate that have taken place over unknown eons of time on Mars, but with much greater uncertainty of course. Final answers will take decades and will require a different approach and new missions, including the return of samples to the Earth.

What is already very clear, from surface measurements with the NASA rovers and other missions including *Mars Express* and *Reconnaissance Orbiter*, is that Mars must have had a thick atmosphere and surface water in the past. But just how warm and wet it was, when, and for how long, remain mysterious, as does the reason why such large changes took place. Masses of studies have been carried out to discuss ways forward on the problem, and an International Mars Working Group that met frequently all over the world was set up in an attempt to coordinate them all. Manned missions to investigate the geological record, perhaps landing first on the moon, Phobos, will eventually be needed. Before that, advanced rovers that can climb cliffs and drill deep into layered terrain were among the ideas that were on the table and science goals and sequences of missions were worked up into plans and 'roadmaps'.

I attended many of these meetings, especially when it was still unclear if we were ever going to get one of our instruments to Mars. Reproduced here is one of many doodles from my files along those lines that I sketched while my mind weighed up strategies during meetings of all kinds. This one dates from a meeting at the Smith Institute, a breakthrough organization which explored overlapping interests between industry and academia, focussing on mathematics, physics, and computing. It was set up in 1993 to promote the mathematical sciences in universities and improve links with industry. I was a founder member of the Board of Directors for its first six years, and the institute is still flourishing more than 20 years later. Except in my private thoughts, it did not deal with Mars or any of the planets, however, but was one of my more down-to-earth activities.

NASA was in no doubt that they wanted to deploy increasingly large and competent robot geologists on the surface of Mars. ESA considered its plans through its various committees and consultations with the community. Europe already had a small orbiter at Mars at last, 20 years after *Kepler* would have been there, in the form of *Mars Express*, and at the same time made the failed landing attempt with *Beagle 2*. There were plenty of discussions about trying again; I supported this, but suggested that instead of another single lander, we should send a dozen or more, all on the same carrier, and deploy them as a network all over the planet.

A Mars network mission had been a high priority for planetary scientists for several decades, and I contributed to a number of detailed studies at JPL. Some of these came very close to becoming real missions, but for one reason or another they all fell at the post. The objectives embraced meteorology, seismology, geomorphology, and geochemistry, all of which could make huge advances in our understanding of Mars with measurements on the surface from a large number of distributed sites. There are many practical reasons for such a mission, also, for instance, understanding the diversity of the Martian surface and the properties of different regions and terrains before committing expensive rovers or human cargoes to a specific location.

The most ambitious of these studies was a network mission called *Mars Environment Survey*, which was very much alive as a real mission prospect from 1991 to 1994, and as a near certainty for some of that

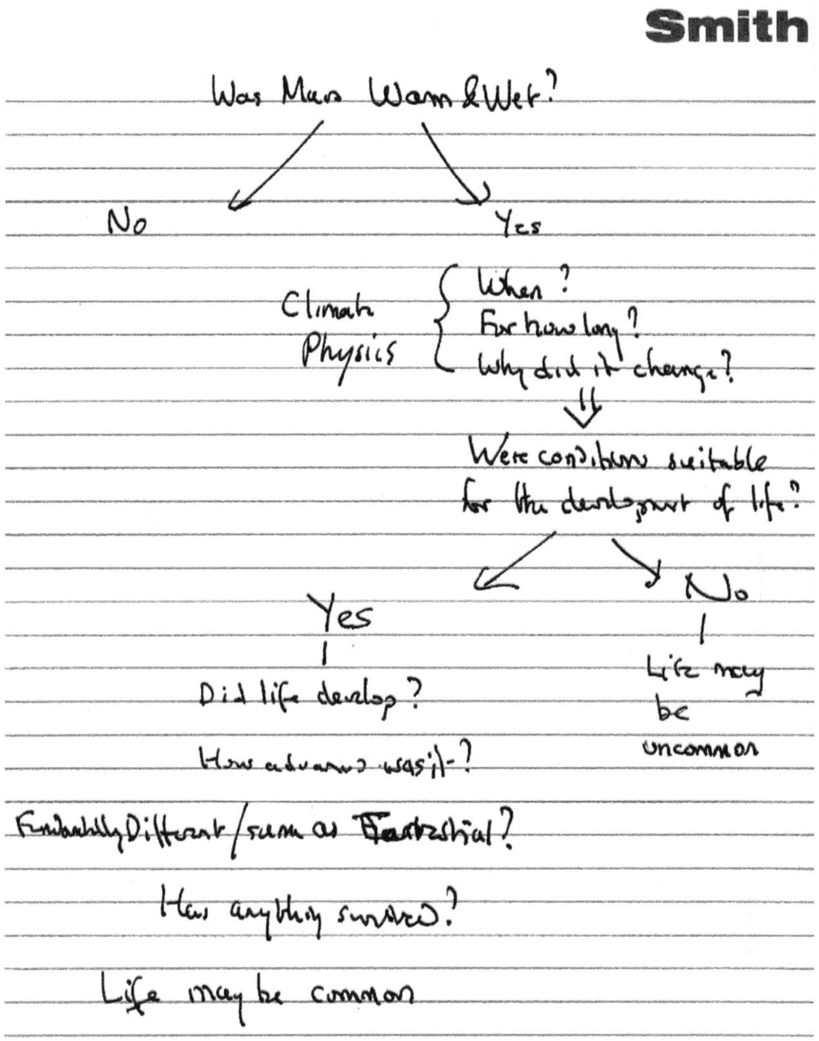

Smith

Was Mars Warm & Wet?

No

Yes

Climate Physics

When?

For how long?

Why did it change?

Were conditions suitable for the development of life?

Yes

Did life develop?

How advanced was it?

Fundamentally Different / same as Terrestrial?

Has anything survived?

Life may be common

No

Life may be uncommon

At a quiet time during a Board meeting of the Smith Institute in the 1990s I jotted down a strategy for a new European Mars exploration programme. By this time we were pretty sure that Mars had been warm and wet and much more Earthlike in the past, a theory that has been amply confirmed by recent missions. We still do not know whether those similarly benign conditions led to life on Mars, as they did here, but the chase is on to find out.

time. This would have provided the necessary global network of surface stations, with a payload I helped to design as part of the study team. It included a weather station, a spectrometer, a seismometer, and cameras and other tools for surveying the environment at each of 16 diverse landing sites scattered over the whole planet. As the study progressed, the mission got more sophisticated and more expensive, as they nearly always do.

For a network to meet its objectives it needs a fairly long lifetime, and it needs to work at night as well. This is a challenge near the poles, where darkness lasts for many months, and temperatures plummet so that electrical heating is also essential to keep the electronics alive. The network approach is much less attractive if high latitudes are not accessible, so the only answer available was nuclear power. This was certainly feasible, using the so-called radioisotope thermal generators that had been developed originally for outer planet missions going far from the Sun, like *Galileo* and *Cassini*. These power sources were fairly compact and almost totally reliable, since they were basically just chunks of plutonium that released heat—lots of it—by radioactive decay, then converted the heat to electricity with solid-state devices based on the thermoelectric effect. However, using plutonium had major safety implications that pushed up the cost.

Then we decided that the fairly sophisticated payload we had in mind called for a soft landing system, which meant using a large heat shield, parachutes, and probably also airbags deployed seconds before landing, all of which added a lot of extra mass. Finally, the stations of the network could not carry powerful enough transmitters to communicate directly back to Earth; a communications orbiter would be required to keep in touch with the widely distributed stations on the surface and relay their data home. As the study progressed and the cost and complexity increased it became clearer and clearer that although it was always attractive scientifically, this was not going to be the cheap mission we had envisaged at first, and eventually NASA cancelled it.

In Europe, various designs for a simpler network mission were circulating, the most successful of which was led, rather surprisingly, by the Finnish Meteorological Service, which had a keen and competent planetary research group imbedded among the usual ranks of Earth modellers and forecasters. I made several visits to Helsinki to help to develop the project and come up with a suitable Oxford involvement. One way to save money would be to use only very rugged instruments that could survive a hard landing, restricting the scientific performance but greatly reducing the mass and complexity.

The landings could be achieved using the torpedo-shaped projectiles called penetrators that had been earlier studied by various groups, including some of which I was a member, but not yet used in anger. These would hit the surface at high speed having been launched from a spacecraft resembling a wine rack that carried a dozen or more of them in orbit. Depending on the nature of the surface they plummeted into, the front part of the penetrator could end up several metres below the surface. Of course, it could also bounce off or be shattered to fragments if it hit hard rock, so you had to have enough of them that you were prepared to lose a few.

Contrary to what most of us would expect, it turns out not to be insuperably difficult to build a vehicle that can hit the surface at 100 metres per second and survive the enormous g-force it then sustains. Obviously

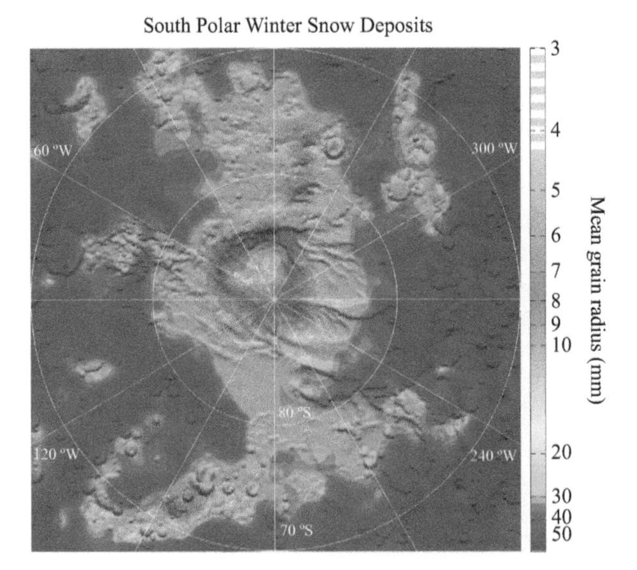

Data from our *Mars Climate Sounder* experiment was used to make this map of the carbon dioxide snowfall over the south pole of Mars in the winter. At the same time, the instrument was measuring the profile of atmospheric temperature and finding a relatively warm layer around fifty kilometres above the surface.

There are many possible approaches to exploring Mars, and I worked on dozens of studies, most of which inevitably end up on the shelf. This sketch illustrates one of them, an unfulfilled (so far) ESA vision of multiple small landers making a network, and multiple small orbiters making a sort of GPS system for communications and navigation on Mars. I studied how the radio signals propagating between the satellites as they set behind the planet could be used to measure temperature profiles in the atmosphere.

the scientific instruments have to be especially robust, but examples have been built and tested that work. The seismometer experts felt sure they could design a solid-state device that could detect tiny Marsquake tremors after surviving the impact of a probe free-falling from space. Some of us were not convinced at the time, and looked at alternatives such as a semi-hard landing, where a small, affordable heat shield and parachute combination reduces the landing velocity at impact, although not to anything close to zero.

If they worked, penetrators would provide some unique opportunities for studying the Martian subsurface, down to perhaps ten metres or so, where we could expect to find water present as permafrost and an environment that had not been sterilized by exposure to the ultraviolet rays from the Sun. This radiation is fierce on Mars, because the planet has no, or very little, ozone layer, and if there are microbes on Mars they must live at some depth in the soil for protection. The probes could search for water with an ice detection device mounted at the front part of the penetrating vehicle body, and also study the thermal conductivity of the soil by measuring the rate at which the probes cool down after impact using temperature sensors mounted on the structure.

There were other advantages. The probes would be designed to have a long lifetime, and the network did not have to be put in place all at once. By starting with a small number of observation posts and adding to it until there were 20 or more, we could use cheaper launch opportunities and spread the costs of building a global grid. If the number eventually got larger still, it could include a dense local network embedded in a region of particular interest from the atmospheric science point of view, for example near the edge of the expanding and retreating polar caps, in one of the large, deep basins like Hellas, or in the deep valley network of the *Valles Marineris*.

Another successful study, successful in the sense that it offered a lot of science for the estimated cost, and involved relatively low risk because of redundancy and simplicity, was a project we called *Marsnet*. This would land a small number of stations at any one time using airbags, which would mean keeping the payload small and light, and it would use solar panels that folded out after landing, much like the ill-fated *Beagle 2*. Both concepts for small landers were developed at about the same time using many of the same people and ideas, and in fact it was seriously discussed at one point that the two might be merged. If they had, the second attempt to use Colin Pillinger's ingenious life-sensing devices would actually involve a network of *Beagles*. A further refinement was a network in space of small satellites surrounding Mars to provide sophisticated positioning information using GPS beacons, as well as a communications network and some additional science experiments. The more I was involved in the studies the more I came to think that a mission like this was a no-brainer for the next European mission to Mars after *Mars Express*.

It was not to be, however. At the key meeting in Birmingham, ESA decided to archive all of the studies on networks of small landers and multiple small orbiters and instead go for broke with a sophisticated surface rover of its own, called *ExoMars*. This would come after half a dozen American rovers had already operated on Mars, and it would be obliged to be much smaller than the latest ones because the NASA capabilities and budgets were well in excess of what ESA felt it could afford. To me, a European rover was a silly choice. Rovers are expensive and risky and the science would be similar to that which the Americans had already addressed a decade earlier. Unless we were very lucky, for example by landing at a site that carried unmistakable signs of past or present life on Mars—fossils, say—the European venture would cost a lot and produce little that was new. And there was an enhanced risk of failure: the relatively simple *Beagle* package had crashed, and now they were going to land a much larger and more delicate rover safely? We were on a hiding to nothing, I thought.

While my colleagues were climbing out on this particular limb and sawing it off I was absent, in bed with a virus, and unable to travel even to Birmingham to make the case for a network as I had planned. Just as with the selection opportunity I had missed with *Kepler*, after the community shot itself in its collective foot I was left wondering if I could have made any difference. Probably not. Then things got more interesting: the European Space Agency contracted to build the rover in England, at Astrium in Stevenage, and they added a precursor, an orbiter mission with a small lander attached, to fly two years sooner. This was in part to answer the criticism being levelled by those of us that pointed out that Europeans had never landed anything successfully on Mars. It also addressed some of the *Netlander* science, by carrying a seismometer and a meteorological station—basically, a network of one.

There were, of course, valid arguments for choosing a rover over a network. The rover was sexier, and the focus on searching for signs of life had much wider appeal than meteorology or geology would ever have. The argument that the Americans had already put several rovers down on Mars, soon to include the very large, nuclear powered *Curiosity* robot geologist, was countered by saying Europe did not have NASA's fear of saying outright that we were searching for life, and that *ExoMars* would carry different kinds of instruments, including a drill that could search six feet below the surface. The *Observer* newspaper quoted me drumming up my positive side by saying: 'Once [Mars] seemed destined to support life. Then we thought it was utterly dead and featureless. Then we discovered—thanks to *Mariner 9*—that it had a landscape through which water had poured. After that, [the *Viking* mission] found that its soil contained no signs of biological material. Since then, we have bounced back and are hopeful life may exist underground.' That's it in a nutshell, I suppose.

ExoMars goes West

ESA knew that by far the best way to land the *ExoMars* rover safely would be to team up with NASA, who had all the expertise in the world when it came to putting machinery down safely on Mars. The Americans were keen on the orbiter: there had been an ambiguous detection of methane in the Martian atmosphere that was potentially crucial for life detection, and they wanted to pin down the source. The two missions, under the collective title of *ExoMars*, would be carried out jointly. ESA and NASA signed an agreement, legally binding but with escape clauses, in Washington in October 2009.

By mid 2010 the new UK Space Agency was in place and one of its first acts was to sign an agreement to cooperate with the USA on Mars exploration and other ventures in space. A senior delegation from NASA flew over to sign the agreement at an event at the Royal Society in London. After the signing, the UK organized a dinner for the VIPs from the States to celebrate this new initiative and the new dawn in space exploration that it promised, and I was invited to be one of the hosts. At the meal, I found myself sitting next to a charming American lady who introduced herself as the wife of one of the astronauts, Mark Kelly, who had come with the Administrator as part of the NASA team. After we discussed a bit about what I did, we talked for a while about her husband's activities as a Space Shuttle Commander.

Then I put my foot in my mouth, or rather I wish I had, if it would have prevented my unconscious chauvinism when I asked 'And what do you do when Mark is off in space?' I was expecting an answer about homemaking, raising a family, and coffee mornings at the church. 'I'm a US Senator' she said, 'for the State of Arizona.' When I got over my embarrassment at being such a klutz, we went on to have a long and interesting conversation about what it's like to be inside the US political scene. Later, I was shocked and saddened to hear on the TV news that Gabrielle Giffords had been shot in the head at point-blank range by a madman in Tucson. Almost miraculously, she survived and made a remarkable recovery.

ExoMars goes East

The *ExoMars* orbiter was an opportunity for our *Mars Climate Sounder* team to propose a new version of our instrument to extend the studies of the atmosphere and also to fix some problems. One of these was that the water mapping channel did not work well; a better selection of operating wavelengths in the far infrared was required. We made the case and were duly selected for flight, 'we' being JPL, Oxford, and our other collaborators but not myself this time. Retirement was looming and I did not want any new long-term commitments, so I withdrew from the team at an early stage.

I had unknowingly saved myself a lot of wasted effort. Work had begun on the new instrument when NASA suddenly pulled out of the whole *ExoMars* collaboration. They cited budgetary problems, although it was clear to those of us on the inside that there was also a darker reason to do with the high-level politics: NASA and ESA were not getting on. ESA seemed undaunted and promptly asked the Russians if they would like to take over the American role, providing rockets to launch the orbiter and the rover, and taking responsibility for the descent and landing systems. As part of the deal, Russian instruments would replace the American ones in the science payload, and that meant that *Climate Sounder Mk. 2* had to go. It was some compensation that the original instrument was still performing, six years after it arrived at Mars.

Neither the Russians nor the Europeans have so far landed successfully on Mars, so the *ExoMars* project will be an exciting one to watch. My role these days now I am not an active participant is limited to chairing a panel for the UK Space Agency that considers proposals for funds for research groups, mainly in universities, to participate in the mission. Handling the funding gives me exposure to plenty of interesting ideas, but is a fraught and difficult business, mainly because the funds available are far smaller than the amount that would be needed to support all of the excellent proposals. There is only enough money to fund around a quarter of them, even after the flawed ones have been eliminated. One has to field complaints from colleagues who are dismayed at being turned down, and cannot understand why. I sympathize, and remember being in that position myself more than once over the years.

The upcoming mission is divided between two launch windows, with the orbiter being dispatched in 2016 and the rover in the following launch window, that is, the next opportunity when the two planets are in the right relative positions in their orbits, which is not until 2018. The *Trace Gas Orbiter* should clarify the exciting mystery that has grown up concerning methane on Mars. The orbiters and landers already at the planet, as well as Earth-based astronomers, have detected traces of the gas in the atmosphere in small and very variable amounts. Methane is often known as 'marsh gas' on Earth, and as this implies it is mainly the product of biological processes such as ruminating cows and rotting vegetation. Presumably there is nothing like that on Mars, so what is going on? Volcanoes are another possible source, and if there are still small volcanic events on Mars that would be almost as exciting as finding cows.

If the orbiter finds localized sources of methane, seeping out from volcanic vents or even from subterranean bacteria, then the rover may be able to use that information to visit these sites and investigate what is going on. This assumes, of course, that we can pinpoint the sites in advance, and that the Russians can put *ExoMars* down on the surface, not only safely but with great precision. Volcanic activity means heat, and heat on an icy world means liquid water. Liquid water may harbour life. Administrating grants is not in the same league as building instruments and sending them there, but it's part of empowering others to do the best science and move the quest forwards. And at least I don't have to travel any further than London to do it.

15
Ice on the Moon

The path less travelled

Yogi Berra once remarked that if you do not know where you are going, you end up somewhere else. Scientific space missions are a bit like that; they are so few and so infrequent that you can expend much effort in planning a trip to one planet and end up at another, just because the opportunity was there. With my focus on atmosphere and climate, I never planned to study the airless Moon, it just showed up and beckoned to our team while we were busy investigating the weather on Mars. What a wonderful opportunity—which of us has not drifted in the moonlight and been seduced by that pale fairy lantern overhead, and asked what is it, where did it come from?

In the early 1960s, if you thought at all about travel into space, it was probably our planet's large satellite, only a quarter of a million miles away, that you imagined as the target. NASA, and the Soviets, were working on simple robotic missions to Mars and Venus, but they were quasi-covert activities and made few headlines. In complete contrast, to the media this was the age of *Apollo*—the first manned missions to another world, the long dreamed-of conquest of the Moon. A disappointingly high proportion of press coverage was concerned with problems, delays, and how much it all cost. 'Wouldn't all this money be better spent on Earth', they cried; where did they think it was being spent? Even the positive coverage tended to be about flippancies like someone's prediction that the lunar surface might be sprinkled with diamonds or suggestions that the astronauts might play a game of golf on the Moon. Others delightfully captured the romance of it all, and the human, technical, and scientific progress it represented; I liked those.

The Inconstant Moon

In the summer of 1966 I had my undergraduate degree in physics and was about to start as a graduate student in the field of atmospheric physics. I chose that because the first scientific satellites of the Earth were exploiting the view from space to improve the science of meteorology and the timing was right to get involved. In Chapter 1, I told how I went to a summer school at the Institute for Space Studies in New York to help prepare for a planned career in space research. This included lectures on what was known about the Moon, and what was mysterious. Towering over all was the question of how a smallish planet like the Earth came to have such a large satellite. The common centre of mass about which the two orbit each other is inside the Earth, about two-thirds of the Earth's radius from its centre, so the pair do not qualify as a double planet system, although they come close. (In the Solar System, only Pluto and its satellite Charon meet this criterion; however, neither qualifies as a proper planet under current rules.) Mercury and Venus have no satellites at all, and Phobos and Deimos of Mars are small, irregular bodies obviously captured from the nearby asteroid belt and not real worlds like Earth's formidable Moon.

The samples from the Moon that *Apollo* would collect and bring back to Earth were expected to provide the data needed to understand how our planet and its satellite formed. (Today we think that the composition of the Moon is similar to the outer layers of the Earth, and that a giant collision stripped this material away from the young Earth to form its large satellite.) Before that could happen, missions were needed to photograph the Moon and identify possible landing sites for the *Apollo* astronauts. These were going on as we studied in New York, and when the formal part of the course was over we got to travel to Cape Canaveral to witness the launch of the first in a series of *Lunar Orbiter* missions, designed for this purpose.

When I arrived at JPL in 1970, there were people there who had worked on the earlier *Lunar Ranger* programme that had preceded *Lunar Orbiter*. The plan for the *Ranger* was simply to fly straight to the Moon and crash on the surface, frantically transmitting images as it zoomed in. A replica of the spacecraft and some of the close photos of the surface were on display in the museum near the von Karman auditorium where the JPL engineers and managers had gathered to watch the data come in. The first six *Rangers* had failed, and it was only the success of *Ranger 7* on 31 July 1964 that saved JPL's reputation and even its existence. The trauma could still be felt nearly a decade later.

The *Lunar Orbiters* which followed were a great success. The one we watched lifting off at the Cape on 10 August 1966 was the first of five, which collectively mapped 99% of the lunar surface with a resolution of about 50 metres. The technology in those days did not include digital imaging, of course. The pictures of the surface taken from the mapping mission's orbit around the Moon were taken on fine-grained Kodak film, like that used by spy planes in the Cold War, and then scanned by a sort of compact fax machine to produce signals for transmission back to Earth. NASA recently re-released the whole set of pictures, after reprocessing them using modern techniques. They look very

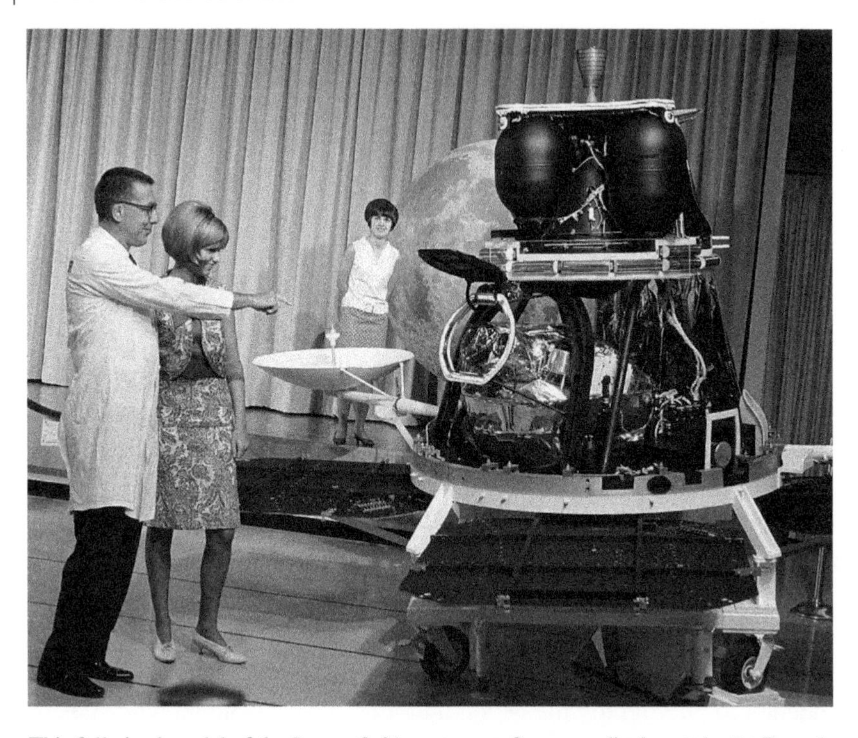

This full-sized model of the *Lunar Orbiter* spacecraft was on display at the Jet Propulsion Laboratory in 1966, at about the same time as I was watching the real thing launch from Cape Canaveral. I was in Florida on a field trip as part of a summer school at the Goddard Institute for Space Studies in New York. I would join the pioneers at JPL just four years later, and become part of a team exploring the Moon another forty years on.

impressive, although of course by now they are all far superseded by subsequent missions, including the *Apollo* flights themselves.

The summer school also included a trip to Grumman Aerospace Corporation on Long Island to see the work being done on the lunar landing module that Neil Armstrong would, three years later, steer down onto the Sea of Tranquillity. We got to climb inside a prototype and feel what it was like to be so confined on such a dangerous journey. The astronauts were true heroes; it still seems incredible to me that they succeeded. They didn't know, because nobody did then, that apart from all the engineering risks they were in danger from solar mass ejections, events on the Sun that fired streams of deadly particles into space during solar storms that took place every few months. These vary in intensity, but the stronger ones would have fried the lunar astronauts in their flimsy spacecraft. Fortunately, and entirely by chance, all of the *Apollo* flights took place in between major ejections.

I was still a student at Oxford on 21 July 1969 when Armstrong stepped onto the dusty surface of our cosmic companion. Doris and I watched from the garden of our friend's house on Boars' Hill, with the TV just inside the patio doors and the huge disc of the Moon itself

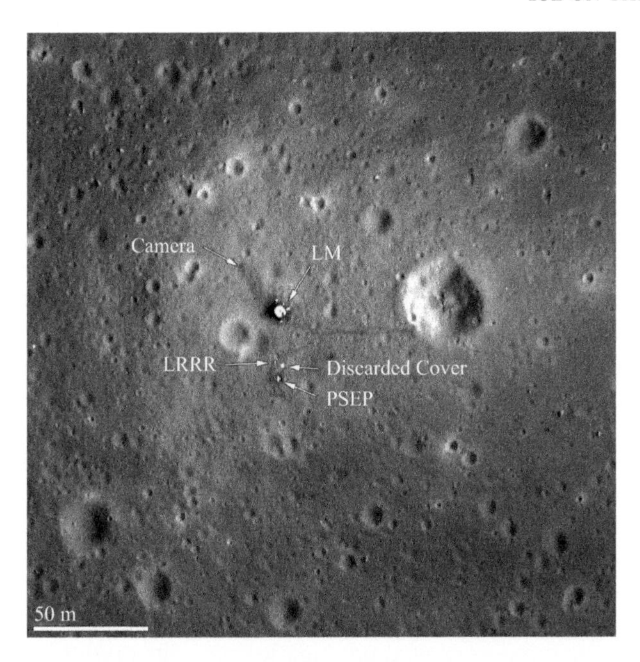

Our *Lunar Diviner* team was making temperature maps of the Moon's surface from the *Lunar Reconnaissance Orbiter* in March 2012 as the camera on the same spacecraft took this picture of the *Apollo 11* landing module (LM) sitting abandoned on the lunar surface. The resolution is so high that individual pieces of scientific equipment can be seen, and even Neil Armstrong's footprints where he walked to Little West crater to the right of the lander, nearly 43 years earlier.

hanging in the night sky. 'They're up there', said Rick, as we stared as if hoping to see them as dots on the big white globe overhead (the Moon is actually chocolate brown in colour, the whiteness we see when we look at it from Earth is an optical illusion). I didn't imagine at that time that 40 years later I would be involved in a mission that would photograph the module as such a speck, seen from an orbit just 15 miles above the surface of the Moon.

A ride to the Moon

Did I stress how hard it is to win participation on missions, especially those going into 'deep' space, that is, beyond Earth orbit, to the Moon and the planets? Such missions are few in number, and each one carries only a few, up to perhaps ten, experiments. The competition to decide who gets to put their instrument on board is extremely keen, and the whole process often takes years. Long-term planning is essential since it definitely helps to become an investigator on a mission if you have been involved in helping to plan and design the whole project from the outset. From such a position you can also help the space agency to win its own battles with the relevant people in government for financial support.

However, there are rare exceptions to this long grind. The chance to explore the Moon that I describe in this chapter is one that I got almost without trying. I had never taken much professional interest in the Earth's large satellite, mainly because it has no atmosphere, and so no climate in the usual sense. Mercury is the same but, as described in an earlier chapter, it suddenly became interesting as an objective partly because I had been asked to help to design a mission to land there, and couldn't resist the challenge, and partly because the discovery of large deposits of ice in the polar regions made it relevant to understanding the history and origins of water in the whole Solar System. That is climate research writ large.

The discovery of ice on Mercury raises the question of whether we should expect the same thing on the Moon. The two bodies are fairly similar in size and appearance, and both have heavily cratered surfaces that have extensive areas near the poles that are permanently shielded from the Sun. The fact that the Moon is cooler overall should make it easier for water to survive than on the scorched surface of Mercury. Why hadn't the radars picked up thick ice deposits on the Moon? Suddenly this question became not just curious but a real political hot potato, when in 2004 the United States announced a grand plan to return to the Moon and set up at least one manned base there. Not only that, they were then going to go on to Mars, using the Moon as a platform for the launch.

This idea came straight from the top, with President George W. Bush tasking NASA with the getting the job done. NASA was not entirely thrilled: the President had not provided enough new money, and some funds would have to come from its existing carefully planned programme, which had little to spare. The Space Shuttle would have to be scrapped, and replaced with something that could go to the Moon. And contrary to what may seem obvious to most of us, the engineers did not think that the Moon was the best place from which to launch a mission to Mars. It would need a very sophisticated facility, and a lot of heavy lifting of supplies, to service flights to Mars from the Moon. But this was politics. Work began on the new launcher and on plans for the manned lunar base.

Lunar Reconnaissance Orbiter

The first step of the planned new era in lunar exploration would be a detailed study of the Moon's surface to decide where the first base would be established. Most of the terrain had not been mapped in the kind of detail NASA wanted before any more landings were attempted. The goal now was a resolution of about half a metre, or around a hundred times

Lunar Reconnaissance Orbiter, with Diviner on board, sits on top of an Atlas V launch vehicle on the crawler taking them to the launch pad at Cape Canaveral in June 2009.

better than the pre-*Apollo* unmanned *Lunar Orbiters* reconnaissance missions of the 1960s achieved, which was still the best that we had.

The new spacecraft that would achieve this ambitious goal was called *Lunar Reconnaissance Orbiter*. The similarity of this name to the Mars mission that was then being built, and would carry our *Climate Sounder* instrument, was not a coincidence; the new lunar mapper was to use a copy of the *Mars Reconnaissance Orbiter* spacecraft and

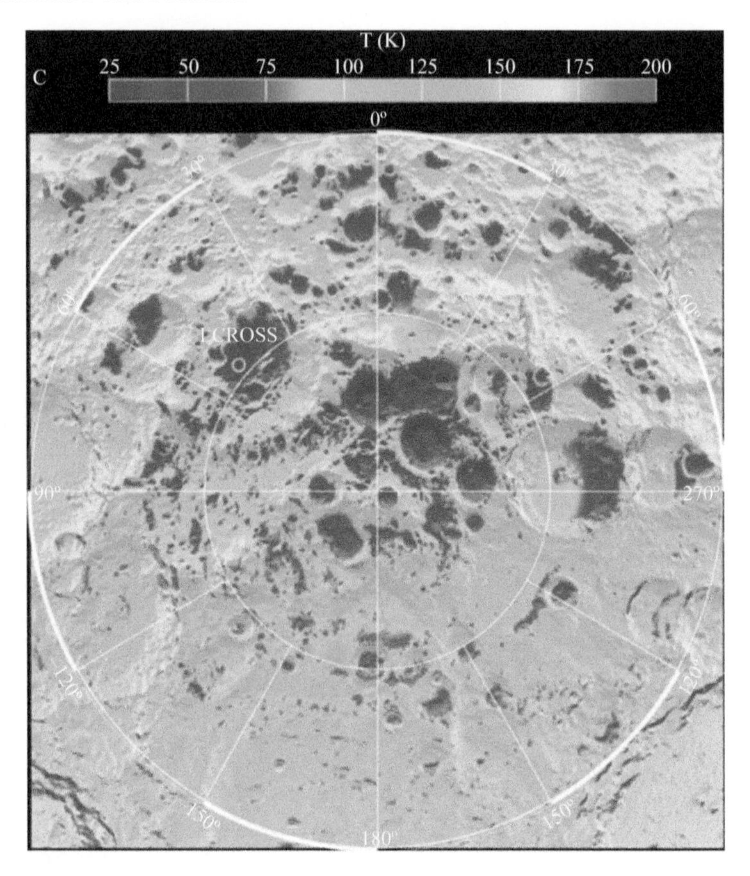

This map by *Lunar Diviner* of the temperatures at the south pole of the Moon shows that the dark places in the polar craters are very cold indeed, less than 30 degrees above absolute zero. At such a low temperature not just water but many other frozen volatile species like carbon monoxide, methane and ammonia would be stable for billions of years.

as many of the same instruments as possible, to save time and money. One of the requirements was to map the temperature of the surface and its variations very accurately and our *Mars Climate Sounder* instrument was just the thing to do this, with some relatively simple modifications to the wavelength coverage and a change of name. We had to go through the formality of submitting a proposal, but the work involved was relatively small, and when it came to the competitive selection our *Lunar Diviner Radiometer* was a shoo-in.

Lunar Diviner got its name because it was designed to find cold traps on the Moon where water ice might have accumulated, providing a valuable resource for future astronauts manning a permanent base at one of the lunar poles. The engineering team that had built the Mars instrument swung into action at JPL and Oxford with very little involvement from me. I didn't even attend science meetings for the most part; as I approached retirement, reducing travel was a priority for me, and anyway most of the discussion was about lunar surface science, a very dry subject in every sense. The hard work was done by the rest of the science team, mainly at JPL, with our Mars team member David Paige from the Los Angeles campus of the University of California as Principal Investigator. I had a new academic colleague at Oxford, my former doctoral student Neil Bowles, and he was keen on surface science. Neil took over my role on the team while I just kept a smidgen of involvement out of interest in what we might learn about the ice question.

The big spacecraft launched on 18 June 2009 and reached the Moon four days later. With no atmospheric drag to worry about, it could orbit very low above the lunar landscape and get great detail on the features below. Soon, it would capture images of the *Apollo* landers and their equipment, including the buggies the astronauts rode, lying abandoned on the silent lunar surface. Their tracks in the dust, and even footprints from walkabouts, show up clearly from a height of just 20 kilometres in the high-resolution photographs. Infrared data is always less sharp, but even our radiometer could see quite small features, less than 100 metres across. The cratered poles, with their shaded regions, were soon revealed.

The plot thickens

Before *Diviner* started its work it was already apparent from the Earth-based radar measurements that there doesn't seem to be anything like as much ice on the Moon as there is on Mercury. The reason is not very clear; we might expect it to be the other way around given the intense heat from the Sun at Mercury. It might be due to differences in the terrain. Small differences in the shape of the polar craters and the outcrops of high rock can easily mean scattered sunlight gets rid of the ice faster than it can form. Or ice might form faster on Mercury, if there is more water being exhaled from a hot interior for instance.

Our goal was to investigate the temperature environment and see where the conditions were right for ice. We were mainly interested in places where ice deposited millions or even billions of years ago would still be present. This means temperatures far below zero centigrade: the actual number, based on lab measurements of the vapour pressure above ice at different temperatures, works out as about –160 °C. Is there anywhere on the Moon as cold as that? The answer is not just yes, but, astonishingly, there are areas as large as Manhattan Island where the temperature is below 30 K, a staggering –250 °C.

At that temperature, water ice would evaporate so slowly that, once captured, it would remain stable for hundreds of billions of years, much longer then the time the Moon has been in existence. It would be trapped permanently, in other words, and could continue to accumulate as long as there was a source of some kind. That might be icy particles from space, small comets for example, or (less likely on the Moon than on Mercury) water might seep out of the interior, a remnant of past volcanism. There is also the possibility that interior warmth reduces the lifetime of the ice, or more probably that a rain of micrometeorites from space erodes the deposits to an unknown degree.

That's not all, however. These crater bottoms are so cold they would trap not just water but a whole range of much more volatile species. These could include ices of ammonia, carbon dioxide, sulphur dioxide, and methane, all of which are common in the outer Solar System and in comets. Inside the asteroid belt, these ices boil away quickly if they are exposed to sunlight. At the poles of the Moon, however, they can be preserved essentially forever, along with anything else that may have come along during the history of our satellite, like simple, or indeed even complex, organic molecules. Everything may be layered like the geological deposits that we use to tell the history of the Earth and if so it will provide a timeline, forming a textbook history of the Moon and the rest of the inner Solar System.

So if future astronauts set up a base in the polar regions on the Moon, they will not only have a source of water, but a unique and exciting chance to drill ice cores that have captured the volatile history of the Moon and the Earth. It's tragic that NASA cancelled the trip; I would give a lot to know what they find when eventually it happens. We'll have to wait for the Chinese, or the Indians. They have the appetite for exploration in space that we seem to be losing in the West.

I saw an example of that when I helped to host a visit to Oxford by a party of high-level delegates from the Chinese Space Agency some years back. After a good lunch with wine (in Christ Church, where the visitors were quick to pick up on the Harry Potter connection without any prompting from us), I asked the senior official sitting next to me whether they have plans to put a Chinese astronaut on the Moon. Oh yes, he said. When? I asked. Impossible to say for sure, was the reply, as (rather like the Japanese) their funding, and therefore progress, depends on the budget, which is set annually. And what if it goes really well? I pressed. By 2020 in the most optimistic case, was the answer, and realistically probably not too long after that. Wow! I predict that they will find a treasure trove of information if they go to the icebound craters at the poles.

16
A Beginner's Guide to the End of the World

Speculative fiction has long been a particular interest of mine. We can never know the future, but we can predict it, often for practical reasons (such as planning a long journey to get to an important meeting on time), but also for entertainment and wonder. In the latter category, true speculative fiction differs from fantasy and what is commonly called science fiction in that it never violates the known laws of physics, and, although it may postulate new laws that we have not yet uncovered, they have to seem reasonable and not too outrageous. H.G. Wells's *Time Machine* is a classic example, known to everyone. So is the following introduction to a chapter in a Batman story (*Year One*, by Frank Miller): 'He's wounded. They've got him cornered. They've got him outnumbered. They've got him trapped. They're in trouble. . . .' Batman is not invulnerable and has no superpowers; he relies on technology, skill, and a great deal of luck to achieve his incredible performance. *Year One* is speculative fiction; swop it for a Superman comic and you are into fantasy. Someone once said you can learn a lot about a person's psyche by asking them which of the two heros they prefer; almost everyone has a definite opinion. I'm with Batman because he doesn't play fast and loose with the laws of physics.

Climate forecasting is a particular form of speculative fiction. We want it to be correct, and we certainly do not want to violate any physical laws, but still we are talking about predicting events in the relatively distant future where there are many unknowns. As such, it takes us well outside the comfort zone of predicting tomorrow's sunrise, the date at which a comet will return on its journey around the Sun, or even the next day's weather, which is largely a matter of extrapolating the behaviour of organized systems that already exist in the atmosphere, like fronts and cyclones. Fifty years from now, will the mean conditions be the same? Unlikely, but it is not easy to say what the differences will be. However, we must make forecasts and they must be as realistic and reasonable as possible, with good estimates of the uncertainties that give rise to a range of possible answers.

The Climate Conundrum

The stability of the Earth's climate has been an interesting topic of debate for a long time. It is important, of course, for the quality of life on our planet, and even the very continuation of life as we know it would be threatened if there should be a large change in things like sea level or rainfall. We know climate can, and does, vary a lot; on Earth the last major ice age was only 12,000 years ago, and our modern society will not last long if anything as severe as that comes again. Also, we now have evidence of massive shifts in climate on Mars, and probably Venus also, that would have changed their habitability greatly, too, if there had been anything alive there to care.

The climate debate used to be confined almost entirely to scientific circles, with politicians standing aloof and the general public largely ignorant or indifferent. That has completely changed over the 50 years or so of my career in planetary and atmospheric science, with today's politicians almost universally buying in to arguments that human industrial activity is changing the climate to a degree that may soon become serious. The general public is taking a while longer. Working in the field that should know the facts, at least to the extent that any predictions about the future can be conflated with facts, I have watched the debate grow and take many forms.

Climate change is a conundrum. According to the *Oxford English Dictionary*, the word used to be a late sixteenth century term of abuse for a crank or pedant, later coming to denote a whim, fancy, or pun, morphing in the late seventeenth century into the current sense of a confusing and difficult problem or question. Any of those definitions can be found among the myriad opinions that swamp the conventional scientific debate with their blogs, arguments, and letters to the newspapers. For some it is a sort of game, for many an annoying deception, for others an amusing fantasy, like a comic book story. A small but increasing number believe it is a crisis.

No universal consensus exists. For instance, in the *Daily Telegraph* on the day I sat down to draft this chapter (9 February 2015) there is a poll in which 94% of 95,148 respondents think that temperature data has been, or may have been, manipulated to show a warming where none really exists. The other 6% are probably the scientists, the ones who should be in the know—but are they? The article has 17,000 comments and the number is growing by the second, nearly all of them hostile to the global warming 'alarmists'. Just a few reiterate the scientists' concern about sea level rise and ecosystem damage due to temperature increases forced by coal and oil combustion that is driving up the level of carbon dioxide in the atmosphere.

Democratically, we could conclude from this overwhelming vote, and others like it, that the much-hyped panic is just contrived by people who stand to benefit, like scientists whose funding may depend on it, or politicians who want to control human behaviour and waste money rather than serve their constituents. Many seem to think that common sense dictates that carbon dioxide cannot possibly affect climate because it is less than one part in a thousand of the atmosphere. Looked at that way, the increase since the industrial revolution of about one third in the abundance of that minor constituent can seem trivial.

Is it not obvious, some say, that manmade climate change is not a threat to civilization, because global temperatures are stable and always have been, within certain modest limits where the fluctuations are due to natural causes anyway? Lower in the denial ratings are those who say OK, so maybe there is a problem, but can we not fix it with 'clean' energy and a few other relatively painless changes? At the bottom of that scale are the alarmists, a small and not particularly vocal minority that fears that humankind may be doomed as a result of its own behaviour. Although relatively few in number, this group includes some, perhaps even most, of the scientists with the specialized knowledge and experience needed to really understand the problem. I am, or ought to be, one of them.

Looking for answers

So I get asked whether the global threat from climate change is real, and if so how bad will it get, and when? The honest answer, of course, is that I don't know, and you should not expect anyone to know. Would you ask a professor of economics to tell you what the exchange rate will be between the dollar and the euro in 50 years' time? Of course not. No one can say how a complex system is going to behave in the future.

Forecasts can be made, however. The question of whether we face serious climate change on the Earth due to anthropogenic (human) activity dominates much of the 'applications' side of atmospheric science these days. When I entered that field, the biggest challenge it confronted was weather forecasting, looking a few days ahead; now, 50 years later, we are pressured to say something about what might happen ten, 50, or 100 years hence.

Climate forecasting is a specialized business. It is undertaken by large teams in specialized institutes with huge computers and budgets to match; there are several of these around the world. The example closest to me is the Hadley Centre for Climate Prediction and Research at the UK Meteorological Office in Exeter. This was founded by my predecessor John Houghton when he moved, in the 1980s, from Oxford to Bracknell to head up the Met Office at its former location. John called it the *Hadley Centre for Climate Research and Prediction* when he set it up, but later the name was inscrutably revised to become the *Hadley Centre for Climate Prediction and Research.*

The Hadley Centre joins with its equivalents in other countries under the banner of the United Nations to form the Intergovernmental

Panel on Climate Change. This enormous venture, involving thousands of scientists and programmers, compares and combines independent forecasts and publishes a consensus, updated regularly every few years. This includes a 'Summary for Policymakers', which drives government policy the world over. The latest report begins:

> Human influence on the climate system is clear, and recent anthropogenic emissions of greenhouse gases are the highest in history. Recent climate changes have had widespread impacts on human and natural systems.
>
> Warming of the climate system is unequivocal, and since the 1950s, many of the observed changes are unprecedented over decades to millennia. The atmosphere and ocean have warmed, the amounts of snow and ice have diminished, and sea level has risen.
>
> Intergovernmental Panel on Climate Change (IPCC), 2015

No wonder the army of sceptics is roused—these are fighting words.

Playing a part

The IPCC is a vast endeavour and any individual participant in its work has only a small part to play in saying what the forecast shall be. I never joined the IPCC effort, although many of my colleagues did. I was involved in refereeing several of their regular 'assessment reports', which were published as books by Cambridge University Press. IPCC Assessment Reports appear about every five years and have become the standard 'bible' for politicians and journalists, as well as well-read members of the public, indeed for anyone seeking the most authoritative position on climate change and its threats, and indeed they led to the award of the Nobel Peace Prize, no less, in 2007.

The Press was, of course, very keen to publish such an important book on such a hot topic, but was required, by its statutes, to obtain a supportive review by at least one independent expert first. In my first review I commented that 'This is the latest, and by far the most authoritative and balanced, statement on a very important, topical and rapidly-changing subject of enormous popular and scientific interest. The IPCC is well-organised and, for such a large organisation, works well to schedule but maintains a consistently high standard scientifically and ethically.' Five years later, I described the next report as representing 'incremental rather than revolutionary progress' but 'It is still the case that the conclusions of the IPCC represent the definitive world opinion on the vital and urgent subject of climate change and its likely effects.'

The fact that I support the IPCC and think they do an outstanding and essential job does not mean that I have no reservations at all, nor that I would make quite such bold summary statements as theirs appearing above. The forecasts depend absolutely on models: large and sophisticated computer programs formulated using current physics and data, and tuned to match the current and past climate behaviour. The models produced independently, often in different countries, are compared to each other and the differences analysed to eliminate downright errors of execution. They can not eliminate any misconceptions or erroneous inputs they may have in common, of course, but still, the predictions they make are undoubtedly the best we have and must be taken seriously, even though they are not and never can be perfect.

The key question, of course, is how good or how bad is 'not perfect'? We don't know. There is some physics, especially that involving clouds and aerosols, that we know is either not very well understood, or is difficult to program into models, or more often both. The dynamical behaviour of the atmosphere and the oceans is likewise represented in models only in a smoothed sense; their real behaviour can include unexpected, steep non-linearities (the 'butterfly effect') which might cause serious shifts in climate that the models know nothing about. Such changes have occurred in the relatively recent past, the largest being the ice ages. In these and various other ways, the IPCC forecasts are the best we have but they may not be right.

Given the 'known unknowns' and the 'unknown unknowns', logic often goes out of the window, especially in the kind of general debates one finds in the newspapers or on television. What we get is:

- *The data that show warming is underway have been manipulated by the scientists involved and therefore are false and designed to fake warming for reasons like getting more funding.*

All data from all experiments in any field are 'manipulated', they have to be; manipulated is not 'fiddled'. And the oft repeated allegation of manoeuvring dishonestly for funding (on such a large and concerted scale—I'm not saying it never, ever, happens) is stupid and offensive.

- *The IPCC forecasts may not be right, therefore they may be wrong, therefore global warming is not 'real' and they are scaremongering.*

'May not be right' does not mean 'probably wrong'.

- *Warmings, and coolings, occur naturally and there is plenty of evidence for such trends long before humans had any significant impact. Therefore any changes we see now are natural.*

This is perhaps the most pernicious misconception—the premise is right, but the deduction is the opposite of that which should be made. If the climate is sensitive to small changes in the Sun, for example, then it is also likely to be sensitive to human perturbations as well.

The current era is characterized by explosive population growth and industrialization. Every car, every factory, every paddy field, is stoking up the trace gases in the air and the concentrations of some of them have more than doubled during the last century. Carbon dioxide is the easiest to measure, because it mixes uniformly and so the value is about the same everywhere away from polluted cities, and that value has increased by about half. No one who understands the basic physics—see below for a discussion of that, where I discuss teaching the subject—has any doubt that the greenhouse gases warm the Earth substantially and made it habitable for our pre-industrial ancestors. There is also no doubt that these gases have increased their presence in the air substantially, nor that this will continue unless checked. Looked at this way, a warming of the whole Earth is relentless and inevitable.

As a species, we are obsessed with climate change, and quite rightly so; it is often described as the greatest threat humanity has faced in its relatively short span on a planet that is barely halfway through its own lifetime. Creatures recognizable as human beings have been around for about five million of the four and a half billion years since the formation of the Earth, that is, only about 0.1% of the planet's history. The Sun is thought to be stable enough in something like its present form to possibly support life on Earth for about another five billion years. Who knows what that will bring? The best books about the future are still those by H.G. Wells, and he was writing those around a century ago.

I'm in danger of getting off-message now; the present book is meant to be a memoir not a climate diatribe. I do have on my shelf a partially completed draft of a book that attempts to explain climate change in non-technical terms; its title is the same as that of this chapter. I spent too long over drafting it, with so many other priorities pushing it to the back of the queue, until about ten years ago I shelved it because the publishers I was talking to had by then put out similar books by other people. There are more and more of these books all the time, and some of them—although by no means all—are very good.

Most of the existing literature, from bestselling books to newspaper reports, is written by professional writers and journalists of one kind or another who do not understand the underlying science terribly well themselves. Of course, there are books written by scientists who *do* know what is going on in the climate 'system' (by which they mean the Sun, the atmosphere, the oceans, icecaps, flora, and fauna, and their mutual relationships), at least as well as anyone does, but these are usually aimed at specialists and anyone else reading them in an earnest attempt to save the world, or at least to understand why that is going to be difficult, may have their work cut out.

The IPCC reports tend to be in the latter category. Their approach is highly statistical, leading easily to summaries that are designed to appeal to politicians. For the general reader, including students, there is a danger that an initial passion for getting to grips with the most implacable threat to life, liberty, and the pursuit of happiness on the current human agenda, may be reduced to boredom by a dreary account of the deliberations of a faceless panel spun off by the United Nations. The experts can seem obsessed with the finer points, such as tables of worldwide greenhouse gas emissions or the molecular symmetry properties at the quantum level of the carbon dioxide molecule. Few are the books at an intermediate level that are fully authoritative, easy and enjoyable to read, and that leave the reader with a warm glow that he or she has grasped what all the fuss is about. That's what I wanted my book to be, and perhaps I will try again some day to finish it.

A Beginner's Guide
to the
End of the World

F.W. Taylor

This is a draft of the cover of my unpublished book on climate change. The book was inspired in part by one on a completely different topic: 'Think' by Simon Blackburn explained Philosophy to the uninitiated in such clear and interesting terms that I was motivated to try to do the same for climate science.

The title slide from my opening talk at the 'International Conference on Comparative Planetology' organized by the European Space Agency in 2009. As presenter of the 'keynote' talk my job was to introduce the topic, and I took an ambitious approach. Using simple models, I showed holistically how the climate might have evolved on all three Earthlike planets, and from this predicted whether to expect to find life (on Venus, no; on Earth and Mars, yes).

Education and outreach

A lifetime of work on atmospheric observations and processes should add up to a modest contribution to understanding the climate. I started investigating how to improve weather forecasting using the first crude data from space, went on to some early studies of weather-related phenomena on the planets, and found them relevant to an increasingly crowded and depleted Earth in trouble. Looking back, everything could arguably be collected under the banner of climate science, even studies of ice on the airless Moon.

Over the years I tried to promote the importance of studying Earthlike climates on other planets, especially Mars, Venus, and Titan, whether in conference talks, when applying for research funds, or when competing for participation in space missions. A year or so before retirement I was invited to give the 'keynote' talk at a big conference at the European Space Science and Technology Centre at Noordwijk near Amsterdam. 'Keynotes' are meant to give a sweeping and stimulating introduction and overview of the rest of the conference, which lasted several days. Rather recklessly I chose the title 'Comparative Planetology, Climatology and Biology of Venus, Earth and Mars'. Including a reference to life was especially daring, but I decided to have a shot at it.

I began 'Spacecraft studies of the three terrestrial planets with atmospheres have made it possible to make meaningful comparisons that shed light on their common origin and divergent evolutionary paths. Early in their histories, all three apparently had oceans and extensive volcanism, Mars and Earth, at least, had magnetic fields, and Earth, at least, had life. All three currently have climates determined by energy balance relationships involving carbon dioxide, water and aerosols, regulated by solar energy deposition, atmospheric and ocean circulation, composition, and cloud physics and chemistry. I will address the extent to which current knowledge allows us to explain the observed state of each planet, its planetology, climatology and biology, within a common framework. Areas of ignorance and mysteries are explored, and prospects for advances in resolving these with missions within the present planning horizon of the space agencies are considered and assessed.'

I did not make a good fist of this talk. I had given so many talks by then that I was overconfident and my discourse overreached itself to the point where it became too lightweight, almost flippant. It was less successful than a similar lecture I had done a few years earlier to an even larger audience, the massive annual gathering of the American Geophysical Society in San Francisco. There, at the customary (for the USA) early starting time of 8am I delivered the Shoemaker Lecture on 'The Atmosphere and Climate of Venus'. In spite of the one-planet title, and the new results just coming in from *Venus Express*, I couldn't resist making comparisons. I began: 'Venus is like another Earth, orbiting closer to the Sun. Its cloudy atmosphere reflects so much of the Sun's heat that, on simple energy balance grounds, we might expect Venus to be cooler than Earth. In fact, its surface glows with a dull red heat and some metals would be molten there. This fantastic case of global warming is due to a greenhouse effect very similar, though stronger, than that which is now threatening to overheat the Earth. What lessons should we draw from this?'

I went on, 'Can we explain the evolutionary tracks of Venus, Earth and Mars with a common paradigm? What key investigations should we be making, particularly in terms of space missions? It is more than twenty years since a space mission to Venus focussed on the planet's enigmatic atmosphere. During that time, little progress has been made on understanding the reasons for the extreme climate on the planet, although there has been increased interest in understanding why it should be so different from the Earth, and from *a priori* expectations. The European Space Agency responded recently to calls from the scientific community there for a new mission to Venus, and produced the *Venus Express* orbiter. This has now been operating at Venus since April, and early results are beginning to flow. I will discuss the interesting questions about Venus that *Venus Express* is addressing and discuss current and expected progress and remaining mysteries.'

I kept a list for a while of all the lectures and media interviews I gave, but it got so long I failed to keep it up. Glancing at it now, I see Friends of the Earth in London; the Ashmolean Christmas Lecture in Oxford, attended by children as young as ten; the Cheltenham Literary Festival; university groups as far afield as France, Germany, Finland, Spain, Poland, Australia, and New Zealand, and dozens of student societies and amateur astronomical societies all over the country and sometimes overseas.

Back at the more mundane level of university teaching, by the 1990s the topic of climate had achieved such a high public profile that I decided the time was right to propose to my colleagues in the Department of Physics that we include a short course on climate physics in the undergraduate syllabus. Changing the content of the Physics degree is always contentious and difficult, made worse by the fact that the subject has grown to such a degree that it has become impossible to include everything. The natural urge to keep up to date with new fields like biophysics or quantum optics vies with the equally natural wish to make

The cover of my textbook for the undergraduate course on Climate Physics that I introduced at Oxford in 2003. Doris designed the cover and drew all of the figures.

sure the basics such as optics and thermodynamics are not squeezed out. I fought my corner and after a lot of discussion and some opposition my proposal went ahead. I was going to teach the course myself and there wasn't a textbook that matched my ideas, and my promises, so I decided to write my own. The guiding principle was to describe climate science starting from basic physics and linking everything with the rest of the syllabus.

The course, and the textbook, gathered up the material from areas such as thermodynamics, quantum mechanics, and optics that are relevant to the key question: what are the mechanisms that maintain the climate, and how may it change? I introduced this with a short discussion of the popular background: is global warming real, and, if so, is it a threat to human survival and wellbeing? The short answers, I said, are yes, and probably.

I was speaking to the young people who were going to have to grapple with the problem before their lives were through. In a lecture theatre full of the nation's brightest students, prone to thinking more about their upcoming exams than the fate of the world, these short answers cannot be justified by saying that any given conclusion must be accepted because it is the prediction of a large computer model. That may be good for the politicians, but the students know, as I do, that no one, not even the people that program and run it, completely understands

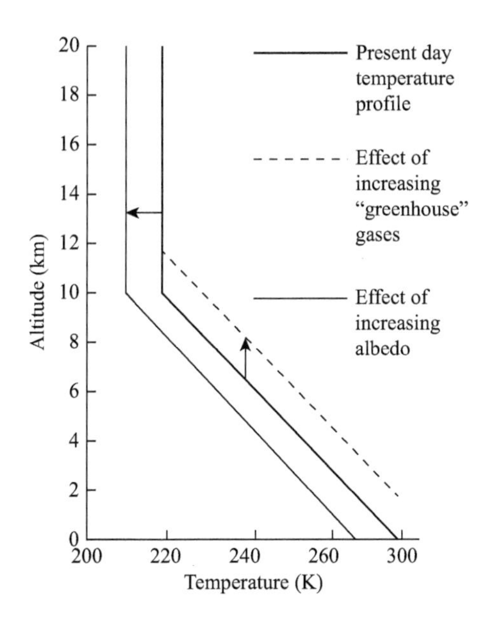

A simple climate model, illustrating global warming (by increasing greenhouse gases like carbon dioxide) and cooling (by increasing refection of sunlight by cloud or aerosol), that uses just basic physics and no computer. In my lectures, I used this to introduce the Physics undergraduates to the global warming conundrum. From *Elementary Climate Physics*.

the model. Instead, I begin the course with a short summary of the processes that go on inside the Sun and the resulting energy flux as radiant heat. The Earth's surface is warmed, not only by the rays of the Sun, but also by the infrared (invisible heat) radiation from the atmosphere. A simple energy balance calculation shows that these contributions are in roughly equal amounts.

Remarkably, and this is where the less well informed sceptics often get into trouble, the atmospheric radiant energy flux does not come from the gases that account for most of its bulk. Instead, relatively small traces of water vapour, carbon dioxide, and methane—the 'greenhouse gases'—do the emitting. The reason has to do with the fact that nitrogen and oxygen are symmetric diatomic molecules and when we look at the quantum mechanical properties of these we find that we should not expect them to absorb or emit infrared radiation. The important ones are triatomics like H_2O and CO_2 which, although they are symmetric in the sense that the basic molecular structure is H-O-H and O-C-O, can break that symmetry by bending and stretching. Try that with N-N or O-O and it stays symmetric.

The important greenhouse gases are present naturally: water vapour from evaporation, mainly from the ocean; carbon dioxide from burning and respiration; and methane from decomposing vegetable matter in swamps and the guts of animals. Although the proportions are relatively small, the 'natural' levels of these gases keep us all more than 30 °C warmer than we would be without them. This can seem astonishing, not least because we wouldn't otherwise notice their absence, but it is a fairly simple matter to make the calculation from first principles and show that it is so. In a world without greenhouse gases the surface would be covered with a shell of ice and, unless we had fled underground and figured out how to keep ourselves warm and fed, none of us would be here. The planet Mars seems to have lost most of its greenhouse effect over the aeons, and the simple lifeforms that might once have dwelt in its erstwhile seas, now frozen and covered over with thick layers of windblown dust, could be huddling underground in the residual warmth of near-extinct volcanoes, and if they are we may soon find them.

Once we understand the role of 'greenhouse' warming in the natural world, the next stage is to ask what basic physics has to say about changing amounts of carbon dioxide and the other active gases on the thing we care about most, the temperature at the surface. The temperature as a function of height in the lowest 20 kilometres of the atmosphere can be worked out using simple thermodynamics and energy balance calculations that can literally be done on the back of an envelope. The brightness of the Sun, or the amount of carbon dioxide or cloud, can be varied to see the effect on surface temperature.

When the proportion of greenhouse gas in the atmosphere goes up, the surface temperature of the Earth also goes up. It can't help itself, and there's no denying that, these are the laws of thermodynamics and energy conservation at work and you're in trouble if you ignore those cornerstones of science. So, we know greenhouse gases are increasing, we know humans are responsible, mostly, for that, and we know that surface warming must follow. Is that the end of the story? Not quite.

Hard evidence

What about measurements, what do they show? Changes in things like bird migration patterns and sea level rises consistent with a warmer planet are being reported along with the temperature data itself. However, most of us have not noticed much of a change, either in our lifetimes or the recorded patterns of life in the two centuries after the industrial revolution. The evidence for warming is there, but it is a bit marginal, and the trends measured over the last 50 years are not all that much larger than the pre-industrial variations deduced from weather records, tree rings, ice cores, and so forth.

We might naively expect that an increase over the natural level of greenhouse gases by 50% would warm the planet in proportion, i.e. by 15 °C or so. We would certainly notice that—the last ice age was a shift of only about five degrees in the other direction. The reasons we haven't felt anything that bad are twofold. Firstly, the relationship isn't linear: doubling the cause doesn't double the effect. Secondly, some of the pollution produces global *cooling*.

The first reason is down to quantum mechanics again and best taken on trust for now. The second has to do with clouds and haze, which float in the atmosphere like giant mirrors and reflect sunlight back to space, keeping the Earth cooler than it would be without them. Cars and factories produce not only greenhouse gases but also smoke, much of it invisible because the particles are so small. Water accumulates on these tiny particles, leading to cloud formation, and they also form the thin 'aerosol' hazes we see everywhere and these reflect sunlight in their own right.

So, that's good, isn't it? Suppose we just relax and hope that global warming, and what has become known as 'global chilling', will cancel out. There are even those who think the Earth will work out a balance on its own, which it might, although if left to itself the solution found by the planet could involve killing off most of the humans, including the Gaia-worshippers. The experts say cancellation is unlikely; almost any set of assumptions they make about future trends in pollution emissions has warming gradually pulling ahead, until we have five degrees of warming—a reverse ice age—possibly in less than 100 years from now. Of course, we can and should argue about how reliable this number is. They can't get the weather forecast right, and they expect us to believe the world is coming to an end? Or, it might only be one or two degrees, we could live with that, right? Life might even, on balance, be better with a bit more warmth, crops will grow faster and so on.

This sort of debate is reasonable and indeed essential, in contrast to those that say there is no greenhouse effect at all, who are just dumb. Not even smart people want to give up their cars on the say-so of a few computer models. But we have to remember that, since the effect is definitely real—it's only the size and speed of it that we can fight over—the warming in 2100 is as likely to be ten or eleven degrees as it is to be one or two. Then we would really be in trouble—relentless sea level rises eventually measured in tens of metres (again we can argue about how many and how soon) that displace coastal populations are on the cards, and that is just the beginning. Water supplies, agriculture, ecosystems, will go to pot, and fragile political stabilities will crumble. What sort of fools would we be to risk that?

Taking action

It was fashionable for a time among the concerned and the more laid-back alike to blame successive American presidents for the whole climate mess because they are too slow to sign up to protocols to limit emissions, the lion's share of which, for the time being, originate in that country (and China). The counter argument is basically that sufficiently useful reductions cannot be made worldwide by legislation, unless it is so draconian as to seriously damage wealth and wellbeing, and stunt progress. A better solution, the ungreen argument goes, is to look for advanced technical solutions that permit an attractive lifestyle and low emissions at the same time. Fossil fuels might be replaced by fusion reactors, really efficient batteries, compact fuel cells, and transport systems using electric propulsion or clean-burning hydrogen. Fine if we can do it, as long as the developing world gets the new technology quickly and cheaply from the rich inventors and licensees. Perhaps a mixed approach is the best answer—some restraint to tide us over, while we figure out how to make fusion power generators work or how to use the abundant energy from the Sun more efficiently.

Deus ex machina is never a comfortable solution to a seriously posed dilemma. What can we really do straight away? Turning down the thermostat and trading the Aston Martin for a Mini or a bike suits the hair shirt brigade, but on its own is too little too late. Reduce the population back to the levels of around 1800 or so? Switch completely from fossil fuels to nuclear power, despite its persistent (and largely undeserved) nasty image? I can't see either of these happening, and neither, it seems, can the politicians. More likely, we're going to behave like humans have always done and fiddle with the problem around the edges until it gets out of hand and then go to war with each other over the residual land, water, and food. There has to be a better way.

The sleeping giant awakens

The wake-up call to world government came mainly through the Inter-Governmental Panel on Climate Change. In Chapter 8, I recounted how Mrs Thatcher's regime in the UK took the lead in organizing international action on ozone depletion, and eventually the nations of the world successfully contained the problem. However, ozone loss was soon replaced by global warming as the worst environmental disaster seen to be threatening mankind. This time there was no obvious step, equivalent to banning fluorocarbons, that governments could agree to take: banning carbon dioxide emissions would involve shutting down most of human activities and a return to the Dark Ages. In the USA, George Bush notoriously declined to acknowledge there was much of a problem and indirectly encouraged climate sceptics and the oil industry to hope the issues would go away. The British government took the opposite tack and, having worked out that a 20% reduction in carbon dioxide would be an equitable share for the UK to aim at in the next 20 years, generously decided to go for 40%, more than any other country. Not only that, but they enshrined it in law, binding themselves (or rather, their successors) to an achievement on which they are certain to default.

Much has changed since then and, while scepticism has by no means gone away, as the *Daily Telegraph* poll shows, most world leaders now accept the scientific arguments and the need to address the looming problem. Along with erecting enormous windmills in beauty spots all over, the main effect of which is to destroy the environment now rather than wait for climate change to do it later, one of the consequences has been a big increase in funding for improving climate predictions and finding better means of ameliorating what are now seen as inevitable serious changes in the world's weather. My own small contribution through writing, speaking, and teaching, and through research to better understand the processes involved by studying them using instruments in space, is mentioned in various places in this book. Most of this deals with the 'big picture', but there is also a lot of work these days to address more specific, local problems. The European Union has encouraged and supported many of these.

We're all Europeans now Part 2

The EU is a very large and very rich organization. From time to time it announces new tranches of funding for research in specific areas, including climate-related problems. When it does, the amounts on offer are staggering, even by the standards of someone used to large grants to build space instruments, totalling in some cases more than ten billion euros. Penurious university groups cannot afford to ignore such largesse, and one of my tasks when running things at Oxford always involved trying to unlock the European treasure chest. In this I had very mixed success. The procedures for applying are labyrinthine, often obscure, and sometimes impossible. For instance, a proposal I planned in the 1980s was blocked because it required me to submit, inter alia, a copy of the documentation which formally created the organization to which I belonged, something it is not easy to find at Oxford University which, it is said, has origins that can be traced back to the time of Alfred the Great.

Eventually that first attempt was in fact approved for funding. The rules said that the EEC, as it then was, would cover half of the cost of an approved project. We had a small but talented group at Oxford which had created one of the first computer models of the atmosphere, the goal being to make simulations that could be compared and contrasted with the data we were acquiring with our satellite instruments. I worked out that we needed about £100k to support this work for the next couple of years, half of which we could probably get from our usual sources in the UK, so I applied to the EEC for the other £50k (these are approximate numbers). They agreed, but then applied the 50% rule so that the award letter offered only £25k. I protested, but the die was cast, and £25k was better than nothing.

However, when the cheque came, someone further down the chain had applied it again, and the amount I received in the end was only £12,500. Trying to sort this out in the days before email and with people on the phone who spoke only idiosyncratic English was a nightmare best forgotten, as was a later application for a much larger sum that took weeks to prepare but was dismissed with a single line of explanation saying the research proposed was too ‘open ended’. I puzzled over this for a while, but this time did not bother to pick up the phone.

Many happier European experiences came later, for instance through collaborating with Ilias Vardavas at the University of Crete. Ilias showed considerable skill in negotiating for funds from the EU and he built a large international team, myself included, to work on problems of the radiation balance in the atmosphere and its effects on the local climate. We met frequently for a while in Heraklion and came up with some new ways of using the data from those meteorological stations that monitored the solar flux as well as the usual temperature and rainfall data. The idea was to investigate cause and effect in a way that would be useful, eventually, to the local agriculture and economy in Crete and elsewhere. My initial contention that space data would add value to this work turned out not to be very practical, which rather limited my contribution, and I withdrew from the consortium after a year or two. But this was not before Ilias and I had laid the foundations for the textbook on *Radiation and Climate* we wrote to support teaching and research in the field.

I also wrote a textbook with Manuel Lopez-Puertas of the Institute for Astrophysics in Granada, Spain. We had a common interest in a phenomenon called *Non-Local Thermodynamic Equilibrium*, a mouthful that is universally shortened to non-LTE. This has to do with the way energy, in the form of heating from the Sun or cooling by the Earth, is moved around by radiation in the atmosphere. It is a common assumption that the molecules of gas in the atmosphere are in equilibrium with each other and with the photons streaming through them, but this is actually false. The discrepancy is small near the surface but gets larger and larger as the pressure drops with height, as collisions between molecules become fewer. Without collisions, any given molecule doesn't know what the others are doing and so its quantum state is determined by interactions with the photons in the radiation field, rather than the Maxwellian velocity distribution dictated by equilibrium thermodynamics. Getting the sums right is essential for the proper interpretation of satellite measurements and for theoretical climate models whenever the upper air is involved, which is much of the time.

The physics and mathematics of calculating non-LTE are extremely complicated, and Manuel devoted his career to it, becoming probably the world's leading expert. I was more interested in the applications side, so together we were able to produce a mighty tome that is still the bible on the subject (*Non-local Thermodynamic Equilibrium in Atmospheres*, World Scientific Publishing, 2002). Creating this involved several long visits to Granada, where I stayed in a hotel on the hill beside the Alhambra and walked each day through the beautiful old town to the institute in the suburbs to write in its placid and friendly surroundings.

Heraklion too, of course, was a magnificent and historic place in which to spend time on authorship, and so were Meudon and Paris when I worked with Athena on our books about Titan. All three added legendary hospitality to an agreeable place to work, and occasionally colleagues in cold, wet England would express suspicion about my motives in selecting partners for writing and research. But it was all driven by our compatible interests in the science, truly.

Avoiding Armageddon

Understanding the physics of a problem like climate change is obviously an essential step on the way to addressing it, and much progress has been made. Whether that means we can make solid predictions is another matter. I again quote Richard Goody, who wrote in his autobiographical notes in 2002 as follows. 'My reading of the literature suggests that we do not understand with any certainty the relative contribution to the observed warming of natural variability, industrial gases, natural and industrial aerosols, land use, solar irradiation, or something else yet to be recognised [I would add ocean circulation variations]. Couple this to undeniable uncertainties in the climate models themselves (particularly with respect to oceans and clouds) and it is very difficult to justify a claim to any quantitative knowledge of climate fifty years from now—except perhaps that anthropogenic activity will lead to changes, although we may be surprised what those changes are.' This is still true today, in my view.

However, like Goody I agree that we can nevertheless infer a need for a different style of industry, one in which safe disposal of all by-products is part of the process. If this step is not taken, and it can be very difficult and expensive, the risks are high. The best models we have, after decades of brilliant work to refine them and ensure that they explain, or at least give consistency to, the increasingly detailed observations, persist in predicting a future so dire that we are left almost in despair. Towards the end of my career in space science I started to feel a benign satisfaction that I had managed to understand and demystify the lure of the planets that had preoccupied me in my youth. But, since my focus had been on atmospheres and climate, the dark side was that everything I had learned pointed towards a bleak future for mankind on Earth.

Perhaps it was time to put what energy I had left not so much on more exploration, but on making whatever contribution was possible to a better future for mankind. Was there anything an ageing professor who thinks he now understands climate physics could do to reduce the chance that future Earth will tend towards a frozen desert like Mars, or more greenhouse heating like Venus? The odds at the moment are on the latter, with accompanying crises like massive sea level rises.

Recently, I was reading some of the latest forecasts by my colleagues and one of them was that the mean global sea level could rise by 90 metres in the next 100 years unless checked by drastic reductions in greenhouse gas emissions. Google Earth tells me that my home on Headington Hill, looking down on the spires of Oxford, is 91 metres above sea level. I know that the River Cherwell, now meandering peacefully towards the Thames at the bottom of the hill, was once part of a massive sea that lapped over our garden, because when I dig I find clamshells and other marine artefacts. If that sort of environment is coming back it is serious.

But what to do? Give up on space instruments and design better windmills and solar farms? Not in one lifetime, and besides I was still committed to space projects, some of which had yet to reach the launch pad. However, something intriguing did come along.

The transport internet

At one of the cocktail parties that regularly punctuate Oxford life I met a local businessman, Noel Hodson. He had for some years managed a company that built special-bodied Bentley cars, and since I now drove a Bentley we talked about this common interest for a while. When he found out what I did, Noel asked if I would be interested in helping with a 'blue skies' project he was championing. His idea was to use capsules in underground pipelines to transport food and other goods rapidly around the country, thus relieving the road system of thousands of lorry journeys and the accompanying congestion, pollution, and expense.

One of the arguments put forth in favour was the fact that the capsules could be driven with 'clean' electricity and much greenhouse gas emission thereby averted. Noel wanted me to say something about the reduction in carbon dioxide, and hence global warming, that might result if his 'transport internet' became established worldwide. It surely could deliver a large part of the government's legal commitment to huge reductions in the nation's carbon dioxide emissions.

We drew up a scheme in which the thousands of lorries on British roads that service supermarket chains like Sainsbury and Tesco were replaced by a network of tubes, which we called FoodTubes, since food (and drink) is the main commodity they carry. Noel worked out the finances and showed that the huge cost of installing the network and running it was offset by the even huger savings on lorries and fuel, such that the initial investment would be paid off in a finite time and large profits made thereafter. And of course there were many other benefits,

I helped to develop the 'FoodTubes' concept, which replace lorries with pipelines enclosing electrically-driven capsules to carry goods to supermarkets and stores. It could be part of the solution to the global warming threat by greatly reducing carbon dioxide emissions. The technology required is not rocket science, and the business model suggest that in addition to the environmental benefits, it could be highly profitable. (Picture courtesy of Noel Hodson).

over and above a reduction in carbon dioxide and other emissions, including relief for congested motorways and faster and more reliable delivery of perishable goods.

When I arrived the technical design was not very advanced and I suggested several improvements including simplifying the capsules, putting them on rails, and pushing them along at high speeds using electromagnets installed in the rail mountings. Sketching this up as a computer-controlled network it becomes a highly efficient and economical system that makes the current hordes of smoking lorries on our roads look like something from the Stone Age. Completely sold on the idea, I joined the consortium of experts and entrepreneurs that Noel had assembled to push the project forwards. We met a number of executives from companies interested in building or using the proposed network and everything was as positive as could be. We tried different points in the labyrinthine government bureaucracy for start-up support and were stonewalled.

Recently, however, it seemed that things were looking up when the government set up a 'Catapult' centre with a specific brief to encourage new initiatives in transportation. In preliminary talks with them, we discovered that their plans for innovation seem to involve making better use of existing roads, rather than augmenting the road and rail networks with expensive new infrastructure. They have developed predictions in which the recent trends among young people to make fewer car journeys, and lorries carrying freight using computerized load management and satnav to follow more efficient routes, are extrapolated to show road usage actually falling as the population and the economy grows. In such a world, we may not need FoodTubes.

The art and science of modelling future traffic behaviour turns out to have a lot in common with climate prediction. Both involve manifold interactions between processes that are quite well understood in isolation, but which become complex, unpredictable, and even unstable when linked together. Much depends, too, on the assumptions made as inputs. Things like population level increases and the rate at which efficient, large-scale clean energy production comes on line must be estimated. If these are seriously wrong we experience the process that used to be irreverently referred to as 'garbage in, garbage out', before it became so important to take the best possible model predictions seriously, even knowing they are surely wrong (but hopefully not *too* wrong). It may be that in 50 years' time road and rail will be so well integrated and so efficient, clean, and uncluttered, that goods and people will travel quickly and cheaply between towns and villages set in idyllic countryside with atmospheric carbon dioxide and other greenhouse gases falling year on year. It's certainly something to aim for, but as a species and a civilisation we're not doing very well at the moment and the trends are worrying.

17
Standing at the edge of time

You must now endure my going hence, even as you once had to endure my coming hither.

George Bernard Shaw, on his retirement
from the Fabian Society *c.* 1940

I was Head of Atmospheric, Oceanic and Planetary Physics for 21 years, during which time a great deal changed. Broadly stated, when I started in 1980 I ran the department in the traditional Oxford manner, which was as a small fiefdom with very little interference from anybody. By the new millennium, the five physics departments were unified and in the process of building up today's network of bureaucracy, while the university as a whole was adding administrators at a dizzying rate. A scheme was introduced where any academic could apply for the title of Professor, without change of job description, workload, or remuneration. The senior administration headed by the Vice-Chancellor had discovered that this was an efficient way of rewarding, motivating, and retaining academics, and this at no cost to the University! It was argued that competing for funding would be more effective if the applicant was badged as a professor, and other universities, and even government laboratories, soon followed suit until nearly everyone leading research anywhere in the country carried the title of professor by the time they reached middle age.

A (possibly) unintended consequence of this innovative bit of grade inflation was that those of us who were 'real' professors vanished into a sea of similarly titled individuals and were no longer distinctive and all-powerful. This, of course, was not entirely (or perhaps at all) a bad thing. Not all brilliant scholars and researchers are good administrators, and anyway once you cease to be in charge you no longer have the burden of responsibility for much of the running of the organization. The administrators do that, finding it necessary to relentlessly increase their numbers to do so. Of course, academics still have to raise money (a lot more than before, with overhead charges on grants and contracts shooting up) and must respond to a flood of rules, schemes, and excessively meticulous interpretations of the laws of the land (and the European Union), with their unending streams of unintended consequences.

The statutory professors like me were those who held a traditional chair with broad responsibility for their subject area but light teaching duties in return for the administrative load this would entail. The position exists in perpetuity, so an election is triggered to find a new occupant for the chair when the incumbent retires. The new-style 'titular' professors, on the other hand, hold their titles on an *ad hominem* basis while they carry on with the duties of a university lecturer until they go, and then the title vanishes. Perhaps to mollify the real professors for their loss of exclusivity, the administration suggested a distinction in which all chairs would be 'named' after a distinguished predecessor or donor. Most of the statutory positions were named already, this being the old tradition, whereas the new titular positions obviously were not. Thus today we have the Savilian Professor of Astronomy as the statutory chairholder alongside several ad-hoc Professors of Astrophysics, or of Cosmology, Solar Physics, and so on. As Reader in Atmospheric Physics, I had morphed into Professor of Atmospheric Physics, and under the new system that wouldn't do. I was invited to propose a name for my chair, which would then have to go through the committees all the way to Congregation for approval.

Choosing a name was not hard. In my subject area, which is one that has been around in Oxford for longer than most, especially outside the Humanities, there were some really distinguished forebears who were not yet honoured by having chairs named after them. Head and shoulders above the rest stood Edmund Halley, who was a contemporary of Newton in the seventeenth century and comparable to him in genius and achievement. Halley's biographer noted that he pioneered the 'firm belief that Nature can be understood rationally and that from what we know we may predict what is yet to be observed.' Halley applied that principle to predicting the transits of Venus that would occur after his death, and the return of the great comet we now know by his name, again something he would not live to see. The idea that numbers and formulae ruled the behaviour of the heavens was revolutionary at the time. As well as astronomy, he was active in geophysics, meteorology, and oceanography. He is the perfect icon for our department in the twenty-first century.

Cambridge venerates its most famous scientist, Newton, in many ways, but even after the naming of the chair there was little comparable fame for the equally great Halley at Oxford. People who knew only about the comet, if that, would ask about the choice. I made a small display that visitors would see when they came into our building, that explained that 'Modern planetary science at Oxford can be traced back to the work of Edmund Halley in the late seventeenth and early eighteenth centuries. In addition to being a pioneering geophysicist and oceanographer, Halley made observations of Mars, Venus and Mercury as well as his famous studies of the orbits of comets. Today, we are involved in instrumentation, analysis and modelling in connection with European and American space missions to all of the planets known in Halley's

When professorial titles began to become universal in Oxford and other universities, I was asked to 'name' the statutory chair of which I was the current occupant. After a brief deliberation I settled on the pioneering atmospheric, oceanic and planetary scientist Edmund Halley (1656–1742), seen here in a bust now at Greenwich Observatory.

time.' The times being what they are, it goes on rather prosaically to add 'This research is supported by a consolidated rolling grant from the Particle Physics and Astronomy Research Council.' Although the notice stopped short of saying so, this grant was worth about a million pounds a year to us. It covered our basic costs; to actually undertake a major project, and build an instrument, we had to apply and win an additional, and usually even larger, sum.

Moving On (a little)

In 1979 I had been appointed Head of Department for life, as was the practice back then. Now, with a big pool of professors available this could be changed, and much mayhem avoided when, as not infrequently happened when there was only one of them, 'the Prof' degenerated in old age into boredom and eccentricity. After much discussion, a sensible compromise was agreed in which the headship of a department was assigned for five years at a time, with no limit to the number of terms one person could have, provided he was willing to continue and was still acceptable to his colleagues and the administration. I had done more than ten years as head when this scheme came in, and went on to add two of the new five-year terms. At the end of the second of these I noticed that every one of my fellow heads in the other parts of Physics had already passed on the job, and had done so at the earliest opportunity. Checking around, I found that this was the trend all over the university; as far as I could tell, at the age of 56 I was the longest-serving head of department in all of Oxford. It was no fun any more, just a tour of duty.

My colleague David Andrews, an atmospheric dynamicist whom I had brought in from Cambridge to one of our academic posts many years earlier, was keen to take over as Head of Subdepartment following a successful stint as Chairman of the Physics Sub-Faculty, the body that organized the teaching and examining side of the subject. With his appointment to the job I had done for so long I was freed of a considerable burden, and could consider how to spend the time thus gained. David was a theoretician, trained as an applied mathematician and an expert in fluid dynamics; he was unlikely to drive the department's programme of space experiments as I had done. However, John Barnett was leading our work on Earth observation experiments now, as Principal Investigator for the HIRDLS experiment, and my erstwhile research student Simon Calcutt was doing a good job on our various planetary instruments. They didn't need me in charge any more. I was already well along on my plan to become a co-investigator on many experiments, savouring the exciting bits of each, instead of being lashed to the oar as Principal Investigator on just one major commitment. A Co-I can focus on his favourite piece of science with far less of the hassle that comes from leading the whole project as PI. I found I could manage as many as six co-investigatorships simultaneously; the stories of those are in the other chapters of this book.

I did consider moving on from Oxford altogether, as my predecessor had done before me, to somewhere where I could paint on a larger canvas for my last decade or so before retirement. Over the years, offers and opportunities for new positions came along from time to time, but I was always busily and (usually) happily engaged on my work and travel with no wish to change horses in mid stream. It is in any case hard to relinquish Oxford, with its valuable reputation and its exceptionally agreeable lifestyle. Three possibilities did give me more than a moment's thought, however. The Space Science Centre at the University of Wisconsin enquired whether I was interested in becoming successor to the formidable Werner Suomi, an American space sciences pioneer with whom I had worked on the *Pioneer Venus* project. But it was not long since I had made the hard decision to move from America, and the thought of returning now, even to take over a thriving outfit in a beautiful part of the country, seemed too retrograde. The University of New York at Stony Brook, at that time home to a thriving planetary group that included my distinguished colleague and occasional collaborator

Toby Owen, wrote to say it had a chair available, and that might have been hard to turn down under different circumstances.

An invitation came to apply for the Directorship of the Max Planck Institute for Solar System Studies in Germany, and another for the Director General position at the British National Space Centre. I was not a fan of the BNSC, which had not yet blossomed into become the UK Space Agency, and in any case the civil service rulebook said I was already too old. The Max Planck opportunity came along soon after I stepped down as head of department, the best time for a move if there was to be one, although to tell the truth I was more than content with life at Oxford, with most of my administrative burden now unloaded. Still, I went to Gottingen and spent some time at the Institute in nearby Lindau to check out our mutual interest. It had a great location, near the Hartz Mountains, with superb staff and laboratories, and Max Planck directors get a huge discretionary budget, but it was a job for life and would have meant resigning at Oxford and transplanting to Germany for the rest of my career at least. The University allows its professors to take leave for up to five years, and if the Max Planck Gesellschaft had accepted my proposal to come for just that much time I would probably have gone.

Retirement and Redux

Oxford professors traditionally retire at age 67, two years later than most academics and the general population. This extra time to continue working was considered a privilege. I reached the age of maturity in September 2011, but knew that retirement doesn't necessarily mean coming to a screeching halt, particularly in academia. Several of the larger space experiments on my co-investigator list were either currently delivering exciting data or in the pipeline to do so on a reasonable timescale. Two of them, the *BepiColombo* mission to orbit Mercury, and *Rosetta*, designed to orbit and land on a comet, were then still in the fairly early stages and yet to produce any data. *Rosetta* was in flight and approaching Churyumov–Gerasimenko, ready for the encounter that has since taken place, as already related (Chapter 11). At the time of writing, *BepiColombo* has not even launched; next year, maybe, but it has been delayed so often any forecast would be wrong by the time you read this. Anyway, as a team member for these two missions I look forward to brand new data and certainly have no intention of letting go just because the day job has changed. *Cassini, Venus Express, Mars Reconnaissance Orbiter*, and *Lunar Reconnaissance Orbiter* were all continuing to send back interesting data and to generate new excitement, and there was plenty to do there, too, thinking, interpreting, and writing about each of them. The younger team members do the hard work that I once did, but need the guidance and advice of their wise and experienced older colleagues from time to time (well, I think so).

The best way to do this is to keep attending the team meetings, held typically about twice a year. The problem with this is that the meetings are often in distant locations. As I write, the *Lunar Diviner* team is meeting in Honolulu, a long 24-hour trip from Oxford. Participating in meetings like that for six teams would mean an average of one foreign trip every month. In addition, there are at least six important conferences every year, and if you want to write books and papers you have to visit your co-authors which in my case are scattered around the world. All told, a foreign trip every fortnight had been the norm for long stretches of my career. As I write this paragraph, the inbox pinged as invitations arrived to a conference in Chania, Crete, which I declined, and one at Charterhouse School, an hour's drive from Oxford, which I accepted.

The sort of frequency of travel I used to take in my stride was no longer sustainable in retirement. The trips to California and Australia that used to seem so exciting are now just incredibly uncomfortable and tedious. Whether it is the increased crowding, queuing, and security, or just an older person's perspective is moot: it's both, probably. My regular trips to Los Angeles to visit JPL required obtaining a work visa, since I was legally employed by Caltech while I was there. Renewing the visa used to be simple, but after 9/11 involved frequent, incredibly frustrating, and sometimes humiliating visits to the Embassy in London, often lasting all day. On arrival at LAX, the slightest anomaly in the paperwork or in the labrythine workings of the US Department of Homeland Security led to a long wait in a back room at the airport, tired and hungry, watching fat smirking cops with massive weapon belts pass by for hours before one of them told me I could go, usually with no explanation of why I had been held in the first place.

And of course these trips cost money: to attend a team meeting or conference you usually have to raise the readies by making successful grant applications. Grovelling for money, as a distinguished, older Oxford colleague once described it, is even more tedious than queuing for hours to have your bags searched or your visa verified and your passport stamped. Invitations to conferences used to come with all expenses paid, sometimes, especially if it involved giving a 'keynote' talk or a prize lecture or something like that. These days, that is rare. Funding applications, too, have got more complicated and more impersonal, and success more unpredictable, over the years and I was glad when retirement provided an excuse to stop. If I had no funding I couldn't travel, and it was time to stop travelling so much.

The other big responsibility I decided to shed was teaching in its various aspects. This of course is a cornerstone of university life, involving not only training graduate students but also lecturing to undergraduates, interviewing for admissions, and examining. As a statutory professor I had a fairly light undergraduate teaching load, my time being taken up by running the department and the big research projects, but I had created two undergraduate courses, *Planetary Atmospheres* and *Elementary Climate Physics*, from scratch and written the text books, delivered the lectures, and set and marked the exams.

I supervised and examined the work of research students not only individually but for a time collectively as Director of Graduate Studies (taking on a stint in this role was known as 'going to the DoGS'). Some of my own students have gone on to significant positions in laboratories in the UK and abroad, with (as is normal) quite a few ending up not only beyond atmospheric physics but sometimes outside science altogether. Christine Rice is perhaps the most outlying example. After taking an MSc degree, at which she showed such ability that she could easily have gone on to a doctorate, she decided to follow her first love and is now a successful opera singer on the international stage. She said much later that she found research 'lonely', an interesting way of putting it but I can certainly see what she meant. It doesn't suit everyone, but I was sorry to lose her.

The responsibilities and workload involved in examining graduate students pale next to finals examining for the undergraduate Physics degree. After three years of study, for the students everything depends on their performance at the end of their final term. Over a couple of weeks they take a set of three-hour papers, suitably clad in the required *sub fusc* uniform, that is, dark suit, gown, and white bow tie (or the female equivalent, often imaginatively expressed). The papers are set, invigilated, and marked by a small cadre of academics who take on the responsibility for three, or sometimes five, years around mid career and then by convention are free of it thereafter. With nearly 200 undergraduates each year taking a dozen or more three-hour papers each, the workload is formidable, and examiners are (or were, the system is changing) expected to have a grip on the entire physics syllabus, right across the board.

I had done my statutory three-year stint as a finals examiner for the Physics degree as soon as I shed my duties as head of department in 2000, having managed to stave it off until I had the time to cope. It was actually quite fun, but only if you don't try to do anything else during the month or so when the papers are set, and the other month or so when they are marked. Sprinkled around that major commitment there are the regular requests to examine graduate students who have completed their theses, which are then examined orally before they can receive the title of Doctor. Such requests come not just from one's own department—one of the examiners has to be 'internal'—but also from other universities in the UK and abroad, where special arrangements may be needed to deal with language difficulties. Reading doctoral theses and conducting oral exams is usually more fun than undergraduate examining, because the timescale is less intense and because the subject matter is original research, usually in a topic of no small interest to the examiner.

Yet another dimension to examining is the role of external assessor at another university than one's own. This involves not examining per se, but rather overseeing the process to ensure fairness and consistency. I did two long stints of this kind of work, first at the University of Edinburgh then later at University College, London. In each case the course was for the Master's degree, involving both taught classes and written exams, and some research leading to a short thesis. Each student was orally examined by me, and then I would scrutinize the class list and attend a meeting to sign off on the awarding of grades. Anecdotes are not appropriate because of the confidential nature of the process, so suffice it to say even this quite restricted sample of attitudes and approaches provided wonderful examples of human diversity. The high point for me was when the professor at UCL who organized their Master's course told me about a post-mortem session he had held with the students. How did they feel about the vivas that the external examiner had conducted? 'Ooh, he's so *nice*' said one of the girls reportedly, and the rest nodded vigorously. I thought I had been pretty tough, but I was pleased with the compliment.

All of this was behind me, I thought, as I settled into my final year before retirement. I had applied for my last research grant, cut back on travel, and was quietly contemplating various agreeable projects (including writing this book) for the post-retirement period. However, as it happened, one of that year's Finals examiners fell seriously ill just as the torrid period of frantic activity was about to begin. The Chairman of Examiners had to search around for a substitute, preferably someone experienced, as there was no time to spare. The Chairman was someone I knew well, and to whom I owed a favour. Furthermore, the field that needed to be covered was mainly thermodynamics and quantum radiation, two of my specialities (for obvious reasons, since these are the basis, along with fluid dynamics, for atmospheric physics). Suddenly I was a Finals examiner again.

Some of my older colleagues had sought ways to continue to teach after retirement (which in Oxford with its tutorial system, is not difficult) but as I signed off on the class list in the summer of 2011 I knew I was not going to be one of those.

Emeritus

Retirement came for me at the beginning of Michaelmas term in 2011 and I was henceforth an Emeritus Professor in the department and an Emeritus Fellow in Jesus College. Speaking at my retirement dinner in college, I quoted Sir John Habbakuk, a former Principal of Jesus, on the meaning of the term Emeritus. At his own retirement celebration, which I had attended years earlier, he had explained to the assembled fellowship that in Latin 'e' means you're out, and 'meritus' means you deserve it. At the corresponding event organized by the department,

I borrowed Shaw's epigram quoted above, itself borrowed from *King Lear*, and added that I was not going far, however.

While most of my administrative duties in the university became history and melted away, what was left was membership of external committees of various kinds. There are the learned societies, for example, which are run by a Council of unpaid devotees, and the committees of the research councils, which disperse funds, and there are conferences to be organized, attended, and run. Membership of these bodies can take up a lot of time, but is generally considered a privilege, and of course one gains a certain amount of influence and information thereby that can prove useful. Committee work was part of my job, over the years; I decided I could do some of this sort of thing in retirement, as a good way to keep in touch.

I had already been Vice-President of the Royal Meteorological Society, and served on the Council of the Royal Astronomical Society, and have to confess both were excruciatingly dull activities. There are few things worse than being a powerless figurehead, endlessly obliged to show up for things (sometimes overseas; as President of the International Commission on Planetary Atmospheres and their Evolution, I had travelled the world for eight years on its business to, I fear, little useful purpose), or write papers to drum up support for vague and often already lost causes (creating a British Space Agency comes to mind, although that did come good in the end). More of that was not on my retirement agenda, although I still enjoy the exclusive dining clubs that meet monthly in London to take wine, eat, and talk. Election to these clubs was something that usually came in late career and it is rare to see the same grey heads that quaff champagne at the Athenaeum in the evening actually at the scientific sessions of the society, although they would have been leading lights in a more youthful phase.

From the point of view of the organizers, professors who have just retired are good prospects for national and international committees: they have the time, the experience, and they no longer have vested interests in the award of a sum of money or a new project or facility, at least not to the same extent. Before long I had several offers, those of interest being from HEFCE, the *Higher Education Funding Committee for England and Wales*, the new *UK Space Agency*, and the Government's *Technology Strategy Board*, since renamed *Innovate UK*. Most of these involve offering advice or awarding funding in the broad area of space science, technology, and strategy, with a typical recent focus on a surge of interest in Mars exploration leading to the *Aurora* programme to put a European rover on the surface of the red planet.

For HEFCE, however, the remit was broader, and had to do with the distribution of billions of pounds of government funding to universities. This is achieved through an exercise that takes place every five or six years, called the *Research Excellence Framework*. Every academic, every department, and every university in the country has to make a detailed submission of their work and its impact to be rated by the body of which I was now a member. The available funds are then distributed according to this rating. Success is vital to the universities, not just because hard cash is a *sine qua non* in modern higher education, but also for the prestige that comes from the rating itself. I had toiled like everyone else under this yoke as a professor; now I had vaulted the fence and gone from poacher to gamekeeper. It was still a massive amount of work, but curiously what had seemed nothing but a pain before now seemed necessary and worthwhile. It is also a good way to keep in touch with what is happening in the field once one becomes less active oneself.

For long periods, the Rutherford Appleton Laboratory had been like a second home to me as we grafted together on large space projects. I am fortunate to be invited by the Director to an annual Space Conference, at which engineers and scientists rub shoulders with royalty, senior politicians, and captains of industry to discuss the nation's current endeavours in space, before retiring in the evening for a sumptuous dinner. It was at one of these conferences in 2009 that the Minister for Science, then Lord Drayson, announced that the UK was to have an executive space agency at last. This would replace the toothless organization that was the British National Space Centre, which specialized mainly in marketing the UK space industry abroad. When it was set up in 1985, the BNSC was meant to run Britain's affairs in space by controlling a partnership of no less than ten government departments and agencies, and for a while (including during the ISAMS crisis related earlier) was actually in charge of university funding in space-related disciplines. However, most of the government bodies (which included the military) wanted to keep their own budgets and the arrangement had collapsed. It took more than 20 years of campaigning, by bodies like the Royal Society Space Research Committee, of which I had been a member, to get the UK back to something like the kind of representation in space that other major nations took for granted.

The management of what used to be the *British Aircraft Corporation*, which now after several rejigs is part of the international consortium *Airbus Defence and Space*, had sponsored me at Oxford with a scholarship in 1966. Their successors still kindly invite me to their corporate Christmas party each year in the sumptuous setting of Lancaster House in London. There I quaff champagne and chat to colleagues and collaborators about their new projects, surrounded by a fine art collection and a decor that puts Buckingham Palace to shame (as Queen Victoria once remarked). Even greater pleasure comes in retirement from hearing the speeches and savouring the plans and achievements of the UK space industry, no longer with the burdensome responsibility for putting any part of them into practice. After a few hours I decant onto the Mall in front of Buckingham Palace, sponsored as a newt, and head for the coach to go home.

Back to Howick

When I can I like to go back to Northumberland, staying sometimes with my friend Tony from schooldays at his home in Tweedmouth or, if Doris is with me, in Howick village where I grew up. The Old Rectory, a fine building in its own grounds dating from the 1830s, now offers excellent bed and breakfast accommodation. From some of the rooms in the Rectory my old home in the Schoolhouse can be seen across the former paddock and orchard, now just a meadow. In the other direction is a glimpse of the North Sea and some of the most beautiful coastline in the country.

The school itself had been converted into a holiday let after it closed in 1961. I had a mind to commemorate my mother's tenure as the last head teacher there with a plaque, which led to correspondence with Lord Howick, on whose estate the village stands. He suggested planting a tree in front of the school, with an engraved stone to carry the message. This we duly did in a small ceremony in the summer of 1994. The plan went awry, however, when the tree, an English cherry, refused to grow. Year after year I would come back and check up on it, only to find that what had apparently been a fine healthy specimen of young treehood, professionally planted in apparently fertile ground, was still a stunted mockery.

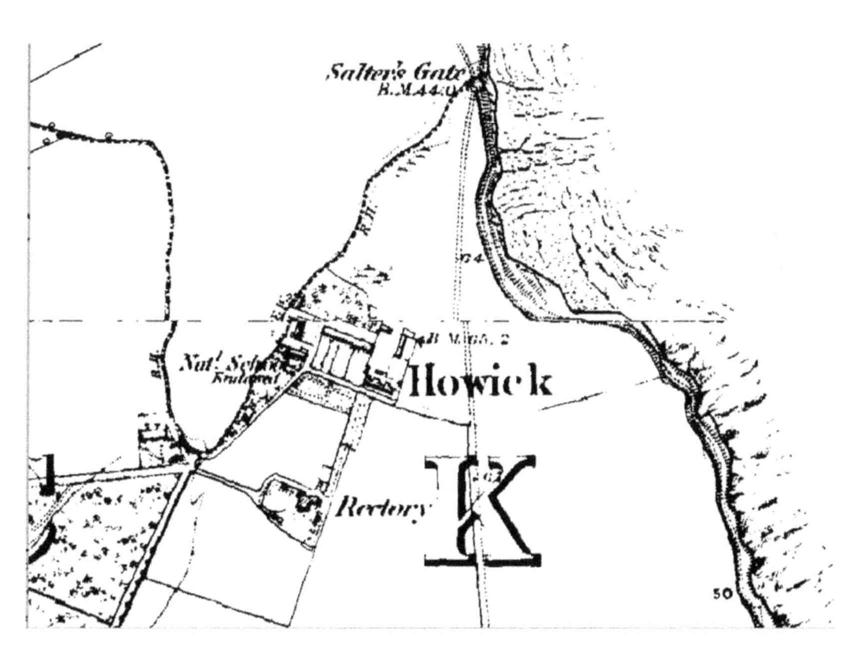

This old map of the village of Howick, where I grew up, validates my theory that the stream we know as Howick burn was diverted away from the houses in about 1820 to make a boating lake for the local aristocracy. It now debouches into the North Sea several miles south of its original path to the small sandy bay known as Salter's Gate. Researching the history of the area has become a hobby in retirement as I wind down my involvement in space experiments and classic cars.

This was distressing, and I searched for a reason, finding one quite by accident. Lord Howick had mentioned that the entire village had been moved about two miles to the north in the early 1800s by his forbear because it had interfered with the view from the Hall. From the date, the ancestor in question would have been none other than the famous 2nd Earl Grey, of Reform Bill (and Tea) fame, who had extensively revamped the entire estate after inheriting it in 1807. This intrigued me and I did some research into the history of the village, quickly discovering that the estate records are all now in an archive in Durham, managed by the university. I contacted the librarian there and established that I could have access, and obtained a general index to the archive, which apparently is mostly untouched and uncatalogued, but very, very large. It would be a major project to unearth the relevant records, and I haven't yet, but God willing I will some day.

What I did find was an old map that shows the village in its present location but with some of the old houses still unbuilt. On this map the stream that nowadays runs along the south side of the village follows a different route, one that takes it right past the school and into the sea at the bay known (mysteriously) as Salter's Gate, where as children my pals and I used to bathe. Going map in hand to look at the actual lay of the land, the bed of the river is still clearly visible in the topography, with a deep cutting where the flow crossed the cliffs, and in fact the bay would have been its mouth. Excellent and unusual specimens of fossilized dinosaur bones have been found nearby.

The present course of Howick Burn runs dead straight from the point where it was diverted to feed the Lily Pond, a large and still picturesque lake that has the decayed remains of a boathouse once used for the entertainment of Earl Grey and his children. So, in around 1820 the Earl had diverted the small river to make the pond, which on close inspection is clearly artificial, although time has mellowed it to a lovely bird and wildlife sanctuary. Moving the river made it possible to build the rest of the village on the recovered land, including a couple of terraced houses and a village hall full of books, completed in 1883. The old river bed must be stony and sterile, and that was where we had planted the tree; no wonder it would not grow.

Back to Liverpool

Ten years before retiring, I had received an unexpected request from the Vice-Chancellor's office asking me to represent Oxford on the Court of the University of Liverpool. The duty to supply a representative extended back to the original charter of my Alma Mater, when the court was set up to monitor the governance of the university. It met just once each year, principally to receive a 'State of the Union' address from the Vice-Chancellor, and to nod through various top-level appointments, including the VC himself when the incumbent changed. We were also invited to attend Degree Day in the Philharmonic Hall, with appropriate pomp and circumstance. Needless to say, I enjoyed this re-introduction at the highest level to the institution I had entered at the lowest nearly 40 years earlier, and carried on into retirement, when it was even more pleasurable because the reduced pressure from my own work allowed me the time to get more engaged at Liverpool.

There was a double blow coming, however. At the meeting in January, 2013, the Vice-Chancellor announced that the university had reviewed its governance and decided to set up a modern management structure, which no longer had any need for the Court. I knew instantly what this entailed, since Oxford had done the same thing a decade earlier. It meant layer upon layer of new administrative posts, costing a lot of money, and making academic life a misery by enforcing with enthusiastic rigour every possible shade of interpretation of European Law and Health and Safety legislation. Surely the Court was there to prevent such folly? As my mind raced with ideas for what I might say that could possibly make a difference, the VC moved on to his next and even bigger bombshell.

Rathbone Hall was to be closed and demolished. This was, he said, because it was past its design lifetime and could no longer be maintained economically. The future of the Court forgotten, I pointed out that Jesus College is more than 400 years old and still works well as a building, it's just a matter of maintenance. Ah, but we've done the sums, he said, and by the way, it is nothing to do with the fact that the older halls stand on large parcels of valuable land that can more profitably be sold these days for housing. The penny dropped and I knew then that it was no use to argue.

As a footnote to this, a year on from the abolition of the court and the death sentence for my beloved Rathbone Hall, I received an invitation to become a Governor of Liverpool John Moores University. Located just yards from its older sibling, John Moores is the former Polytechnic that was elevated to university status in the Australian-style reforms of the early 1990s. Reading the literature, I found it to be a thriving institution that was actually larger than its more venerable neighbour and larger, in terms of student numbers, than Oxford as well. What a great way to resume my love affair with Liverpool! Alas, I then read about the duties and responsibilities of a governor and realized that this was no sinecure, like the Court had been, and would have involved far more time and commitment than I thought I could raise in my eightieth decade, with a lot of other things I wanted to do.

Another false start, at about the same time, had to do with editing scientific journals. This is a common academic activity, and I had done quite a lot of it over the years as an editorial board member for journals including *Planetary and Space Science, Atmospheric Chemistry, Quantitative Spectroscopy and Radiative Transfer*, and even *Idojaras*, a Hungarian journal of meteorology. Just as I was contemplating retirement, the editorship of the Planetary Science volume of the *Journal of Geophysical Research* became vacant. JGR, published in America, is probably the top journal in my field and the chance to take this on seemed heaven-sent, in retirement when I thought I would have time to read hundreds of papers and adjudicate many often bitter quarrels between authors and referees. But just like the governorship, when I looked into the details it became clear that I would be worked into the ground if I tried to do this along with keeping involved in my favourite space projects and the other things I had planned.

The Jesus Chair

Life in college is a joy, especially once one no longer has to attend ponderous meetings of governing body or undertake other worthy duties. However, one seeks to make a contribution in ways other than merely shedding ones illustrious presence and hopefully continuing to mentor the young, if not as a teacher any more then at least as 'the wise old sage who knows his onions'. I had a fascinating job as the chairman of a small group charged with determining the future of the Fellows' Library, a venerable presence in the college dating from Elizabethan times and with a renowned collection of rare books. It had fallen into disuse, except by the occasional scholar, usually from overseas, who came to find and consult a particular artefact. Extra books, old paintings, and various bric-a-brac had been stored there over the years, giving it the appearance of Tutankhamun's tomb when you used the enormous iron key to open the massive old door and looked inside.

Now, the roof was leaking and water, insects, and uncontrolled temperatures were taking toll of the books, and causing millions of pounds worth of damage. The floor was rotten and the balcony, installed 200 years ago when it was brought from an old church in Wales that faced demolition, was unsafe. It would cost a huge amount to renovate the whole thing, and almost no one was using it. Some of the Fellows thought the books should be sold and the space used for student rooms, a coffee bar, or even (I'm not sure this was serious, but there are iconoclasts in every ancient order) a bowling alley.

Our committee, a majority of Fellows, and all of the old members who replied to a survey, were in favour of restoring and keeping the Library. Accordingly, we embarked on a massive programme of building work that began with specialists to wrap each book and store them all at a remote site before the stacks, the floor, and the roof, could be removed. An enormous marquee was erected over the whole wing of the College to keep out the elements while rebuilding took place; the floor could not be saved, but neither, in a Grade 1 listed building, could it be replaced with modern wood. We had to fit seventeenth-century wood

As a Fellow of Jesus College, Oxford I helped to rescue and restore the near-derelict Elizabethan Fellows' Library with its ancient tomes and wonderful scholarly atmosphere. One aftermath of this was the introduction, at my behest, of this comfortable leather armchair for quiet contemplation and meditation. The Bursar named it Professor Taylor's Chair.

from a company that specialized in reclaiming material from demolished stately homes; it was surprisingly inexpensive.

A key part of our plan was to furnish the bays between the book stacks each with a modern desk and chair with lighting, power points, and Wi-Fi. The essential point was that, while the ancient surroundings were preserved, the room would become a *modern* Fellows' Library in which we would actually sit and work, whether or not we were using the old books for any purpose other than as an inspiring backdrop.

However, the furniture part of this was rejected by the Governing Body in favour of old chairs from an antique dealer. It is virtually impossible to get Elizabethan chairs, and those that do still exist are extremely uncomfortable. My colleagues voted to choose instead, to my horror, basic Victorian chairs of the kind that used to be found in every village hall around the land, before folding or stacking chairs took over. These looked like they had been recovered from a dump somewhere and extortionately priced in an antique shop in case some mug wanted them. Apart from being expensive, uncomfortable, and ugly, they were (I thought) neither appropriate in the Elizabethan setting of our library, nor did they provided the 'ancient meets modern' ethos I was striving to achieve in its resurrection. I lost that fight, but did manage to have one bay fitted with a sumptuous leather armchair, for reading, meditation, and occasionally even sleeping.

I also mildly vented my spleen by posting in the Senior Common Room a segment from a book by Kingsley Amis, *Jake's Thing*, in which the governance of an imaginary Oxford college is satirised by having fellows argue endlessly about furniture while once-in-a-lifetime decisions about the sale to developers of vast tracts of valuable rural college land slip through without discussion in moments of bored neglect. This echoed what had really happened at Jesus, with the proposed selling off of the sports fields finding lamentably little resistance while we spent hours discussing the chairs for the library. Amis was probably aware of earlier occurrences, since he had been at Oxford and most good satire is based on real events. And, after all, there is a funny side to (almost) everything.

The great and the good

Oxford is a good place for stumbling across famous people, and this is often enhanced by having a modest reputation as an expert on some topic of current interest to the chattering classes, such as climate change. I have already described how this took me to 10 Downing Street and tea with Mrs Thatcher in the 1980s. Around that time, I also met the Lord Chancellor, Lord Hailsham, formerly Quentin Hogg, at a cocktail party held in Oxford by our Vice-Chancellor. Academics then would wear gowns in the presence of the VC and so I was recognizable as a local; the jolly figure of his lordship approached me and enquired where there might be a loo. We were outside, as it was summer, and I wasn't familiar with the plumbing in the nearby seventeenth-century buildings, but I said I would go and enquire. 'No, don't bother,' said the Lord Chancellor. 'I will hold on until this is over', adding ruminatively 'Tempus Pissit'.

An even more stellar encounter came when I was giving an invited talk at a conference in Sheffield in the summer of 1989. Sheffield is not far from Chatsworth House, the seat of the Duke of Devonshire, whose family name was Cavendish. The Cavendishes have been at Chatsworth since 1549, and the 7th Duke endowed the physics laboratory at Cambridge, later named (by Maxwell) in honour of the great scientist Henry

Cavendish, another relative. Among many other achievements, Henry had discovered the element hydrogen and obtained the first good estimate of the mass of the Earth. It was said that some of his equipment was archived at Chatsworth and a few of us were given the chance to go and see it. I was not sure what to expect, certainly not to be greeted by the Duke himself nor to be served tea by the Duchess in person, as actually happened. She was famous in her own right as Deborah, the youngest of the glamorous Mitford sisters, but like her husband she was charming and friendly to our small and rather scruffy group of visiting scientists.

I had been to Chatsworth before, as a tourist, and had seen nothing of Henry Cavendish's artefacts then. However, Chatsworth is enormous and not all of it is open to the public so perhaps that was not surprising. What was startling, however, was the secret panel in a large wall painting that swung back to reveal the entrance to an eighteenth-century laboratory, laid out with Cavendish's apparatus. The Duke explained that Henry had actually worked mostly in London, at his house in Clapham, but that his artefacts had been collected and sympathetically set up at Chatsworth after his death. They were rarely shown to anyone. We were awestruck.

In July 2003 I encountered a different kind of distinguished family. The occasion was a dinner for departing Rhodes scholars and their partners, held in Hall in my college. As hosts, Doris and I would sit with one of the soon-to-be-departing student couples. I consulted the seating plan on the way in and saw that my companion was to be Miss C. Clinton. I knew that Chelsea was in Oxford but not that her boyfriend was a Rhodes scholar, which is what brought them both to the dinner. She was soon to join the management consultants McKinsey & Co in New York, and was planning to focus in part on the global challenges faced by companies and populations from the threat of climate change. We talked about that for hours. I also told her that I had unwittingly lived next door to her father when we were both students; she said that he was currently in Oxford for a few days as we spoke. Again by chance, I encountered the ex-President in the street the next day as we travelled in opposite directions down Parks Road. I was startled but managed a smile as I wove past the tall, very fit man with a tell-tale bulge under his arm that stood between me and Mr Clinton, and I carried on up the street. Too late I realized that he was on a walkabout and wanted to meet locals; his body language made it clear he wanted to talk to me. I could have told him that we had been neighbours and that I had had dinner with his daughter the night before, and who knows what else. Damn!

At another dinner I was seated next to Richard Dawkins. After some small talk and quite a lot of wine I decided to tackle him on some of his views about religion, as expressed in his recent best-selling book *The God Delusion*. I assumed that was what he would like and was a courteous thing to do, and anyway I thought he was too extreme and could benefit from a little of the Unitarian Universalist philosophy I had embraced years ago while living in California. However, I had barely started when his brow turned thunderous and he got up and left, the meal unfinished. I had had a similar experience a couple of years earlier when a colleague brought in a cabinet minister (who shall be nameless, and actually I have forgotten his name) as a guest. He, too, fled before the end of the meal in the face of high-table banter. I now know that many, if not all, famous people do not relish having their views challenged in this way, at least not in the way that I do it.

One of my duties was to invite distinguished speakers to give the Halley Lecture, one of the 'named' lectures that are minor highlights of the university calendar. When that came due in 1988 I had just come back from a routine meeting in the USA, and in the airport at Washington I followed my usual practice and picked up a book to read on the flight home. On the 'just published' shelf was a book by a Cambridge astrophysicist whose main claim to fame then was a brave fight against disability. The blurb on the cover claimed that he had taken cosmology, a subject which most people find incomprehensible, and, by expressing deep insights in simple language, made it accessible to 'the interested layman'. Communicating science is a skill that deeply interests me, and that I have often tried to acquire myself (this memoir is my latest and last attempt) so I bought it and started reading immediately while waiting for the flight to board.

I enjoyed the book but was not convinced that anyone without a degree in physics would be able to get anywhere with it. Its sales would therefore be very limited. However, the topic was perfect for the Halley Lecture, which was attended mainly by scientists wanting an authoritative summary of a topical and interesting area of astronomy. It was said that Stephen Hawking—for it was he—could give a good lecture despite having to speak through a synthesizer, and that this added drama to the occasion.

I therefore invited Hawking to come over from Cambridge, with some misgivings about whether he would attract the sort of audience the occasion demanded given his specialized fame in what was still a narrow field, and that he could only speak slowly through a machine. There was little sign at first of his book catching on—I think in fact it was published a few months later in the UK than in the USA—and anyway, I did not think it would. But I need not have worried. In the nine-month interval between the invitation and the lecture *A Brief History of Time* became a runaway best-seller and everyone wanted to hear Hawking speak about it. We booked the biggest lecture theatre in the university, which at that time was in the new Zoology building, and when the time came I headed over there to introduce the speaker to what I now knew would be a capacity audience.

It was much more than that, however. The entire building was surrounded by people who could not get in, including myself. Hawking and his entourage had been sneaked in earlier via a fire door at the back, which was now sealed, and the security staff were busy trying to bar entry at the front and also get all the surplus people sitting in the corridors and aisles out. It was not quite Beatlemania, since there was no screaming, but nevertheless there was no chance of me getting inside any time soon. In the end I went back to my own building and just hoped for the best. I was told later that the lecture went very well.

I had done quite a lot of interviews on radio and television for one reason or another, usually to do with climate, or the latest venture in space, but I thought it was a highlight when the BBC called to invite me to appear live on breakfast television. Fired up by this new experience, I rose very early and drove down from Oxford to London, only to get stuck in stationary traffic a mile or so away from the television centre at White City. When I finally arrived there were only minutes to go before the scheduled air time, so I ignored signs about parking and drove right up to the front entrance and abandoned my car there. With a hurried explanation to the security man on the door I was allowed in and hustled upstairs to the studio where the familiar figures of the presenters were busy talking to a minor celebrity about her latest movie or something.

Crouched behind the cameras, I explained to the director in a loud whisper designed to penetrate his headphones that I was due on next. He looked puzzled, consulted a clipboard, and called over an assistant who beckoned me away. I ended up in another studio, in the basement, where another set of presenters were busy with news and interviews. The producer there filled me in: they were running a prototype of the 24-hour BBC news channel that was soon to launch worldwide. For now, however, although the cameras were active and everything looked real, it wasn't being broadcast at all, just developed and rehearsed. This was what they wanted me for. The producer swore they had told me my venue wasn't the real thing, but they hadn't, and it was obvious to me that they knew I would not have turned up if I knew it was only a dress rehearsal. We did the interview and I tried not to seem annoyed, although I'm pretty sure that I did not succeed. As soon as it was over I retrieved my car and drove rapidly home.

Onward and upward

I was enjoying the ongoing space missions for which I was a team member at the time of my retirement—*Venus Express, Mars Reconnaissance Orbiter, Lunar Reconnaissance Orbiter, Cassini, Rosetta*, and *BepiColombo*. I had no desire to give any of them up, although close involvement would no longer be possible without resources for students, post docs, and the vital but incessant travel to meetings. This did not mean it was necessary to resign, especially if you have had involvement in a mission since its conception and put in your time, so to speak. A bit of the knowledge that only comes from long experience is also essential when it comes to interpreting the findings and crafting the publications containing on the results.

There are also always a few key science questions that one feels have not been satisfactorily resolved despite a lifetime of effort. These can burn a hole in your soul if you don't do something about them while you can. Finding some time to think about them, talking to younger people who are working on them too, making contributions to planning new missions that might provide more complete answers, all of these things can continue in an active retirement. What those key questions actually are is specific to the individual, with their own background and history, and their own quirks. I do not think I could have rested happily had we not got the new, detailed data from *Venus Express* on the weirdly configured polar vortices that we glimpsed but did not understand when my first instrument got to Venus 35 years ago. However, *Venus Express* also left me with the sort of unsolved puzzle that can still keep me awake at night: what is the role of active volcanoes in maintaining the extreme climate on Venus? How I wish I knew.

Volcanoes probably do explain long-term climate change on Mars, but in ways we still haven't fully fathomed. The new rovers on the surface are digging into the geological record now and that will surely help, with important advances already and more to come soon. And, although it is a saga that will run forever, I expect that there will be an initial breakthrough on the search for life in the next decade. And manned landings? Don't hold your breathe, unless the Chinese go for it, which they may well once they have stepped on the Moon. I hope they, or the Indians, will send astronauts or robots to drill cores in the icy deposits at the poles of our satellite and find out what has been trapped there over the aeons. Someone needs to do that at Mercury, too, but it will take a little longer and few of us will be there to watch.

We will see what comes from the Mercury orbiter mission *BepiColombo*, however. I hope the answer is not nothing; it is a long and difficult journey with many pitfalls. I retain faith in the skill and ingenuity of the European engineers who have been so impressive in the past, landing on Titan, and on the nucleus of a comet, for example. Although it is not really designed to address my favourite goals—the heat flux from the interior, and the icy Mercurian ski slopes—I will be glad if our German colleagues still remember I'm on their team for the MERTIS instrument come New Year's day, 2024 (or whenever it finally gets there), so I can get an insider's look at the data that might shed light on those conundrums. I'm already delighted to be steeped in data from *Rosetta*, coming from a comet that is stranger than we imagined, and which is currently busy blowing itself apart as the frozen gases in its interior vaporize while we watch and wonder.

Don Hunten, my friend and mentor from Arizona, who was one of the first to recognize that Saturn's planet-sized moon Titan had an Earth-like atmosphere of nitrogen, was one of the prime movers who created the *Cassini* mission. He used to say that he would never see the data from the mission he had done so much to bring into being. Whether it was just to ward off the grim reaper or whether he actually believed it I'm not sure, but he kept up this line all the time the spacecraft was on its long trek to Saturn. In the end he did see a lot of the data, which started flowing in 2005, well before Don left us in 2010. I hope to see the mission out until it plunges into Saturn's atmosphere in 2017, and still have time to write the third edition of my book about Titan with Athena Coustenis.

Athena is young enough to be working actively on ideas for new journeys to this bizarre world, as I did too for a while. The next mission to Titan could involve balloon-borne platforms investigating the winds and the clouds, and doing a photographic survey of the terrain below. Mobile wheeled and floating craft for deployment on the surface are also being studied. However, it is extremely unlikely that any of these will get there before 2035 or even later. When I worked out the timescale, I realized I would be nearly a centurion by the time we had analysed the data, even without the usual delays. At that point I politely declined to become further involved.

18
Epilogue

He travels on, and in his face, his step,
His gait, is one expression; every limb,
His look and bending figure, all bespeak
A man who does not move with pain, but moves
With thought – He is insensibly subdued
To settled quiet: he is one by whom
All effort seems forgotten, one to whom
Long patience has such mild composure given,
That patience now doth seem a thing, of which
He hath no need. He is by nature led
To peace so perfect, that the young behold
With envy, what the old man hardly feels.

William Wordsworth, 1798

Rudyard Kipling described his autobiography *Something of Myself* as dealing with 'his life from the point of view of his work', and I have sought to do the same. This means leaving out most of the details of one's life history, and so diverging from the path followed by most biographies and memoirs. Those employ painstaking finesse to describe the personal lives of people known for specific achievements, such as famous authors, whether or not they were interesting or special in other ways. Many of them were not: we could contrast Philip Larkin, who led a quite boring life but wrote exquisite poetry, with the unknown soldier who wrote an account of the battle of Waterloo because he was there on one of the most momentous days in modern history and could tell us what it was like. Kipling, although undeniably and deservedly famous, classified himself for the purpose of his memoir with those of us lesser mortals who are interesting mainly because they experienced and wrote about something unique and exciting.

My history might mean something to a young person thinking about getting into a career in science, possibly even space research, or someone older who didn't but rather wishes they had, or just someone who wonders what went on behind the scenes. Reading other people's memoirs has taught me that many authors believe that they should be about lessons learned, as well as interesting history from an individual perspective. James D. Watson, of DNA fame, wrote a memoir that was specifically about listing nuggets of advice (the title, *Avoid Boring People*, has a double meaning that was meant to be taken both ways). His goal was to convey 'life' lessons: what do special people, or people who witness special circumstances, have to say that can teach those who come along later to think, do, or feel? This was not on my agenda until one of the reviewers appointed by the Press to advise on my book proposal suggested I should include my attempt at this, which made me realize that most memoirs do, and for many like Watson and Medawar it is their first priority.

The following is an apposite quote from the American author George Saunders, who was delivering a valedictory speech to students at Syracuse University in the USA:

> 'Down through the ages, a traditional form has evolved for this type of speech, which is: Some old fart, his best years behind him, who, over the course of his life, has made a series of dreadful mistakes (*that would be me*), gives heartfelt advice to a group of shining, energetic young people, with all of their best years ahead of them (*that would be you*).' Saunders goes on to say (now I paraphrase) that his advice is, as a goal in life, *Try to be kinder.* When young we are anxious and ambitious and we tend to prioritize our own needs over the needs of others, even though what we really want, in our hearts, is to be less selfish and more aware of the present moment. We get there through education and art; frank talking with friends; and establishing ourselves in some kind of spiritual tradition, recognizing that there have been countless really smart people before us who have asked these same questions and left behind answers for us.

But any philosophy of life has to be reconciled with the dog-eat-dog business that is the competition for funds for research, for senior posts in research institutes and universities, or for team memberships on pioneering space missions. This is easier if you know what you want, and so are not drawn by opportunities that may be open doors but which don't further your dream or help your friends, colleagues, and students. Winning a large grant or a prestigious role in a project is not always a good thing: once gained, funds have to be spent properly, and that means committing a big slice of your life to objectives that may not even have seemed like a good idea at the time.

Honesty is a vital and under-rated virtue. When, once upon a time, a director of a research council said tartly that I should 'take what's on offer' when a sum was offered for organizing an annual summer school in geophysical fluid dynamics, something that was about as attractive to me as a sojourn in Wormwood Scrubs; or when a manager at the Met Office wrote that they had decided the joint institute headed by me should change direction and what I thought about it was irrelevant; or when an arrogant dispenser of taxpayers' funds insulted an audience of professors with 'you can get academics to do anything by trailing the money', they were not ignoring conventional truths. However, academics still have a rare measure of freedom if we can find the courage to use it; so turn your back when you have to and be true to yourself.

Try to be methodical: don't be like the man who 'jumped on his horse and rode off in all directions'. I learned this while taking a course in transcendental meditation during the three months that I lived alone in Oxford after moving from the USA, while Doris remained in Pasadena looking after house and job transitions. The technique is still useful but best of all is if you can train yourself to put *First Things First.* There is a book with that title by Stephen R. Covey, but even better is *Zen and the Art of Motorcycle Maintenance* by Robert Pirsig. I read that over again every few years. Pirsig is not one of the world's greatest philosophers, but he hit on a simple message and a magical way of telling it that few others have achieved. Basically, it is that sanity consists of doing things in the right order.

Try to trust people, unless you have clear and understandable reasons not to. It's just so much more efficient, even if you do get burned sometimes. When you do get a measure of what you want, appreciate it and be grateful. Don't rush to the next challenge until you are sure you are ready, and keenly interested in it. It is all too easy to rush up a mountain just because it is put in front of you or, as Covey puts it, to climb a ladder all the way to the top before realizing that it is leaning against the wrong wall.

The Best Job in the World

There is a splendid organization called the Planetary Society, which has the goal of communicating the excitement of planetary exploration to the general public. It has over 40,000 subscribers, including 50 founder members of which I am one. The Planetary Society was started in 1980 by Bruce Murray, then the Director of JPL and my boss until I moved back to Oxford; Louis Friedman, a senior JPL colleague who ran the show as Executive Director, and Carl Sagan, who at the time was at the peak of his fame as a popularizer of astronomy, especially planetary science, with his television series *Cosmos.*

The UK branch of the society organized a Carl Sagan Lecture in his honour in 1998 and I was asked to be the lecturer. I duly spoke on 'Planets and their Atmospheres: Carl Sagan's Vision Today' at the event in Birmingham Civic Centre, working in some of the details of my brief encounters with the great man over the years. The last time had been in 1995, some months before Carl died, when we were both at a planetary sciences conference in Hawaii. He had been ill for some time with some ghastly variant of leukaemia, and looked very weak, although he claimed to be cured.

Carl was scheduled to give a keynote lecture, and duly did, speaking about the future of planetary exploration. I was in the front row, and when he asked for questions I spoke up and said that I was surprised he had put so much emphasis on the outer planets and so little on Venus. I was sure it was his favourite planet, as it was mine. Did he not think we should be planning new missions to Venus, in order to address the mysterious climatic conditions there? This was a loaded question, because I was working on proposals for a mission to Venus at that time, and I knew of course that Carl was an expert on the subject. Indeed, it was he more than anyone who had shown that the very high temperature at the surface of Venus was due to the 'greenhouse' effect, and he had to be at least as curious as I was about the processes at work there.

I was disappointed with his answer. The outer planets, and comets, were top of his political agenda then and he just said Venus is interesting but we have to pick and choose our priorities. Although I was deflated by this, in the main body of his talk he had articulated several ideas that I would eventually work into our successful proposal, back in Europe, for *Venus Express*, so all was well in the end. I also decanted them from the scribbled notes I had made in Kona onto a slide for my talk in Birmingham. This is what those notes said:

> The space program used to be driven by the cold war.
>
> Now we don't have that, will we still have a planetary exploration programme?
>
> Yes, if we convince people to want it. But why should they?
>
> It turns out there are two basic things about it that ordinary people care about. The first is connected with origins; the second has something to do with survival.
>
> ORIGINS: Where did the universe come from? The Solar System, the Earth? Are there other Earths? Life? Intelligence?
>
> SURVIVAL: Understanding and living with environments, especially our own. Dealing with threats to our survival and our wellbeing.

Carl certainly had a rare gift for putting things in a nutshell.

After the Sagan tribute lecture was over, the organizers took me out for a meal. We chatted all evening about what I did and the various space missions in which I had some involvement. At the end of the evening, as he drove me back to my hotel, the Chairman said solemnly 'You know, you have the best job in the world.' I had never thought of it like that, but I didn't disagree. A year or two later I recalled this remark when I was being interviewed by the *Sunday Times*.

I had helped them with a special edition of their Magazine, one that covered recent advances in science, and they came back and asked me to give them some words for a biographical sketch that they wanted to include in a box to go with the relevant article about space. Somehow the discussion got onto the problems of trying to carry out these difficult and expensive space projects (we had just crashed on Mars without getting any scientific data, for the second time). At the end the correspondent said it sounded a bit agonizing and was it really the best job in the world? 'No', I said, 'I'd rather be the manager of Manchester United.' They printed that.

Awards and inspiration

You cannot plan your way to awards and honours so don't let them divert you from your goals. Their targeting is capricious and their impact is fleeting; they will just distract and disappoint you. One is grateful for them when they come, of course, and some sort of affirmation helps for achieving closure, especially when it comes as recognition for making it through the rain on ambitious projects that were painful at the time but fruitful in the end. I have a few that were like the balm of Gilead, including the NASA Exceptional Scientific Achievement medal for *Pioneer Venus*, the Prize for Opto-Electronics awarded by the Rank Foundation for the pressure modulator radiometer and its several applications at Earth, Venus, and Mars; a special award of the Scientific Instrument Makers Guild for the *Improved Stratospheric and Mesospheric Sounder*, the Bates medal of the European Geophysical Union for 'outstanding contributions to planetary and solar system science', and the Arthur C. Clarke Lifetime Achievement Award, for everything.

The NASA award is a real 'gong' on a ribbon, with separate coloured bars that you can wear, military style, when not actually wearing the medal itself, although I never have, not even at the awards ceremony where the Administrator stopped short of actually pinning it on. I might have got another one of these 15 years later for ISAMS, since all of my colleagues who were Principal Investigators for other instruments on the *Upper Atmosphere Research Satellite* did. At the time I wondered why they had left me out. Perhaps it was because the instrument did not last as long as it should have, perhaps it was just that the awards came right after 9/11 when America was not big on foreigners of any kind, even allies, perhaps it was something I did, or something I didn't do. It just shows why it's not worth worrying about baubles.

The Rank Foundation is a bequest of J. Arthur Rank, who made his fortune from flour with Rank-Hovis-McDougall, and then from cinema (including many classic movies, which began with a muscular individual striking a huge gong). It makes its awards at a grand event in London each year. One of the winners at about the same time as ours had developed a slide projector that, instead of transparencies and lenses, used tiny mirrors to project an image pixel by pixel. I thought this a stunning achievement but totally impractical for everyday use; now, of course, there is one in every lecture theatre and every cinema. The Rank Prize came with a cheque, which helped me not to mind when I took it back to Oxford to find that some wag had posted anonymously on the departmental noticeboard an altered copy of the citation that said the award was for 'Rank Incompetence'.

Four of us were awarded the Rank Prize for Innovation in Opto-electronics in 1989 for the development of the pressure-modulator radiometer and its applications. At the awards ceremony at the Royal Society are, from left, Clive Rodgers, Sir George Porter (President of the Royal Society), myself, Sir John Davis (Chairman of the Rank Trustees), John Houghton and Guy Peskett.

The Worshipful Company of Scientific Instrument Makers is a Guild of the City of London, and as part of an ancient tradition their presentation was at an even more splendid dinner in Mansion House. Uncomfortably encased in white tie and tails, I explained briefly to the assembled company what an Improved Stratospheric and Mesospheric Sounder is and what it does, although I don't think I got very much across. My wing collar, along with the rest of the outfit rented from Walters in Oxford, was becoming detached and it was all a bit of a farce, but the place and the event reeked of history and the evening was a delight.

The European medal was named after Sir David Bates, an Ulsterman who became one of the legends of atmospheric science. I had once heard him speak when he was very old and I was still a student, and in my acceptance speech at the award event in Vienna I passed on the advice he gave us young people at the time: 'Never send a letter that you enjoyed writing.' It's an even more useful principle now, in the age of email and tweets, although I can't say I have always stuck by it. It was gratifying when the *Oxford University Gazette* picked up on the award, and reported it in glowing terms, citing '. . . Professor Taylor's outstanding work in atmospheric physics, planetary sciences, molecular spectroscopy, and infrared physics.' These kind words were somewhat offset by the fact that they illustrated them with a picture of me fast asleep in full academic dress at a formal event in 2005.

Some kind words in the University Gazette were accompanied by a picture of me fast asleep in the warm sun during a break at a degree ceremony in 2005. When the reporter asked for a picture to go with her article I sent it to them as a joke, along with the one they were supposed to use; obviously, the Gazette office has a sense of humour too.

One Saturday morning in August 2000 I was paging through the previous day's *Oxford Times* in bed and discovered a special colour supplement. To mark the Millennium, they were celebrating 'Oxfordshire's Top 100 Sons and Daughters'. I looked to see whom they had chosen and was gobsmacked to find myself in there. I had done some interviews with one of their senior reporters about our Mars adventures. He had found it interesting and responded with a half-page article in the paper. But this? There I was on the same page as Jack Straw, the Home Secretary, and the marvellous actress Dame Maggie Smith. Wonders never cease.

The Arthur C. Clarke award was a surprise, too. This was announced at the annual UK Space Conference, which was held that year, 2009, at Charterhouse School. I had given a talk entitled 'The Climate Problem from Space' earlier on the final day of the conference and, as a speaker, was invited to the banquet in the Great Hall in the evening that closed off the week-long proceedings.

From the supplement to the Oxford Times, August 2000, 'Oxfordshire's Top 100 Sons and Daughters'. I am holding a picture of the doomed *Mars Climate Orbiter*; the caption said 'Prof Taylor is out of this world.' Photo by Jon Lewis, by kind permission of Oxford Mail/The Oxford Times (Newsquest Oxfordshire).

I arrived for dinner in the battered business suit that I wore on such occasions to find everyone else in evening dress. I had actually been told that the dinner was formal, but completely forgot; this was very embarrassing. I thought I found a solution when I noticed one other guest in mufti, and together we located an empty table hidden at the back of the hall. I introduced myself to the other sartorial offender, who was a friendly German called Tibor Pacher, and we took ourselves off to the remote location where we had a jolly time drinking the wine that had been provided in expectation of twelve, not just two, people at the table.

A picture from the BBC web site, captioned: 'Professor Fred Taylor, from Oxford University, received the Lifetime Achievement award. Professor Taylor has worked on a range of planetary missions, including ESA's *Venus Express* mission and on NASA's *Galileo* spacecraft sent to Jupiter. 'They kindly did not mention that I was inadvertently underdressed for the occasion, which was the awards banquet at the UK Space Conference in 1999.

This went fine until the awards ceremony at the end of the meal. The Arthurs claim to be the 'Space equivalent of the Oscars' and are conducted in similar fashion, with a sealed envelope opened with great ceremony as the citation for the relevant achievement is read out in a sonorous tone. When, last of all, they came to the Lifetime Achievement award, it began to penetrate my well-soused brain that they were talking about me. I had to go on stage, lamentably underdressed, and make an acceptance speech with no preparation or competence to do so. I cannot remember what I said, but I was delighted to receive the prize, which takes the form of an engraved monolith with the same (relative) dimensions as the Sentinel in the film *2001, A Space Odyssey.*

It was well after midnight when I walked alone across the vast, darkling playing fields of Charterhouse, back to the chastely distant women's dorm where I was staying (the women all being gone; this was the long vacation), clutching the monolith to my chest. As I walked I woke Doris up with a call on my mobile to tell her about the award, knowing full well that if I had earned it then so had she for always being there.

Valediction

For the last word I turn to Oliver Goldsmith, author of the well-known play *She Stoops to Conquer*. His biographer said 'Goldsmith wrote with confidence, if slight accuracy, on all branches of knowledge.' Goldsmith also wrote fine poetry, most famously *The Vicar of Wakefield.* Just before he died in 1774, he penned the following inspiring lines:

> O, blest retirement! Friend to life's decline -
> How blest is he who crowns, in shades like these,
> A youth of labour with an age of ease!

However, it turns out that Goldsmith was only 45 when he died, as a result of excessive debauchery occasioned by his new-found wealth. Of course, you can drink a lot and still reach a ripe old age. The once-popular newspaper columnist J.B. Morton, according to his biographer, became cantankerous in his dotage in the 1950s and

> "he was forever pounding his thick blackthorn stick on the bars of pubs and bellowing for ale".

It's not quite that bad yet, but that will suit me too.

Glossary

Ames Ames Research Center, a NASA facility in Mountain View near San Francisco, California.

AOPP Atmospheric, Oceanic and Planetary Physics, the department at Oxford University that I headed from 1979 to 2000.

Astrium was the successor to British Aerospace, with the space work now based mainly in Stevenage.

ATMOS Atmospheric Molecular Spectroscopy, an instrument developed at JPL to study atmospheric pollution from a platform on the Space Shuttle.

BAE British Aerospace, a UK industrial conglomerate with its main bases for space work at Stevenage and Bristol.

BepiColombo is a European mission to Mercury scheduled to launch in 2017. It is named after an Italian astronomer who studied the planet's complex orbit.

BNSC British National Space Centre, the precursor to the UK Space Agency, responsible for funding university research in space.

Caltech The California Institute of Technology, a leading US university located in Pasadena, near Los Angeles.

Cassini a joint US- European mission to orbit Saturn and deliver the Huygens probe to the surface of Saturn's largest moon, Titan.

CFC chlorofluorocarbons, a family of chemicals that pollute the stratosphere and damage the ozone layer.

CIRS Composite Infrared Spectrometer, an instrument on the Cassini Saturn Orbiter mission.

Co-I Co-investigator, an elected member of an experiment team working with the PI on a space mission.

COMPLEX Committee on Planetary and Lunar Exploration, a body of the US National Academy of Sciences.

COSPAR Committee on Space Research, an international body that held a huge Congress each year at various locations worldwide.

DPS Division for Planetary Sciences of the American Astronomical Society, whose annual meeting was the top venue for planetary sciences.

ESA The European Space Agency, which has its administrative headquarters in Paris, but its main operating base at ESTEC, in the Netherlands.

ESRO the European Space Research Organisation, the precursor to ESA.

ESTEC The European Space Technology Centre in Noordwijk, near Amsterdam, the largest facility belonging to the European Space Agency.

FRAM Fine Resolution Antarctic (Ocean) Model, the first realistic high resolution ocean model run in the UK.

Galileo named after the great Italian astronomer who discovered the four large moon of Jupiter, this was a mission to study them close-up from orbit around the giant planet. The spacecraft carried our NIMS instrument.

GISS Goddard Institute for Space Studies, an offshoot of GSFC located on the campus of Columbia University in New York City and engaged mainly in theoretical work.

GSFC Goddard Space Flight Center, a large NASA facility in Greenbelt, Maryland, near Washington DC.

HIRDLS The High-Resolution Dynamics Limb Sounder, a space instrument developed at Oxford and NCAR for NASA's *Earth Observing System.*

IRIS Infrared Interferometer Spectrometer, a satellite instrument and precursor to CIRS.

ISAMS The Improved Stratospheric and Mesospheric Sounder, a large space instrument developed at Oxford, RAL and BAe, for flight on NASA's UARS spacecraft.

IUS Inertial Upper Stage, a booster used to send payloads into deep space after they had been lifted into low Earth orbit by the Space Shuttle. *Galileo* was sent to Jupiter in this way.

JPL Jet Propulsion Laboratory, a NASA facility in Pasadena California operated by the California Institute of Technology (Caltech) and the lead centre for the American planetary programme.

JTH Sir John Theodore Houghton, Head of Atmospheric Physics at Oxford until 1979 and later the Director-General of the Met Office.

LRO Lunar Reconnaissance Orbiter, a mission launched in 2009 to map the surface of the Moon prior to a renewed programme of manned landings.

Lunar Diviner was named for its role in searching for water ice on the Moon as part of the *Lunar Reconnaissance Orbiter* mission.

MCS Mars Climate Sounder, an instrument to measure temperature, humidity and dust on Mars developed by JPL and Oxford and flown on the successful *Mars Reconnaissance Orbiter* spacecraft in 2006.

MERTIS Mercury Thermal Infrared Spectrometer, an instrument on the BepiColombo Mercury Orbiter spacecraft.

Met Office The British Meteorological Office, headquartered at Bracknell for most of the time covered in this book, but since 2003 at Exeter.

NASA National Aeronautics and Space Administration, the American space agency.

NCAR National Center for Atmospheric Research in Boulder, Colorado.

NERC Natural Environment Research Council. The UK agency responsible for funding university research in atmospheric and climate science.

Nimbus satellites, the name means 'cloud' in Latin, were launched by NASA from 1964 to 1978 to provide a test bed for innovative instruments designed to make meteorological observations form space.

NOx refers to the oxides of nitrogen that pollute the stratosphere and damage the ozone layer.

NIMS Near Infrared Mapping Spectrometer, an instrument on the Galileo Jupiter Orbiter spacecraft.

OSD Oxford Scientific Design, a company formed to spin off the research and development work of the Oxford department to industry and other users.

PI Principal Investigator, the leader of a team carrying out an experiment on a space mission.

PMIRR The Pressure-Modulator Infrared Radiometer, an instrument to measure temperature, humidity and dust on Mars developed by JPL and Oxford and flown on the ill-fated *Mars Observer* and *Mars Climate Orbiter* spacecraft in the 1990s.

PPARC Particle Physics and Astronomy Research Council. A UK agency responsible for funding university research in planetary science.

RAL Rutherford Appleton Laboratory, a large government centre at Chilton, 10 miles south of Oxford, specialising in high technology work including space research.

Rosetta Named after the Rosetta Stone, a European spacecraft that intercepted comet 67/P in 2014.

RHI Robert Hooke Institute for Collaborative Atmospheric Research, a joint venture at Oxford by the University, the Met Office, and the Natural Environment Research Council.

SAMS The Stratospheric and Mesospheric Sounder, a space instrument developed at Oxford for the American weather satellite *Nimbus 7*.

SERC Science and Engineering Research Council, followed SSSC and preceded PPARC as the UK agency responsible for funding university research in planetary science.

SSSC Solar System Science Committee, a fund-awarding and policy planning body ion the UK.

SSU Stratospheric Sounding Unit, an instrument built by the Meteorological office, based on designs developed at Oxford, for deployment on operational weather satellites.

TES Tropospheric Emission Spectrometer, an instrument developed at JPL to study pollution near Earth's surface from Space.

TIROS Television and Infrared Observation Satellite. The first successful weather satellite, launched by the United States in 1960, and a precursor to *Nimbus*.

TOGA Tropical Ocean, Global Atmosphere. A large international research programme designed to improve long-range weather forecasts by a better understanding of the role of the oceans.

UARS The Upper Atmospheric Research Satellite, a six-ton orbiting laboratory developed by NASA to study the stratospheric ozone layer.

UKSA United Kingdom Space Agency, created in 2010, based in Swindon.

Viking a NASA mission to Mars, which successfully placed the first robotic landers on the planet in 1976.

VIRTIS Visible Infrared Imaging Spectrometer, an instrument developed for the *Rosetta* comet rendezvous mission, with a new version used on *Venus Express*.

VORTEX Venus Orbiter Radiometric Temperature-sounding Experiment, an instrument on the *Pioneer Venus Orbiter* spacecraft launched in 1979.

Voyager a NASA mission to the outer planets Jupiter, Saturn, Uranus, and Neptune.

WMO World Meteorological Organisation, based in Geneva.

Index

G

H

K

L

M

U

V

W